The New York Times

BOOK OF

Physics and Astronomy

The New York Times

BOOK OF

Physics and
Astronomy

MORE THAN 100 YEARS OF COVERING
THE EXPANDING UNIVERSE

Edited by

CORNELIA DEAN

Foreword by

NEIL DEGRASSE TYSON

STERLING

New York

STERLING

New York

An Imprint of Sterling Publishing
387 Park Avenue South
New York, NY 10016

ISBN 978-1-4027-9320-2

Distributed in Canada by Sterling Publishing
c/o Canadian Manda Group, 165 Dufferin Street
Toronto, Ontario, Canada M6K 3H6
Distributed in the United Kingdom by GMC Distribution Services
Castle Place, 166 High Street, Lewes, East Sussex, England BN7 1XU
Distributed in Australia by Capricorn Link (Australia) Pty. Ltd.
P.O. Box 704, Windsor, NSW 2756, Australia

For information about custom editions, special sales,
and premium and corporate purchases, please contact Sterling Special Sales
at 800-805-5489 or specialsales@sterlingpublishing.com.

Manufactured in the United States of America

2 4 6 8 10 9 7 5 3 1

www.sterlingpublishing.com

CONTENTS

CHAPTER 1

The Nature of Matter

CHAPTER 3

The Fate of the Universe

FOREWORD

The relationship between journalists and scientists has not always been harmonious. And in many circles today it still isn't. Most fields of science use a lexicon that hardly anybody from the public understands. In fact, most scientists can't even communicate with fellow scientists from other disciplines for the very same reason. So when the time comes for a journalist to report on a discovery in which the ideas might be obscure, the language intractable, and the discoverer inexperienced with public communication, the chances of error for the journalist are high. When combined with the journalistic tradition to not show your final draft to the person you interviewed, you've got a recipe for rampant misconception.

Meanwhile, the journalistic ethos to report "both sides" of a story implies an honesty of coverage intended to respect all views, no matter how much they differ from one another. In politics and religion, of course, people fight wars to declare who is right, and it's not the job of the dispassionate journalist to judge this in advance. In science, a journalist can either report on a discovery, or report on an emergent controversy. But when two scientists disagree, it's because they're in need of more—or better—data to show that one of them is wrong, or both of them are wrong. In the face of compelling evidence, scientists generally agree and move on, never seeing the conflict as an occasion to have killed one another.

For the journalist, however, it's tempting to allocate column space to fringe ideas, offering them as the "opposite" view to the mainstream. Who wouldn't want to cheer the underdog? But journalists don't always see the limits of this exercise. They know to not interview the person who is sure Earth is flat—just to get an opposing view. But at some point, awareness of the published scientific literature can inform these choices. If it's not a scientific controversy it shouldn't be a journalistic one. Moreover, urges are strong to write phrases such as, "these new discoveries will challenge cherished theories" or "scientists must now return to their drawing boards." Fact is, research scientists are always at the drawing board. And no theory is cherished, at least not in the way a pious person might revere a religious artifact. In science, all ideas, all hypotheses, all theories, are fair game to be overthrown. And some of our highest honors—including Nobel Prizes—go to those who succeed. Any resistance the scientific community might offer to ideas that conflict with our prevailing understanding of the universe will be proportional to the accumulated evidence that supports our understanding in the first place. In fewer words: Extraordinary ideas require extraordinary evidence.

Apart from these subtleties, most scientific discoveries overthrow no previous idea at all. They simply emerge from our ignorance, filling gaps in our base of cosmos knowledge and wisdom. These are new paradigms, without reference to whether any paradigm has shifted.

One of the great, uncelebrated challenges of science journalism has been to convey this textured, but booby-trapped, landscape on which science discovery unfolds. *The New York Times* has done this better than most. That's because its editorial policy, whether done on purpose or by accident, treats science not simply as intermittent news stories but as an enterprise to be monitored, tracked, reported on, and interpreted. How else could the discovery of the Higgs boson, an important but obscure subatomic particle with no obvious or foreseeable utility to the human condition, garner a banner headline on page one?

A book just as long, or longer, could easily be assembled from articles on advances in medicine, on the environment, or on engineering marvels. But this book explicitly honors the fact that physics lies at the beating heart of the operations of the universe; that physics is the founding science of 20th-century geopolitics; and that astrophysics is the founding science of wonder. Under that journalistic umbrella, science is part of human culture. Science is life.

During middle school and high school, I clipped every article on physics and astrophysics that appeared in the daily *New York Times* and in its Sunday magazine. Where others kept family photos, secure in albums, I mounted frail newsprint. Overall, the *Times'* coverage was quite good—brought about by a new generation of journalists devoted to reporting solely on science stories. There are scientifically literate journalists who nurture relationships with scientists, who review the relevant research literature, and who reach beyond the perfunctory press releases issued by science agencies. What a luxury for a newspaper to support such a staff. Regardless of what they occasionally got wrong, I wouldn't have to look far—in print, radio, or television—to discover how inept everyone else was at this task. So I came to greatly respect the science journalists of *The New York Times*. And I also came to see their telling of this timeless and epic adventure of cosmic discovery as a kind of time-capsule-in-the-making—a chronicle of our species' search for how the universe works and what our place within it might be.

Neil deGrasse Tyson
New York

INTRODUCTION
AN INVITATION TO OUR READERS

S cience is a search for knowledge about the natural world in which we live and a struggle to apply this knowledge in the world's dark corners of ignorance. It is a realm where researchers seek, see, find and ponder, often in a state of surprise or even awe.

This book invites you to enter into this quest. The portal is the archives of *The New York Times,* in particular its coverage of physics and astrophysics, realms that have dominated science and the popular imagination like few others in modern times.

Of course, a collection of articles from *The New York Times* cannot be a comprehensive guide to physics and cosmology—at least not for all of it. In the first place, *The New-York Daily Times* began publication in 1851, so we missed Archimedes, Newton, and other early giants of the field. Plus, for the first few decades of its life, the newspaper, which became *The New York Times* in 1857, was not the model of journalistic rectitude and thoroughness it strives to be today.

That began to change in 1896, when the Chattanooga newspaper publisher Adolph Ochs established a journalistic beachhead in the north by purchasing *The Times.* Ochs, who aspired to create "the model of decent and dignified journalism," especially valued coverage of subjects like science.

But in those days science writing was not well-developed. Reading our early coverage, it is clear we were earnest and trying hard, but all too often reporters and editors struggled to make sense of their arcane subject matter.

Perhaps more important, until the middle of the 20th century, the center of gravity for physics research was in England, Germany and elsewhere in Europe. Americans who engaged in what might loosely be described as physics research—people like Thomas Alva Edison, for example—focused on lightbulbs, phonographs and other practical uses of physics-related technology, rather than the field's theoretical underpinnings and implications.

Until the fascists drove Albert Einstein, Niels Bohr, Enrico Fermi and a host of other luminaries to the New World, the epicenter of theoretical physics was an ocean and more away. Often, important findings were published in European journals, in European languages, like German, that our journalists could not read. So physics was not a field we followed closely.

For example, we had little to say when Albert Michelson and Edward Morley (Americans!) devised an experiment to detect the motion of light relative to the "ether," the substance believed—until then—to fill the seemingly empty vastness of the universe. By failing to detect any difference in the speed of light, regardless of how and where it was moving, they struck a lethal blow to the ether theory.

That was a landmark in physics and it led Michelson to a Nobel Prize. More importantly, it led Einstein to a line of research that culminated in the theory of relativity.

But if you search *The Times* archives, you won't find an on-the-spot report. Here, in its entirety, is the earliest reference to Michelson and Morley, months after they performed their landmark work:

INVESTIGATING LIGHT WAVES

CLEVELAND, Ohio, March 4—Scientists are watching with great interest the joint experiments of Profs. Michelson of Case School of Applied Science and Morley of Adelbert College in an effort to determine the feasibility of making the wave length of light the ultimate standard of accurate measurement. Previous to their experiments, the limit at which interference of light had been secured was 50,000 wave lengths. On Friday they secured such interference at 250,000, and think the ultimate limit not yet reached.

In 1905, Einstein's so-called "miracle year," he produced five major papers, including one on the theory of special relativity and another on the equivalence of matter and energy. Yet a reference to "Albert Einstein" in *The Times* does not turn up until 1919. It occurs in a story with this (unintentionally) hilarious headline:

LIGHTS ALL ASKEW IN THE HEAVENS

The article is about the way light is bent when it travels past massive bodies like planets. The observations confirm "the Einstein theory of light," the story says. But it identifies the great physicist only as "a Swiss citizen, about 50 years of age." And its first paragraph is practically an admission of journalistic incompetence:

*Efforts made to put in words intelligible to the non-scientific
public the Einstein theory of light proved by the eclipse
expedition so far have not been very successful.*

Our inability to understand what is going on doesn't matter, the article
informs us, because the new findings have no practical implications for most
people. It quotes an astronomer: "Only astronomers are affected."

In the early decades of the 20th century, *The Times* persistently misspelled
the first name of the Danish theorist of the atom, Niels Bohr, and in one
memorable burst identified the German researcher Max Planck, the father of
quantum physics, as *Professor Planc* and called Werner Heisenberg, scholar of
quantum uncertainty, *August.*

Even well into the 20th century, the paper appears to catch up with things
months—or even years—after they have been announced to the scientific
community. Sometimes circumstances excused the lapses. For example, William
L. Laurence, a *Times* science writer, was among the first to report on the theoretical
usefulness of uranium in weapons (the experts he quoted pooh-poohed it). But
he had nothing to say about the development of the atomic bomb until after it was
dropped on Japan.

But then he wrote a lot. At the request of the War Department, he had been
embedded with the project and actually observed the first nuclear explosion—the
Trinity test in New Mexico—of a plutonium-fueled bomb. And he flew on the
mission that dropped the second bomb on the Japanese city of Nagasaki and
reported on it.

Laurence is famously (and affectionately) remembered nowadays in the
science department of *The Times* as "Atomic Bill." But then, as now, close ties to
the people we cover raises eyebrows. There are those who say he got too tight with
the War Department and, as a result, played down the horrors of nuclear warfare.
Five years after the war, however, he reported on the possibility that a hydrogen
bomb could be rigged to exterminate the entire world population, or most of it.

The Times overlooked some discoveries or inventions, or failed to discern
their significance. The first article explaining the potential usefulness of the
"transistor" appeared in print two years after the device was invented at Bell Labs,
the research arm of AT&T. Mention of the devices appeared near the bottom of
an article about military research on radios.

Eventually, though, *The Times* got better at being right on top of things. When
Bell Labs researchers discovered what turned out to be microwave background

radiation—echoes of the Big Bang—there are those who say they did not fully grasp its impact until they read Walter Sullivan's description in *The Times*.

When you look seriously at our coverage, something else jumps out: it is practically devoid of women, whose achievements get short shrift, at least until relatively late in the 20th century. Marie Curie is more notable in pages of *The Times* as the wife of the researcher Pierre Curie—and as a principal in a sex scandal after he died—than as the discoverer of radium. Only at her death do we describe her, accurately, as a "martyr to science."

Lise Meitner, whose discoveries about splitting the atom led to development of the atomic bomb, barely makes it into our pages, and then almost parenthetically.

Perhaps that is because, until relatively recently, journalism at *The New York Times* was a man's game. This book contains scores of articles and only a handful are by women.

Happily, women are on the cutting edge of physics and cosmology research and are written about with respect, even though *The Times* is still a bit short of women writing on physics and astrophysics.

Things generally began improving after World War II. By that time the atomic bomb, radar and other wartime advances had drawn attention to physics and, indeed, to science generally. And by then *The Times* had earned a reputation as a source of high-quality science reporting.

From those years forward, the chief challenge in compiling a book like this is choosing what to reprint from the vast wealth of possibilities. Inevitably, readers will disagree with some of the choices and omissions, but in general the shape of the field itself offered some guidance.

The chief accomplishments of modern physics are in the realm of the very small (atoms and subatomic particles and the forces that act on them) and the very large (stars, galaxies, the universe itself). In a sense, the quest of modern physics is to link our understanding of those two realms—something researchers and theorists have yet to achieve.

Their quest is the heart of this book, so most of the articles here describe researchers' probing into the mysteries of the atom and its subatomic components, or the gradual unraveling of the secrets of the universe—what it's made of and what holds it together.

In both realms, the answers are never as simple as they seem.

Chapter 1, called "The Nature of Matter," begins with an assessment of particle physics in 2000, the 100th anniversary of the introduction of the theory of quantum mechanics, which would overturn the field. If you don't know much about physics, this article will put you in the picture, so that you can better enjoy following early efforts to determine what comprises and animates the physical world.

This effort gradually produced what we know today as the Standard Model, the framework of quarks that make up particles like protons and neutrons; leptons like the electron; and the bosons that tie the whole business together. For example, the Higgs boson, discovered with much fanfare in the summer of 2012, endows things with mass.

Chapter 2, "The Practical Atom," follows what happens when theory goes into practice, from sending wireless signals and radar to the atomic bomb, brain surgery without a knife, and working transistors made from single atoms. Finally, in chapter 3, "The Fate of the Universe," you will encounter astronomers' first modern efforts to determine where our planet fits in the cosmos and the gradual embrace of the big bang theory—the idea that the universe and everything in it was created in an instant about 13.5 billion years ago. The working out of that event continues as well.

Here you can read about the discovery of distant galaxies, black holes, quasars and other celestial objects. There is much, of course, that you will not read.

Moon missions, for instance, or the debate over whether Pluto is really a planet (evidently not), or the so-far-disappointing search for intelligent life elsewhere in the universe. You will not read about the prosecution of J. Robert Oppenheimer, who led the United States's atomic bomb effort and whom red-baiters later accused of being a security risk.

And we offer only a hint of physics in communication, medicine, solar panels, computers, the Internet and a host of other useful technologies that are, in their way, triumphs of the engineering of physics. Recounting all of them is simply beyond this book's scope. Even a small sample would overwhelm this collection.

On the other hand, we do devote some of our pages to a kind of engineering, the engineering required to design and build the instruments we use to peer into the tiny interstices of the atom and into the vastness of space. Few moments in physics are more inspiring than when an electron microscope discerns the outline of a single atom, or when an instrument like the Hubble Space Telescope—its faulty vision restored by a mission by space shuttle astronauts—shows us how things look in distant nebulae.

And because the fingerprints of theoretical physics are all over modern life, readers will find discussions of practical matters: atomic physics at war; the uses of physics in energy production; physics in medicine, starting with the early days of X-rays; the physics of concert halls and traffic jams, and what happened when Einstein invented a refrigerator.

Look for articles about so-called "cold fusion" for examples of how the scientific method works. When researchers made the startling announcement that they had achieved energy-producing fusion in a simple tabletop apparatus, it held immense promise for practical implications. But some (including the estimable *Times* reporter Malcolm W. Browne) were skeptical until colleagues could replicate the result. When they did, those other researchers began to come around—and *The Times* put the story on the front page. Then, of course, it turned out the replication was a laboratory "artifact"—a mistake—and the theory went the way of the ether and other bad physics ideas.

Some would say the same fate awaits "string theory," at one time the favorite candidate of deep physics thinkers seeking to explain the nature of physical reality. See what you think.

What comes through from the beginning of our coverage is the passion of the researchers we cover, their excitement at being at the cutting edge of the effort to explain the universe and our place in it, and their complete certainty that, succeed or fail, the work is important.

This collection ends with a look into the future. The fate of the Webb space-based telescope, NASA's successor to Hubble, is uncertain. With the discovery of the Higgs, which took place at a giant particle accelerator in Europe, American physicists worry that once again the center of gravity of physics is shifting to Europe. But despite funding difficulties, occasional political indifference or even opposition to basic science, the enterprise thrives.

When the great theorist Max Planck was a young man, at the end of the 19th century, his teachers discouraged him from a career in physics on the grounds that all the great discoveries had already been made. Needless to say, Planck personally disproved that idea.

Today, even as telescopes peer back to the very beginning of time, even as researchers discover the Higgs, they create opportunities to ask new questions and discern new answers. The more we learn about physics and cosmology, the more there is to find out. So take this journey. You will not be disappointed.

Cornelia Dean

CHAPTER 1

The Nature of Matter

Quantum Theory Tugged,
and All of Physics Unraveled

By DENNIS OVERBYE

They tried to talk Max Planck out of becoming a physicist, on the grounds that there was nothing left to discover. The young Planck didn't mind. A conservative youth from the south of Germany, a descendant of church rectors and professors, he was happy to add to the perfection of what was already known.

Instead, he destroyed it, by discovering what was in effect a loose thread that when tugged would eventually unravel the entire fabric of what had passed for reality.

As a new professor at the University of Berlin, Planck embarked in the fall of 1900 on a mundane-sounding calculation of the spectral characteristics of the glow from a heated object. Physicists had good reason to think the answer would elucidate the relationship between light and matter as well as give German industry a leg up in the electric light business. But the calculation had been plagued with difficulties.

Planck succeeded in finding the right formula, but at a cost, as he reported to the German Physical Society on Dec. 14. In what he called "an act of desperation," he had to assume that atoms could only emit energy in discrete amounts that he later called quanta (from the Latin *quantus* for "how much") rather than in the continuous waves prescribed by electromagnetic theory. Nature seemed to be acting like a fussy bank teller who would not make change, and would not accept it either.

That was the first shot in a revolution. Within a quarter of a century, the common sense laws of science had been overthrown. In their place was a bizarre set of rules known as quantum mechanics, in which causes were not guaranteed to be linked to effects; a subatomic particle like an electron could be in two places at once, everywhere or nowhere until someone measured it; and light could be a wave or a particle.

Niels Bohr, a Danish physicist and leader of this revolution, once said that a person who was not shocked by quantum theory did not understand it.

This week, some 700 physicists and historians are gathering in Berlin, where Planck started it all 100 years ago, to celebrate a theory whose meaning they still do not understand but that is the foundation of modern science. Quantum effects

2

are now invoked to explain everything from the periodic table of the elements to the existence of the universe itself.

Fortunes have been made on quantum weirdness, as it is sometimes called. Transistors and computer chips and lasers run on it. So do CAT scans and PET scans and MRI machines. Some computer scientists call it the future of computing, while some physicists say that computing is the future of quantum theory.

"If everything we understand about the atom stopped working," said Leon Lederman, former director of the Fermi National Accelerator Laboratory, "the GNP would go to zero."

The revolution had an inauspicious start. Planck first regarded the quantum as a bookkeeping device with no physical meaning. In 1905, Albert Einstein, then a patent clerk in Switzerland, took it more seriously. He pointed out that light itself behaved in some respects as if it were composed of little energy bundles he called *lichtquanten*. (A few months later Einstein invented relativity.)

He spent the next decade wondering how to reconcile these quanta with the traditional electromagnetic wave theory of light. "On quantum theory I use up more brain grease than on relativity," he told a friend.

The next great quantum step was taken by Bohr. In 1913, he set forth a model of the atom as a miniature solar system in which the electrons were limited to specific orbits around the nucleus. The model explained why atoms did not just collapse—the lowest orbit was still some slight distance from the nucleus. It also explained why different elements emitted light at characteristic wavelengths—the orbits were like rungs on a ladder and those wavelengths corresponded to the energy released or absorbed by an electron when it jumped between rungs.

But it did not explain why only some orbits were permitted, or where the electron was when it jumped between orbits. Einstein praised Bohr's theory as "musicality in the sphere of thought," but told him later, "If all this is true, then it means the end of physics."

While Bohr's theory worked for hydrogen, the simplest atom, it bogged down when theorists tried to calculate the spectrum of bigger atoms. "The whole system of concepts of physics must be reconstructed from the ground up," Max Born, a physicist at University of Göttingen, wrote in 1923. He termed the as-yet-unborn new physics "quantum mechanics."

Boy's Mechanics

The new physics was born in a paroxysm of debate and discovery from 1925 to 1928 that has been called the second scientific revolution. Wolfgang Pauli, one of its ringleaders, called it "boy's mechanics," because many of the physicists, including himself, then 25, Werner Heisenberg, 24, Paul Dirac, 23, Enrico Fermi, 23, and Pascual Jordan, 23, were so young when it began.

Bohr, who turned 40 in 1925, was their father-confessor and philosopher king. His new institute for theoretical physics in Copenhagen became the center of European science.

The decisive moment came in the fall of 1925 when Heisenberg, who had just returned to the University of Göttingen after a year in Copenhagen, suggested that physicists stop trying to visualize the inside of the atom and instead base physics exclusively on what can be seen and measured. In his "matrix mechanics," various properties of subatomic particles could be computed—but, disturbingly, the answers depended on the order of the calculations.

In fact, according to the uncertainty principle, which Heisenberg enunciated two years later, it was impossible to know both the position and velocity of a particle at once. The act of measuring one necessarily disturbed the other.

Physicists uncomfortable with Heisenberg's abstract mathematics took up with a friendlier version of quantum mechanics based on the familiar mathematics of waves. In 1923, the Frenchman Louis de Broglie had asked in his doctoral thesis, if light could be a particle, then why couldn't particles be waves?

Inspired by de Broglie's ideas, the Austrian Erwin Schrödinger, then at the University of Zurich and, at 38, himself older than the wunderkind, sequestered himself in the Swiss resort of Arosa over the 1925 Christmas holidays with a mysterious woman friend and came back with an equation that would become the yin to Heisenberg's yang.

In Schrödinger's equation, the electron was not a point or a table, but a mathematical entity called a wave function, which extended throughout space. According to Born, this wave represented the probability of finding the electron at some particular place. When it was measured, the particle was usually in the most likely place, but not guaranteed to be, even though the wave function itself could be calculated exactly.

Born's interpretation was rapidly adopted by the quantum gang. It was a pivotal moment because it enshrined chance as an integral part of physics and of nature.

"The motion of particles follows probability laws, but the probability itself propagates according to the law of causality," he explained.

That was not good enough for Einstein. "The theory produces a good deal but hardly brings us closer to the secret of the Old One," Einstein wrote in late 1926. "I am at all events convinced that he does not play dice."

Heisenberg called Schrödinger's theory "disgusting"—but both versions of quantum mechanics were soon found to be mathematically equivalent.

Uncertainty, which added to the metaphysical unease surrounding quantum physics, was followed in turn in 1927 by Bohr's complementarity principle. Ask not whether light was a particle or a wave, said Bohr, asserting that both concepts were necessary to describe nature, but that since they were contradictory, an experimenter could choose to measure one aspect or the other but not both. This was not a paradox, he maintained, because physics was not about things but about the results of experiments.

Complementarity became the cornerstone of the Copenhagen interpretation of quantum mechanics—or as Einstein called it, "the Heisenberg-Bohr tranquilizing philosophy."

A year later, Dirac married quantum mechanics to Einstein's special relativity, in the process predicting the existence of antimatter. (The positron, the antiparticle to the electron, was discovered four years later by Carl Anderson.)

Dirac's version, known as quantum field theory, has been the basis of particle physics ever since, and signifies, in physics histories, the end of the quantum revolution. But the fight over the meaning of the revolution had just barely begun, and it has continued to this day.

Quantum Wars

The first and greatest counterrevolutionary was Einstein, who hoped some deeper theory would rescue God from playing dice. In the fall of 1927 at a meeting in Brussels, Einstein challenged Bohr with a series of *gedanken,* or thought experiments, designed to show that quantum mechanics was inconsistent. Bohr, stumped in the morning, always had an answer by dinner.

Einstein never gave up. A 1935 paper written with Boris Podolsky and Nathan Rosen described the ultimate quantum *gedanken,* in which measuring

5

a particle in one place could instantly affect measurements of the other particle, even if it was millions of miles away. Was this any way to run a universe?

Einstein called it "spooky action at a distance."

Modern physicists who have managed to create this strange situation in the laboratory call it "entanglement."

Einstein's defection from the quantum revolution was a blow to his more conservative colleagues, but he was not alone. Planck also found himself at odds with the direction of the revolution and Schrödinger, another of "the conservative old gentlemen," as Pauli once described them, advanced his cat *gedanken* experiment to illustrate how silly physics had become.

According to the Copenhagen view, it was the act of observation that "collapsed" the wave function of some particle, freezing it into one particular state, a location or velocity. Until then, all the possible states of the particle coexisted, like overlapping waves, in a condition known as quantum superposition.

Schrödinger imagined a cat in a sealed container in which the radioactive decay of an atom would trigger the release of cyanide, killing the cat. By the rules of quantum mechanics the atom was both decayed and not decayed until somebody looked inside, which meant that Schrödinger's poor cat was both alive and dead.

This seemed to be giving an awful lot of power to the "observer." It was definitely no way to run a universe.

Over the years physicists have proposed alternatives to the Copenhagen view. Starting in 1952, when he was at Princeton, the physicist David Bohm, who died in 1982, argued for a version of quantum mechanics in which there was a deeper level, a so-called quantum potential or "implicate order," guiding the apparent unruliness of quantum events.

Another variant is the many-worlds hypothesis developed by Hugh Everett III and John Wheeler, at Princeton in 1957. In this version the wave function does not collapse when a physicist observes an electron or a cat; instead it splits into parallel universes, one for every possible outcome of an experiment or a measurement.

Shut Up and Compute

Most physicists simply ignored the debate about the meaning of quantum theory in favor of using it to probe the world, an attitude known as "shut up and compute."

Pauli's discovery that no two electrons could share the same orbit in an atom led to a new understanding of atoms, the elements and modern chemistry.

Quantum mechanics split the atom and placed humanity on the verge of plausible catastrophe. Engineers learned how to "pump" electrons into the upper energy rungs in large numbers of atoms and then make them all dump their energy all at once, giving rise to the laser. And as Dr. Lederman said in an interview, "The history of transistors is the history of solving Schrödinger's equation in various materials."

Quantum effects were not confined to the small. The uncertainty principle dictates that the energy in a field or in empty space is not constant, but can fluctuate more and more wildly the smaller the period of time that one looks at it. Such quantum fluctuations during the big bang are now thought to be the origin of galaxies.

In some theories, the universe itself is a quantum effect, the result of a fluctuation in some sort of preuniversal nothingness. "So we take a quantum leap from eternity into time," as the Harvard physicist Sidney Coleman once put it.

Where the Weirdness Goes

Bohr ignored Schrödinger's cat, on the basis that a cat was too big to be a quantum object, but the cat cannot be ignored anymore. In the last three decades, the *gedanken* experiments envisioned by Einstein and his friends have become "ungedankened," bringing the issues of their meaning back to the fore.

Last summer, two teams of physicists managed to make currents go in two directions at once around tiny superconducting loops of wire—a feat they compared to the cat. Such feats, said Wojciech Zurek, a theorist at Los Alamos National Laboratory, raise the question of why we live in a classical world at all, rather than in a quantum blur.

Bohr postulated a border between the quantum and classical worlds, but theorists prefer that there be only one world that can somehow supply its own solidity. That is the idea behind a new concept called decoherence, in which the interaction of wave functions with the environment upsets the delicate balance of quantum states and makes a cat alive or dead but not in between.

"We don't need an observer, just some 'thing' watching," Dr. Zurek explained. When we look at something, he said, we take advantage of photons, the carriers of light, which contain information that has been extracted from the object. It is this loss of information into the environment that is enough to crash the wave function, Dr. Zurek said.

Decoherence, as Dr. Zurek noted, takes the observer off a pedestal and relieves quantum theory of some of its mysticism, but there is plenty of weirdness left. Take the quantum computer, which Dr. Lederman refers to as "a kinder, gentler interpretation of quantum spookiness."

Ordinary computers store data and perform computations as a series of "bits," switches that are either on or off, but in a quantum computer, due to the principle of superposition, so-called qubits can be on and off at the same time, enabling them to calculate and store myriads of numbers at a time.

In principle, according to David Deutsch, an Oxford University researcher who is one of quantum computing's more outspoken pioneers, a vast number of computations, "potentially more than there are atoms in the universe," could be superposed inside a quantum computer to solve problems that would take a classical computer longer than the age of the universe.

In the minds of many experts, this kind of computing illuminates the nature of reality itself.

Dr. Deutsch claims that the very theory of a quantum computer forces physicists to take seriously the many-worlds interpretation of quantum theory. The amount of information being processed in these parallel computations, he explains, is more than the universe can hold. Therefore, they must be happening in other parallel universes out in the "multiverse," as it is sometimes called.

"There is no other theory of what is happening," he said. The world is much bigger than it looks, a realization that he thinks will have a psychological impact equivalent to the first photographs of atoms. Indeed, for Dr. Deutsch there seems to be a deep connection between physics and computation. The structure of the quantum computer, he says, consists of many things going on at once, lots or parallel computations. "Any physical process in quantum mechanics," he said. "consists of classical computations going on in parallel."

"The quantum theory of computation is quantum theory," he said.

The Roots of Weirdness

Quantum mechanics is the language in which physicists describe all the phenomena of nature save one, namely gravity, which is explained by Einstein's general theory of relativity. The two theories—one describing a discontinuous "quantized" reality and the other a smoothly curving space-time continuum—are mathematically incompatible, but physicists look to their eventual marriage, a so-called quantum gravity.

"There are different views as to whether quantum theory will encompass gravity or whether both quantum theory and general relativity will have to be modified," said Lee Smolin, a theorist at Penn State.

Some groundwork was laid as far back as the 1960s by Dr. Wheeler, 89, who has argued quantum theory with both Einstein and Bohr. Even space and time, Dr. Wheeler has pointed out, must ultimately pay their dues to the uncertainty principle and become discontinuous, breaking down at very small distances or in the compressed throes of the big bang into a space-time "foam."

Most physicists today put their hope for such a theory in superstrings, an ongoing and mathematically dense effort to understand nature as consisting of tiny strings vibrating in 10-dimensional space.

In a sort of missive from the front, Edward Witten of the Institute for Advanced Study in Princeton, New Jersey, said recently that so far quantum mechanics appeared to hold up in string land exactly as it was described in textbooks. But, he said in an e-mail message, "Quantum mechanics is somehow integrated with geometry in a way that we don't really understand yet."

The quantum is mysterious, he went on, because it goes against intuition. "I am one of those who believes that the quantum will remain mysterious in the sense that if the future brings any changes in the basic formulation of quantum mechanics, I suspect our ordinary intuition will be left even farther behind."

Intuition notwithstanding, some thinkers wonder whether or not quantum weirdness might, in fact, be the simplest way to make a universe. After all, without the uncertainty principle to fuzz the locations of its buzzing inhabitants, the atom would collapse in an electromagnetic heap. Without quantum fluctuations to roil the unholy smoothness of the big bang, there would be no galaxies, stars or friendly warm planets. Without the uncertainty principle to forbid nothingness, there might not even be a universe.

"We will first recognize how simple the universe is," Dr. Wheeler has often said, "when we recognize how strange it is." Einstein often said that the question that really consumed him was whether God had any choice in creating the world. It may be in the end that we find out that for God, the only game in town was a dice game.

—*December 12, 2000*

Investigating Light Waves

Scientists are watching with great interest the joint experiments of Profs. Michelson of Case School of Applied Science and Morley of Adelbert College in an effort to determine the feasibility of making the wavelength of light the ultimate standard of accurate measurement. Previous to their experiments, the limit at which interference of light had been secured was 50,000 wavelengths. On Friday they secured such interference at 250,000, and think the ultimate limit not yet reached.

—March 5, 1888

Prof. Röntgen's X-Rays

The preliminary communication of Prof. Wilhelm Conrad Röntgen to the Würzberg Physico-Medical Society of his discovery of a new form of radiant energy appears this week translated in full in several of the English papers. As the chief interest of men of science is centered in the question of the nature of the rays, those portions of Prof. Röntgen's paper which deal with this aspect of the subject are here reproduced in full.

The name given by Prof. Röntgen to the newly discovered form of radiant energy is X-rays. The translation appended was made by Arthur Stanton, and appears in the current number of *Nature*. After describing his experiments in making shadow photographs of various substances, Prof. Röntgen says:

7. After my experiments on the transparency of increasing thicknesses of different media, I proceeded to investigate whether the X-rays could be deflected by a prism. Investigations with water and carbon bisulphide in mica prisms of 30° showed no deviation either on the photographic or the fluorescent plate. For comparison, light rays were allowed to fall on the prism as the apparatus was set up for the experiment. They were deviated 10mm. and 20 mm. respectively in the case of the two prisms.

With prisms of ebonite and aluminum I have obtained images on the photographic plate which point to a possible deviation. It is, however, uncertain, and at most would point to a refractive index 1.05. No deviation can be observed by means of the fluorescent screen. Investigations with the heavier metals have not as yet led to any result, because of their small transparency and the consequent enfeebling of the transmitted rays.

On account of the importance of the question it is desirable to try in other ways whether the X-rays are susceptible of refraction. Finely powered bodies allow in thick layers but little of the incident light to pass through, in consequence of refraction and reflection. In the case of the X-rays, however, such layers of powder are for equal masses of substance equally transparent with the coherent solid itself. Hence we cannot conclude any regular reflection or refraction of the X-rays. The research was conducted by the aid of finely powdered rock salt, fine electrolytic silver powder, and zinc dust already many times employed in chemical work. In all these cases the result, whether by the fluorescent screen or the photographic method, indicated no difference in transparency between the powder and the coherent solid.

It is, hence, obvious that lenses cannot be looked upon as capable of concentrating the X-rays; in effect, both an ebonite and a glass lens of large size prove to be without action. The shadow photograph of a round rod is darker in the middle than at the edge; the image of a cylinder filled with a body more transparent that its walls exhibits the middle brighter than the edge.

8. The preceding experiments and others which I pass over point to the rays being incapable of regular reflection. It is, however, well to detail an observation which at first sight seemed to lead to an opposite conclusion.

I exposed a plate, protected by a black paper sheath, to the X-rays, so that the glass side lay next to the vacuum tube. The sensitive film was partly covered with star-shaped pieces of platinum, lead, zinc, and aluminum. On the developed negative the star-shaped impression showed dark under platinum, lead, and more markedly under zinc; the aluminum gave no image. It seems, therefore, that these three metals can reflect the X-rays; as, however, another explanation is possible, I repeated the experiment with only this difference, that a film of thin aluminum foil was interposed between the sensitive film and the metal stars. Such an aluminum plate is opaque to ultraviolet rays, but transparent to X-rays. In the result the images appeared as before, this pointing still to the existence of reflection at metal surfaces.

If one considers this observation in connection with others—namely, on the transparency of powders, and on the state of the surface not being effective in altering the passage of the X-rays through a body—it leads to the probable conclusion that regular reflection does not exist, but that bodies behave to the X-rays as turbid media to light.

Since I have obtained no evidence of refraction at the surface of different media, it seems probable that the X-rays move with the same velocity in all bodies, and in a medium which penetrates everything, and in which the molecules of bodies are imbedded. The molecules obstruct the X-rays, the more effectively as the density of the body concerned is greater.

9. It seemed possible that the geometrical arrangement of the molecules might affect the action of a body upon the X-rays, so that, for example, Iceland spar might exhibit different phenomena according to the relation of the surface of the plate to the axis of the crystal. Experiments with quartz and Iceland spar on this point lead to a negative result.

10. It is known that [Philipp Eduard Anton von] Lenard, in his investigations on cathode rays, has shown that they belong to the ether, and can pass through all bodies. Concerning the X-rays the same may be said.

In his latest work, Lenard has investigated the absorption coefficients of various bodies for the cathode rays, including air at atmospheric pressure, which gives 4.10, 3.40, 3.10 for 1 cm., according to the degree of exhaustion of the gas in the discharge tube. To judge from the nature of the discharge, I have worked at about the same pressure, but occasionally at greater or small pressures. I find, using a Weber photometer, that the intensity of the fluorescent light varies nearly as the inverse square of the distance between screen and discharge tube. This result is obtained from three very consistent sets of observations at distances of 100 and 200 mm. Hence, air absorbs the X-rays much less than the cathode rays. This result is in complete agreement with the previously described result, that the fluorescence of the screen can still be observed at two meters from the vacuum tube. In general, other bodies behave like air; they are more transparent for the X-rays than for the cathode rays.

11. A further distinction, and a noteworthy one, results from the action of a magnet. I have not succeeded in observing any deviation of the X-rays even in very strong magnetic fields.

The deviation of cathode rays by the magnet is one of their peculiar characteristics; it has been observed by [Heinrich] Hertz and Lenard that several kinds of cathode rays exist, which differ by their power of exciting phosphorescence, their susceptibility of absorption, and their deviation by the magnet; but a notable deviation has been observed in all cases which have yet been investigated, and I think that such deviation affords a characteristic not to be set aside lightly.

12. As a result of many researches, it appears that the place of most brilliant phosphorescence of the walls of the discharge tube is the chief seat whence the X-rays originate and spread in all directions; that is, the X-rays proceed from the front where the cathode rays strike the glass. If one deviates the cathode rays within the tube by means of a magnet, it is seen that the X-rays proceed from a new point—i.e., again from the end of the cathode rays.

Also for this reason the X-rays, which are not deflected by a magnet, cannot be regarded as cathode rays which have passed through the glass, for that passage cannot, according to Lenard, be the cause of the different deflection of the rays. Hence I conclude that the X-rays are not identical with the cathode rays, but are produced from the cathode rays at the glass surface of the tube.

13. The rays are generated not only in glass. I have obtained them in an apparatus closed by an aluminum plate 2 mm. thick. I propose later to investigate the behavior of other substances.

14. The justification of the term "rays," applied to the phenomena, lies partly in the regular shadow pictures produced by the interposition of a more or less permeable body between the source and a photographic plate or fluorescent screen.

I have observed and photographed many such shadow pictures. Thus I have an outline of part of a door covered with lead paint; the image was produced by placing the discharge tube on one side of the door, and the sensitive plate on the other. I have also a shadow of the bones of the hand, of a wire wound upon a bobbin, of a set of weights in a box, of a compass card and needle completely enclosed in a metal case, of a piece of metal where the rays show the want of homogeneity, and of other things.

For the rectilinear propagation of the rays I have a pinhole photograph of the discharge apparatus, covered with black paper. It is faint, but unmistakable.

15. Researches to investigate whether electrostative forces act on the X-rays are begun, but not yet concluded.

16. If one asks, what then are these X-rays; since they are not cathode rays, one might suppose, from their power of exciting fluorescence and chemical action, them to be due to ultraviolet light. In opposition to this view, a weighty set of considerations presents itself. If X-rays be, indeed, ultraviolet light, then that light must possess the following properties:

(a) It is not refracted in passing from air into water, carbon bisulphide, aluminum, rock salt, glass, or zinc. (b) It is incapable of regular reflection at the surfaces of the above bodies. (c) It cannot be polarized by any ordinary polarizing media. (d) The absorption by various bodies must depend chiefly on their density.

That is to say, these ultraviolet rays must behave quite differently from the visible, infra-red, and hitherto-known ultraviolet rays.

These things appear so unlikely that I have fought for another hypothesis.

A kind of relationship between the new rays and light rays appears to exist; at least the formation of shadows, fluorescence, and the production of chemical action point in this direction. Now, it has been known for a long time that, besides the transverse vibrations which account for the phenomena of light, it is possible that longitudinal vibrations should exist in the ether, and, according to the view of some physicists, must exist. It is granted that their existence has not yet been made clear, and their properties are not experimentally demonstrated. Should not the new rays be ascribed to longitudinal waves in the ether?

I must confess that I have in the course of this research made myself more and more familiar with this thought, and venture to put the opinion forward, while I am quite conscious that the hypothesis advanced still requires a more solid foundation.

The *London Electrician,* in its last number, points out briefly the similarity and difference between the Röntgen rays and the Lenard, or true cathode rays, as follows:

It may not be without interest at the present moment to recall the main points of difference and of similarity between Röntgen rays and Lenard rays— to use two brief and convenient expressions. Röntgen rays are not deflected by a magnet; Lenard rays are. Röntgen rays suffer far less absorption and diffusion than Lenard rays. Lenard found that his cathode rays failed to pass through anything but the thinnest soap films, glass, and aluminum foil, etc.; the Röntgen variety will traverse several centimeters of wood and several millimeters of metal or glass. Röntgen was able to take "shadowgraphs" and detect fluorescence 200cm. away from the discharge tube; 6 cm. or 8 cm. were enough to wipe out Lenard rays in air at atmospheric pressure, and even in hydrogen gas at only 0.0164 mm. pressure, the "radiation length" for cathode rays was only 130 cm.; hydrogen at atmospheric pressure behaving as a decidedly turbid medium. These are, however, rather differences in degree than in kind. Lenard rays emanate, of course, from the cathode itself, but Röntgen rays, according to their discoverer, start from the luminescent spot on the glass wall of the discharge tube, at which the cathode rays terminate.

The points of similarity between Röntgen and Lenard rays are their photographic activity, their rectilinear propagation (as evidenced by the sharp shadows cast), and the fact that in both cases it would seem the total mass of molecules contained in unit volume of any substance practically determines its transparency. All things tend to show that we are on the verge of a great scientific discovery, which may oblige us, *nolens volens,* to "rearrange our ideas."

Even the *Lancet* has succumbed to the photographic evidence presented to it, and says in an editorial paragraph in its current issue:

The application of this remarkable phenomenon to the discovery of bullets and abnormalities in the structure of bone has already been made, with very promising results. It is reported already from Vienna, for instance, that photographic pictures taken by this means showed with the greatest clearness and precision the injuries caused by a revolver shot in the hand of a man and the position of the bullet. In another case, that of a girl, the position and nature of a malformation in the left foot were ascertained. As the conditions of these experiments become perfected with regard to the source of the radiations, there is no doubt that the result will be even more brilliant.

An abstruse extract taken from Lord Kelvin's Baltimore lectures, delivered at Johns Hopkins University in 1884, which is quoted in *Nature*, seems to show that he anticipated some discovery which would prove that ether was compressible like ordinary forms of matter. The point is to show that the new energy behaves more like sound than like light. It is worthy of note that Prof. Sylvanus Thompson has already termed the new form of energy "ultraviolet sound."

—February 5, 1896

Character of the X-Rays

Ogden N. Rood, professor of physics in Columbia College, was asked yesterday by a reporter for *The New-York Times* whether recent experiments with the Röntgen rays had altered his views as to the nature of the phenomena.

It will be remembered that before full details of Röntgen's discovery reached here Prof. Rood tentatively expressed his view that the rays were of the same nature as light waves, only much shorter than any previously known.

Since that time numerous theories have been suggested by experimenters on both sides of the Atlantic. One of these, adapted from William Crookes, is that the rays represent a bombardment of the molecules in a highly rarefied atmosphere, and that this bombardment is communicated to other particles of matter outside the enclosing envelope. Prof. Rood thinks this theory in its entirety is not necessary to account for the Röntgen phenomena, and is extremely improbable in any case.

After glancing through the article by Nikola Tesla in *The Electrical Review*, also printed in part in *The New-York Times* Wednesday last, Prof. Rood dictated the following statement of his views:

> It is never proper to frame a new hypothesis in order to account for discoveries when an old and well-established theory is capable of giving them an explanation. Our knowledge of the waves of light, of radiant heat, and of the ultra violet rays is confined within narrow limits.
>
> Thus far, according to our previous experience, the shorter the length of the wave the more it will be bent, or refracted, by substances in general. It, however, does not at all follow that this will hold true if the length of the wave is shortened to a degree beyond our experience.
>
> On the other hand, it is highly probable that if this shortening process goes on till the length of the wave has become of the same order of magnitude with the molecules of matter that the effect in the matter of refraction will diminish. This would account for the fact that no one has succeeded thus far in refracting or reflecting the Röntgen rays. In my own experiments I have so far found them to act in the same general way with waves of ordinary light, and I am inclined to believe that they are simply waves of light of unprecedented shortness.

This view necessarily excludes all idea of a molecular bombardment outside of the Crookes tube. I believe that it will be eventually demonstrated that the Röntgen rays are capable of reflection and refraction, although it will require more delicate work than has thus far been extended in this direction.

Experiments at Princeton Club

The monthly meeting of the Princeton Club of New-York was held at the Brunswick Hotel last night. President Hugh L. Cole presided, and as soon as the business of the club had been disposed of Prof. W. F. Magie, PhD, delivered a lecture on the development of the Röntgen rays.

The medical possibilities of the rays he illustrated with lantern slides, which threw upon the big sheet enormous pictures of hands, feet, elbows, and joints of the human body, showing bullets, deformities, and malformations. A successful photograph was also taken of a hand. The exposure was twenty minutes, and the negative was developed under the eyes of those present.

"President Cole, perhaps you will give me your hand?" ventured the lecturer.

"With pleasure," said Mr. Cole, advancing, and shaking hands with Prof. Magie, amid general laughter.

"No, I mean to be photographed," explained the professor.

Mr. Cole sat beside the Crookes tube and laid his hand on the plate. There was a ring on his third finger. "Just sit like that for twenty minutes, please," was the order.

"Will someone put a coin in this aluminum box?" asked the professor next.

Job Hedges, who sat in the front row, volunteered. He dropped a dime into the box, which, together with a key and an open pocket knife, were placed upon a second plate.

The current was then turned on in the Crookes tube, which instantly lit up with a greenish light. While the experiment was in progress pictures were thrown upon the big sheet. A hand, a frog, a bat, a fish, and a cat's head were seen in quick succession. The fish showed two white spots.

"Those are the swimming bladders," explained Prof. Magie. "They cast no shadows. That is how a man's brain would look—just the inside of the skull—like a pumpkin. The softer portions of the body throw no shadow, and cannot be photographed by the X-rays."

Then appeared a hand covering a pair of scissors, which showed plainly underneath it; a needle in a thumb, the swollen joints and displaced portions of a mangled hand, a forearm with a diseased elbow, the human foot in a shoe showing the toes plainly, and two masses of plaster held by an iron brace in the case of a broken elbow.

"The value of these rays in watching the progress of discovery in such cases is inestimable," said Prof. Magie.

The lecturer said he hoped the rays would make the doctors even more careful with their bone settings, "for what jury," he asked, "can be won over when it sees the photograph of a badly set arm or leg?"

By this time Mr. Cole's ordeal was over. The plates were taken away into a dark room by L. L. Coe (who photographs subjects for the Rogues' Gallery at Police Headquarters) and were developed. The results were excellent. The bones of Mr. Cole's hand were thrown upon the sheet with great distinctness and amid general cheering. The ring on the third finger was very plain.

The key, the knife, and the dime of Job Hedges in the aluminum box were also clear and unmistakable.

When the lecture and experiments were over, supper was served to the members of the club in an adjoining room.

—March 13, 1896

About X-Ray Photography

While some discoveries of a purely scientific character appeal only to a limited class, others broadly affect the life and happiness of the human race, and thus become of universal importance. The discovery of Prof. [Wilhelm Conrad] Röntgen is unique in that it interests alike the scientific and non-scientific intelligent minds of all countries. To the world of science it suggests new problems as to the constitution of matter and the subtleties of electricity, while to the race at large it opens up a new means of diagnosis and relief of suffering and disease.

But much of the benefit of this marvelous discovery is for the time being in abeyance. Since its announcement there has been a perfect craze for X-ray photography. Many scientists and medical men who have used what has been put into their hands as X-ray apparatus have signally failed in their results. The failure has not seldom been due as much to ignorance of the proper technique of the new photography as to inefficient apparatus. The consequence is there has been much disappointment, and in not a few cases complete subduing of scientific enthusiasm. But the X-ray is not for today only. Its use and development are destined to go beyond anything we can now even conjecture. Before long every physician, surgeon, and dentist will have to rely upon it for a large proportion of the diagnostic and possibly therapeutic purposes within the range of his practice.

A book describing in simple language the best outfit for X-ray photography and the method by which the most effective pictures can be taken is now in the press, and will appear shortly. The author of the book is Dr. William J. Morton. Dr. Morton has long been recognized at home and abroad as the leading electrotherapist in this country. His X-ray photography is as remarkable as his work in other fields. Mr. Thomas Alva Edison recently wrote to a correspondent who applied to him for advice as to how to go about Röntgen ray photography: "Go to Dr. Morton; he is the best X-ray expert in America."

The first part of this book, which is entitled: "The X Ray; or, Photography of the Invisible, and Its Value in Surgery," treats of the various electrical features of the X-ray apparatus. A short chapter is devoted to the fluoroscope, which is not only of practical value in enabling the surgeon instantly to locate foreign bodies, such as bullets, needles, etc. embedded in the flesh, but determines for the operator whether X-rays are being produced or not in the Crookes tube, and if they are produced it indicates their degree of intensity. The value of this method of detection is supreme. It may be explained here that the X-ray is often a most elusive quantity. An expert has expressed the opinion that for all-round aggravation it has few rivals. After hours of labor the green light, which is an assured sign of high efficiency, may show itself for a moment, and then suddenly fade out and leave the exasperated experimenter wearily to go over the ground again. An X-ray, strange to say, will make an impression as quickly on a "slow" as on a "quick" plate, with even less chance of fogging. It is not so much a question of the length of exposure, as of the quality of the rays being developed. The fluoroscope, therefore, saves the operator no end of uncertainty and anxiety. This splendid sequel to the advent of the Röntgen ray throws an interesting sidelight on the tireless quality of temperament which seems inseparable from the successful inventor. Mr. Edison investigated over 1,800 different substances before adopting for the screen of the instrument the tungstate of calcium which, by giving the maximum fluorescence most effectively brings out the shadow of the object under inspection.

The induction coil is a most important part of the apparatus. For obtaining pictures of the hands, arms, feet, and lower portions of the legs, a coil having a four-inch spark will suffice, but for pictures of the shoulder, chest, abdomen, hip, or thighs a coil with an 8- or 10-inch length of spark will be required. A condenser and blower should be used with high-voltage current in the primary. The only other apparatus needed in this connection are the proper rheostats to govern the current respectively supplied to the primary of the induction coil, the motor which drives the break-wheel, and the motor which drives the blower.

The nature of the X-ray is discussed and the opinions of various investigators are given. Edison believes that the X-ray is of the nature of sound waves, a view also entertained by Oliver J. Lodge and J. J. Thomson. A sound wave is a longitudinal wave occurring in our atmosphere, but its counterpart has not yet been discovered in the ether. That the X-ray is such a wave in the ether was Röntgen's original surmise. Mr. Tesla holds that the X radiation is a stream of material particles projected from the cathode capable not only of penetrating the glass walls of the bulb or tube, but also of being projected onward into space, penetrating in greater or less degree some substances, such as flesh, leather, wood, etc., and arrested by other substances, such as metals, bones, etc. According to this view, a "bombardment" may be taking place outside of the tubes similar to that already generally recognized as existing within the tube. Many observers of the French school maintain that the X-ray is of the nature of light—namely (for lack of a better descriptive term), invisible or black light.

Fortunately, these varying speculations are not interfering with the practical utilization of the rays.

One of the great anxieties of the X-ray operator is to maintain the low vacuum of the Crookes tube, which is essential to certain classes of work. Again, the vacuum must be raised, and the process may involve much perturbation of spirit. Dr. Morton's single sentence, "The amateur will often have an exciting contest with his vacuum," is pregnant with meaning. What such a contest may be is suggested in the description of one of its stages, which runs thus: "To accomplish the raising of the vacuum electrically, a moderate current must be passed through the tube continuously, the operator watching its behavior all the time with the fluoroscope, at the same time observing whether a spark jumps across the gap space, and watching the electrodes to see that they do not become too hot. When the vacuum is thus low in a focus tube, the platinum may heat to a red, or even to a white, heat. This should be prevented either by reducing the current still more, or by interrupting its flow for a few moments. In a few minutes, after passing this moderate current through the tube, it will be noticed that a spark will jump across the inch space between the discharging rods; these must now be moved half an inch further apart. After another interval the spark will again jump, and the rods must again be separated until another spark jumps. This process is repeated until a vacuum is reached which will force three, four, six, eight, or more inches of spark to jump across between the discharging rods, rather than pass through the high resistance of the tube." Again, if a lowered vacuum is produced while working with a large amperage, there may be a sudden rush of current through the tube itself, breaking down the glass at its contact with the platinum entering wire, and "the operator sees the characteristic green of his X-ray-producing tube suddenly turn to blue, then to purples and whites, and soon to an arching stream from electrode to electrode. The tube is ruined."

It is often found in practice that the fluoroscope gives a fleeting and indistinct view, and that for the purposes of delicate operations a fixed and permanent record upon a sensitive plate is of much higher value. One of the latest advances is a combination of camera and fluoroscope, whereby an object can be examined in the fluoroscope and be photographed at the same time. As soon as the operator has completed his examination, he presses a button and the image is instantly transferred to the sensitive plate. In this way valuable work has been done in shadowing out and recording tumors, abnormal growths, and various diseased conditions of the larynx and bones of the face and their accessory cavities, as well as of the lungs, with their many complicated ailments. One of the first applications of this convenient instrument secured a picture of a silver tube, which had been placed in the throat and had slipped down out of sight in the trachea. A singular fact confirmatory of experience in other fields of photography is brought out in the use of this instrument. It is well known that the camera will see more than the human eye, and that many of the planets and stars which have been made perfectly visible on a sensitive plate have never been reached by mortal sight. In the same way the picture fixed upon the plate by photo-fluoroscopy has more strength and detail than is evident when it is viewed by the eye in the fluoroscope.

Great as is the interest which has been excited by the X-ray in the scientific laboratories of the world, among electrical engineers, photographers, students, and amateurs, its interest to the physician and surgeon is still more vital, for in its application to surgery lies its highest field of usefulness to humanity. The physician who, in the exploration of the mysteries of the human body, has been wont to employ the ophthalmoscope, the stethoscope, the cystoscope, the percussion hammer, and the probe has now added to these the most valuable means of diagnosis ever known to the science and art of medicine. Conspicuous among the revelations of the X-ray are those relating to normal anatomy. It might be claimed that the bones of the animal body could be studied from prepared skeletons, but such artificial arrangements of the bones can never, in reality, give their exact relation as well as the X-ray picture, nor in any sense afford a correct idea of these relations in the varied postures permitted by the changing position of the bones which compose the joints.

In teaching the anatomy of the blood vessels, the X-ray opens out a new and feasible method. The arteries and veins of dead bodies may be injected with a substance opaque to the X-ray, and thus their distribution may be more accurately followed than by any possible dissection. The feasibility of this method applies equally well to the study of other structures and organs of the dead body. To a certain extent, therefore, X-ray photography may replace both dissection and vivisection. And in the living body the location and size of a hollow organ, as, for instance, the stomach, may be ascertained by causing the subject to drink a harmless fluid, more or less opaque to the X-ray, or an effervescing mixture which will cause distension, and then taking the picture.

For the exhibition of fractures, dislocations, diseases of the bones and deformities, the X-ray is now indispensable, and it is recognized that no hospital in the land can do justice to its patients if it does not possess a complete X-ray outfit. By this means it is possible to detect and diagnose fractures and dislocations, and the very important point which often presents itself, whether the case is a fracture or dislocation, or both. Deformities of the bone are discovered, and even diseases like tuberculosis and cancer, which, in destroying the bone structure, have varied bone density.

In dentistry the X-ray has already become of supreme importance. By it pictures of the living teeth can be taken, showing each wandering fang or root, however deeply embedded in its socket. Children's teeth may be photographed before they have escaped from the gums, and the extent, area, and location of metallic fillings may be sharply delineated, even though concealed from the outer view. The lost end of a broken drill may be found, and, what is most interesting, the fact that even the central cavity of the tooth may be outlined, so that diseases within the tooth may be detected. It is equally obvious that diseases of the bone and other tissue in the neighborhood of the teeth may also be observed.

One of the first applications of Röntgen's discovery was for the detection of foreign objects in the body, and for this purpose it is now being incessantly employed. Bullets and needles are difficult things to find when embedded in flesh, but their discovery is made a certainty by the X-ray. The English War Office has supplied its military expedition up the Nile with complete X-ray outfits, and our own Navy Department is said to contemplate equipping each of its vessels with similar apparatus. If the X-ray had been known in its present form in the time of Garfield, the fatal bullet might have been located, and the life of the president saved.

The medico-legal side of the X-ray is one of great moment. Already court records contain numerous cases in which the Röntgen's ray has rendered valuable testimony. A picture has just been taken which may play a prominent part in a case soon to be brought into court. The patient was thrown down with violence in a trolley-car accident more than a year ago, and has suffered more or less ever since. An exposure was first made of the injured knee only, and no positive evidence of the degree of the injury was afforded. By resorting to the comparative method, a picture of both knees was obtained, which showed that the upper portion of the large bone of the leg below the knee was nearly three-quarters of an inch wider in the injured knee than in the normal one. This was doubtless due to fracture and subsequent growth of bone. Such a picture would be likely to have great weight with a jury.

Of the physiological effect of the X-ray little is yet known. The testimony on the subject is very conflicting, even taking only that of the highest authorities.

Prof. J. J. Thomson, in the Rede lecture at the University of Cambridge, insists that X-rays do not exert any of those deleterious effects on bacteria which are fortunately associated with ultraviolet light. In contradistinction to this, Dr. [William] Shrader of the Missouri State University stated, as the result of his experiments on the effect of the Röntgen rays on disease germs, that in nearly every instance the germs were found to be destroyed by the action of the rays. On this head judgment must be suspended until further data can be brought to light.

In respect to one of the effects of these rays, testimony seems to be unanimous. The flesh of those on whom the rays have been projected has become sunburned, and frequently the skin would strip off. In many cases the passage of the rays through the skull caused the hair to drop out, and active inflammation of the eyelids, upper lips, and of the skin of the face generally has been exhibited in experimenters who have devoted much of their time continuously to X-ray work. Many instances are reported where scientists have been prostrated, and have had to relinquish their Röntgen-ray work for a while. It is said, indeed, that the death of Dr. Shrader, whose investigations have just been referred to, was caused by exhaustion resulting from his unremitting study of the subject.

—September 6, 1896

The Mystery of Radium

Under the heading "The Mystery of Radium," *The London Times,* on Aug. 13, gave prominence to another of those statements regarding the discovery by Prof. and Mme. Curie which are believed to be from the pen of that eminent chemist, Prof. Sir William Crookes.

The article first refers to the announcement last March of the "astonishing fact that radium, in addition to the radioactive properties rendered more or less familiar by the researches of [Antoine Henri] Becquerel on uranium, possesses the property of maintaining its temperature at a point three degrees higher than that of its surroundings, and of continuously emitting heat without any apparent diminution of bulk or alteration of physical constitution." This announcement, it is added, was received with great incredulity. Eminent scientists refused to accept a statement so irreconcilable with scientific experience, and maintained that there must have been somewhere a serious error of observation. The writer goes on to say:

"That radium possessed radioactive properties indefinitely more powerful than those displayed by any other body was a fact of an order to which we were accustomed. These properties, however remarkable, differed only in degree from properties with which the scientific world was familiar. That difference in degree has indeed become sufficiently astonishing in the light of further study, for it has become clear that radium without external stimulus can produce effects hitherto only obtainable by means of the electric discharge in high vacua. It can throw gases into that state of vibration which causes the production of their characteristic spectrum, and it emits at the same time a radiation resembling the Röntgen rays and producing like them marked physical and physiological effects. Superadded to this extraordinary development of powers not unfamiliar in their lower manifestations is the unique and unprecedented power of emission of heat."

The article declares that this power is now established beyond all possibility of question. "That gross physical effect," says the writer, "in addition to the radioactive and physiological effects produced on so large a scale, points to an amazing total output of energy for which no compensation has yet been discovered."

The writer then goes on to state the latest developments in connection with the discovery of radium. He says that of course strenuous efforts have been made to obtain accurate measurements of the heat production and to determine the effect of external conditions in promoting or retarding it. M. Curie, he adds, has just communicated to the French Physical Society a paper stating the results of a recent inquiry.

"It appears," says the article, "that at the time of his [Prof. Curie's] lecture at the Royal Institution in June, the resources of that laboratory in producing and manipulating liquid gases were utilized in a new series of experiments. Prof. William Dewar had already, in 1893, improved the calorimetric use of liquid gases by means of a combination of vacuum vessels, so that heat evolution at the temperature of boiling liquid air or hydrogen could be determined with accuracy. When a sample of radium bromide weighing 0.7 gram was tested in this way it was found to be capable of volatilizing an amount of liquid oxygen and hydrogen equivalent respectively to 6 cc and 73 cc of the gases measured at the ordinary temperature. It seems that through a very wide range of temperature the thermal emission remains unchanged. Whether at the temperature of a summer day or at that of liquid air, the emission of heat goes on without perceptible variation.

"When, however, we make a long downward stride from liquid air to liquid hydrogen, radium shows that it is not always unaffected by external temperature. Within a comparatively short distance of absolute zero a change occurs in the rate of heat emission, but not in the direction that might be anticipated in view of the effect of low temperatures on ordinary chemical action. Instead of being reduced, the emission of heat, so far as present data can be relied on, is augmented at the temperature of liquid hydrogen. Whatever be the nature of this extraordinary phenomenon, it only increases in intensity at a point where all but the most powerful chemical affinities are in abeyance. The evaporation of a liquid gas gives an absolute measurement of the amount of heat given off by radium. Changes in the degree of radioactivity may escape the most careful observer, or may be imagined where they do not exist, but the quantity of liquid hydrogen which a given mass of radium converts into gas in a given time can be easily measured with an accuracy which is beyond trivial objection, and the amount of heat required for the conversion can be ascertained with great precision. Hence there is no longer any doubt either of the quantity of heat evolved by radium or of the fact that the rate of emission is apparently greater in liquid hydrogen than at any temperature from that of liquid air up to that of an ordinary room.

"At the beginning of these experiments on liquid hydrogen a contrary result appeared to emerge when the low-temperature thermal measurements were compared with the early Curie values observed at the temperature of melting ice, as formerly given in *The Times*. This led to the curious discovery that a freshly prepared salt of radium has a comparatively feeble power of giving off heat at all temperatures, but that its power steadily increases with age until about a month from its preparation, when it reaches the maximum activity, which it afterward maintains apparently indefinitely. A solution of a radium salt behaves in exactly the same way; its power of heat emission is at first relatively low, but goes on increasing for about a month, when it becomes equal to that of the solid salt, and so remains. These remarkable results throw no light upon the process by which radium maintains its constant emission of heat and radioactivity, but they will have to be accounted for by any theory that may be constructed."

—August 23, 1903

A Lecture by M. Curie

M. Curie's promised lecture on radium, which was discovered by himself and his talented wife, and secured for them this year a Nobel Prize, drew vast crowds to the Sorbonne one evening this week. Although the hall in which the now world-famed chemist was to speak is built to seat 3,000 persons, numbers who had stood for hours at the door in the Rue des Ecoles were finally turned away, every corner available being occupied by a deeply interested crowd.

When the lecturer appeared upon the platform with Mme. Curie, they received a magnificent ovation, and every word M. Curie spoke was followed with the closest attention. His account of the work done by M. Becquerel upon uranium, which formed the starting point for his own and Mme. Curie's experiments, was very clearly given, and his lecture was illustrated by tests of the properties of the newly discovered metal which is destined to revolutionize the science of chemistry.

In one series of experiments M. Curie demonstrated that radium emits three sorts of rays, two of which give off such powerful electric emanations that they cause phosphorescence and produce perpetual motion. Then, plunging a tube of radium into liquid air, M. Curie demonstrated in the darkness of the hall that it was a light-radiating body, and that it also gave out heat.

Taking into consideration the deeply scientific and necessarily technical nature of the subject, Paris is probably the only city in the world where so large and so representative an audience could be collected for a lecture of the kind.

It was curious to notice the intelligent interest taken by the humbler ranks of the Paris population in the great discovery of M. Curie, when, a few weeks ago, a small atom of radium was exhibited upon the boulevards. Even the artisans on their way to their work stopped to inspect and discuss it.

—March 6, 1904

Atom of Matter Can Be Detected

Prof. Ernest Rutherford, director of the physical laboratories at Manchester University, and one of the world's foremost authorities on radioactivity, details in a communication addressed to *The Scientific Weekly* the nature of certain experiments which, in addition to important results from the point of view of radioactive data, are noteworthy from the fact that during their progress it was for the first time possible to detect a single atom of matter. This can be done in two ways, one electrical and the other optical.

The possibility of the detection of a single atom of matter is due to the great energy of motion of alpha particles, which, as Prof. Rutherford showed in 1903, are veritable atoms of matter, which are ejected from radioactive matter at a speed of about 10,000 miles per second. Prof. Rutherford's more recent experiments show that 136 million alpha particles are expelled every second from one milligram of radium in a radioactive equilibrium.

"From the point of view of modern theory," says Prof. Rutherford, "the appearance of an alpha particle is the sign of a violent atomic explosion, in which the fragment of an atom—an alpha particle—is ejected at high speed. In a majority of known active substances, the expulsion of an alpha particle accomplishes the transformation of one substance into another, and the decrease of the atomic mass consequent upon the loss of an alpha particle at once offers a reasonable explanation of the appearance of an entirely new kind of matter in place of the old."

Prof. Rutherford reverts to his old suggestion, made in 1905, that very probably the alpha particle is an atom of helium, carrying two unit charges, and he refers to the difficulty of proving or disproving experimentally the correctness of this hypothesis, although the settlement of this question has been for the last few years the most important problem in radioactivity, because the proof that the alpha particle is an atom of helium carries numerous consequences of the first importance in its train.

Prof. Rutherford now asserts that his recent experiments have thrown further light on this question and have led to important conclusions in several directions. The description of the experiments is too technical to be gone into here, but Prof. Rutherford considers that they demonstrate the correctness of his theory that the alpha particle must be an atom of helium carrying a double charge, or, in other words, that an alpha particle, when its charge is neutralized, is a helium atom. It must be concluded, he says, that atoms of known radioactive elements are, in part, at least, constituted of helium atoms, which are liberated at definite stages during disintegration.

Prof. Rutherford's conclusion is put with characteristic modesty. "It may be of interest to note," he says, "that the experimental results recorded in this article lead to experimental proof, if proof be needed, of the correctness of the atomic hypothesis with regard to the discrete structure of matter."

This is truly a modest claim to victory in the greatest scientific battle of the century.

In response to a message from the correspondent of *The New York Times*, Prof. Frederick Soddy, lecturer in physical chemistry at Glasgow University and coworker with Prof. Rutherford at McGill University, Montreal, telegraphed as follows:

"I am in entire agreement with Rutherford's conclusions. As to the alpha particle it has hardly seemed possible to doubt it was an atom of helium for many years. My first impressions in the matter, dating from 1904, have proved correct. The experiments are beyond all praise."

Alfred Russel Wallace, who was also asked for an expression of his views, replied: "I know nothing of the subject and have no opinion."

—November 8, 1908

Madame Curie's Genius

Mme. Marie Curie, discoverer of radium, is coming to New York in May to receive the gift of one gram of this very valuable element from American women. The rise of this woman to the position of a leading scientist of the world is without parallel in history. The daughter of a poor but distinguished Polish scientist, she worked her way up in the field of her choice until today she bears the unique distinction of having, together with her husband, received the Nobel Prize award, being the first and only woman professor at the Sorbonne in Paris, of having refused the ribbon of the Legion of Honor because, as she put it, "I don't like decorations," of receiving the English Albert Medal from the Royal Society of Arts, the second to be granted to a woman, Queen Victoria having received the first. There were many more honors and decorations offered, but the response in almost all instances was similar to the one given to the suggestion of the Legion of Honor award.

Comparatively little is known about the private life of Mme. Curie, owing to her modesty and reticence. From scraps gathered here and there, however, one gains an impression of a personality too greatly concerned with the big things of life to waste any more thought or time than is necessary over the little.

Mme. Curie's love of research is an inheritance. The passion for science came to her when she was a tiny child. Her nursery was a laboratory. Instead of dolls, she played with test tubes, retorts, and crucibles while her father, M. [Władysław] Sklodowski, professor of physics at a college in Warsaw, was working at his experiments. As she grew older, she became more and more useful to the professor. She learned the various places in the laboratory for every instrument and something of their meaning and use. She constituted herself "washer" for her father, always cleaning his apparatus after his work was over. Quietly, it is told, the little girl would don a large apron and devote herself to the rinsing and scouring of flasks, beakers, mortars, burettes and pipettes. These were her precious toys, and she handled them reverently.

Professor Sklodowski died, and the little household at Warsaw broke up. There was another daughter besides Marie. The two girls, left without funds, had to take up the struggle for existence by themselves. Political disfavor made it expedient for the family to leave Warsaw. The older sister went to Austria, where she became a well-known doctor of medicine. Mme. Curie came to Paris, there to try to find a place for herself in the scientific world of the city.

She had no resources other than courage, about 50 francs in money and a knowledge of chemistry greater even than she herself appreciated at the time. In an obscure quarter of the city, she found a bare garret, furnished with a cot and chair. Her food was black bread and blue milk. She lived on this diet so long, one account goes, that she had afterward to cultivate a taste for meat and wine.

Her one ambition was to gain admission to a laboratory as a student assistant. After a period the doors were grudgingly opened to her. She was admitted because there was need of a cheap assistant to prepare furnaces and to clean bottles, a kind of expert janitress service.

She had not been there a week before Professor Gabriel Lippmann, who was at the head of the laboratory, discovered that she possessed a knowledge of science and an originality of mind far above the average. Rather tardily it became known that she was the daughter of a scientist of note in her own country and that she had grown up in his laboratory. Someone else was found to wash bottles and arrangements were made to give the Polish girl full facilities for work.

It was at the Sorbonne that she met Pierre Curie, a young instructor, who was making a name for himself by independent research in physics. His privilege was to have one of the young women students to assist him in the laboratory, and it fell to Marie Sklodowski to be this assistant. They worked and studied together, and in 1895 were married. Marie Curie was then about 28 years old.

Announcement of the radium discovery came three years later. The finding of the new element was no laboratory accident. The two young scientists had toiled and experimented through years of depression and poverty. Few who knew them believed in their theories. The Curies were, in fact, a sort of joke in scientific circles, it is stated.

It was Mme. Curie's courage which brought them success. Pierre Curie suffered periods of complete discouragement. He is said often to have doubted his conclusions. But Marie Curie never doubted. She never lost faith. Night and day she worked in their little laboratory at home. When the name of Curie became worldwide, Professor Curie publicly declared that more than half the credit belonged to his wife. The research work was begun before her marriage, it is stated, and it was through her that he became interested in radioactivity.

In 1906, at the very height of their career, Pierre Curie was run over and killed by a cart. There was no scientist to take the place of Pierre Curie as special lecturer at the Sorbonne, except his widow. The most conservative of universities was obliged to break its rules and invite a woman to full professorship. When Mme. Curie began her lectures in chemistry, succeeding her husband, the lecture hall was filled with scientists, students and leaders of the educational world. The ministry was praised in the press for having appointed her. Some idea of what this woman is like and how she works is given in the following extract from an English article printed in 1910:

> With honors heaped upon her, she remains the same unobtrusive, reserved person that I saw in the tumbledown shanty that the City of Paris gave to M. Curie for a laboratory and a school for chemistry. She is a little better dressed now than then, but with extreme plainness. The complexion is still that of one brought up in stove-heated rooms, ashen, and the lusterless hair unchanged in all but a few silver threads. She remains hard to read, a consequence of being brought up at Warsaw under the heel of the Russian boot and the eyes of an officialdom, jealous of all scientific investigation. Mme. Curie spoke of the university in which her father filled the chair of chemistry as having in all its corridors finger posts pointing to Siberia. There was always something sad and wrapped up in her look. On losing M. Curie, she apparently had got to that state in which "joy has no balm and affliction no sting."
>
> As a lecturer, she closely confines herself to statement and demonstration, risking nothing that is unproved, however strong cause she may have for divining inference. She is completely innocent under all circumstances of any wish to dazzle or show off. In this respect Mme. Curie presents a refreshing contrast to some "guinea hens" (pintades) who have opened literary salons and to two or three ladies of risky antecedents who intend at the next elections to stand for divisions of Paris. Her French is grammatical and clear and she is never at a loss for a word; but it is a little stiff.
>
> The employment of laboratory apparatus as she lectures helps to give it more flexibility. Her laboratory is kept in apple pie order and her notebooks show the plain, straightforward and scrupulously exact observation of a good seaman's log. They bristle with notes of interrogation.

You will perhaps say that I describe a scientist of the other sex. Not at all. Mme. Curie is essentially womanly. But she lost her mother early and was brought up at her father's side in his laboratory, and was not warped from her true nature according to any conventional standard of femininity. She evolved from within according to her opportunities and the tender paternal guidance, and became on chemistry an authority in the minds of the university students who came to the laboratory. The suspicious prying of the police taught her how necessary it was to hold her tongue. Reticence in speech became her second nature.

There is nothing in the desire of self-advertisement about Mme. Curie. She is shy and sensitive to a degree. Her day in the laboratory is long and she always hurries to the train for home at the end of it. While always ready to help a fellow worker or to do an act of kindness, Mme. Curie shuns interviews.

"It is not the personalities of people that interest: it is their work," she says.

Mme. Curie shares with most savants the faculty of abstraction which so many men engaged in scientific work possess. Old M. Curie, it is related, used to chaff his daughter-in-law about her abstraction and had one story that he used to tell as an example of her utter indifference to anything not concerned with her investigations when she was engaged upon a difficult piece of research work. He said that once when she was in the middle of an absorbing experiment a servant ran into the laboratory screaming loudly, "Madame, madame, I have swallowed a pin!"

To which Mme. Curie responded soothingly:

"There, there; don't cry. Here is another you may have."

According to the descriptions given of her, Mme. Curie is of medium height, thin, almost angular, with a very high forehead, which gains an unusual prominence by her hair being drawn straight back on her head. She is pale, probably from her continuous work in the laboratory. Perhaps her most prominent features are her eyes, of bluish gray, which have the clearness of youth with the penetration and depth of a profound thinker. They are set very far back in her head. Her attitude is one of reserve and her speech is calm and slow. Mme. Curie has been naturalized as a French citizen.

—*May 1, 1921*

Pictures Electrons Speeding in Atom

The statement of the modern conception of the atom as a miniature solar system with electrons whirling around a nucleus with velocities ranging as high as 93,000 miles a second, was a feature of Sir Ernest Rutherford's address before the British Association on his assumption today of his presidential duties.

He seized on the fact that the association met here in 1918 as affording an opportunity for a review of what physical science had accomplished since that time. The years between, he said, had been aptly termed "the heroic age of physical science," and never before had discoveries of fundamental importance followed one another with such bewildering rapidity. He enumerated some of those known in 1893—the Röntgen rays and the radioactivity of uranium and Guglielmo Marconi's success with wireless—and this led him to speak of the relation between pure and applied science.

"In this age," he declared, "no one can draw any sharp line of distinction between the importance of so-called pure and applied research. Both are equally essential to progress, and we cannot but recognize that without flourishing schools of research on fundamental matters in our universities and scientific institutions, technical research must tend to wither."

Praises Work on Electrons

Sir Ernest then entered on his review and paid tribute to Sir J. J. Thomson for his boldness of ideas and ingenuity in the study of electrons.

"He early took the view," he said, "that the atom must be an electrical structure held together by electrical forces, and showed in a general way the lines of possible explanation of the variation of physical and chemical properties of elements as exemplified in the periodical law."

Next he spoke of the discovery of radium and the emanation of alpha, beta and gamma rays, and declared: "The use of alpha particles as projectiles with which to explore the interior of the atom has definitely exhibited its nuclear structure, has led to the artificial disintegration of certain light atoms and promises to yield more information yet as to the actual structure of the nucleus itself."

This led Sir Ernest to describe the work of Sir William Crookes and C.T.R. Wilson in showing how the radiation of alpha particles worked. He spoke of the fundamental differences between the units of positive and negative electricity.

These were, he declared, of the utmost importance, but he confessed that so far science could not give an explanation of them.

"It is practically established," he said, "that these two are the fundamental and indivisible units which build up our universe, so we may reserve in our mind the possibility that further inquiry may some day show that these units are complex and divisible into even more fundamental entities."

Bombardment of the Atom

Then, taking up the development of ideas concerning the structure of the atom, Sir Ernest spoke of the so-called bombardment of atoms. It was found, he pointed out, that occasionally a swift alpha particle was deflected from its rectilinear path through more than a right angle by an encounter with a single atom. From this "arose the conception of the nuclear atom, now so well known, in which the heart of the atom is supposed to consist of a minute but massive nucleus carrying a positive charge of electricity and surrounded at a distance by the requisite number of electrons to form a neutral atom."

This led to a detailed study of the scattering of alpha particles with the result that "a relation of unexpected simplicity is found to hold between elements. No one could have anticipated that with a few exceptions all atomic numbers between hydrogen 1 and uranium 92 would correspond to known elements." Further study along these lines resulted in a definite conception of what the atom really was. Taking the heaviest, uranium 92, Sir Ernest described it thus:

"At the center is a minute nucleus surrounded by a swirling group of 92 electrons, all in motion in definite orbits and occupying, but by no means filling, a volume very large compared with that of the nucleus. Some of the electrons describe nearly circular orbits round the nucleus; others, orbits of more elliptical shape whose axes rotate rapidly round the nucleus. The motion of the electrons in the different groups is not necessarily confined to a definite region of the atom, but the electrons of one group may penetrate deeply into the region mainly occupied by another group, thus giving a type of interconnection or coupling between the various groups."

Marvelous Speed of Electrons

"The maximum speed of any electron depends on the closeness of its approach to the nucleus, but the outermost electrons will have a minimum speed of more than 1,000 kilometers [about 620 miles] per second, while the innermost electrons have an average speed of more than 150,000 kilometers [about 93,000 miles] per second, or half the speed of light. When we visualize the extraordinary complexity of the electronic system we may be surprised that it has been possible to find any order in the apparent medley of motions."

Sir Ernest thinks it is possibly too soon to express a final opinion on the accuracy of this explanation of the outer structure of the atom, but he considers it a great advance. Thus, it aided in the prediction of the chemical properties of the missing element No. 3. There is, however, an enormous amount of work to be done before anything like a complete picture of even the outer structure of the atom can be formed.

Discusses Nucleus of the Atom

Passing on to consider the nucleus of the atom, Sir Ernest said:

"We know it is a minute compound, and we can with confidence set a maximum to its dimensions. It is possible for us to lay it down, that the proton and the electron are the ultimate units which take part in the building up of all nuclei and to deduce with some certainty the number of protons and electrons in the nuclei of all atoms: but we have little, if any information on the distribution of these units in the atom or on the nature of the forces that hold them in equilibrium."

It might be, he said, that new and unexpected forces might come into importance at the very small distances separating the protons and electrons in the nucleus. Until more information was gained on the nature and law of variation of the forces inside the nucleus, further progress toward a knowledge of the detailed structure of the nucleus might be difficult. At the same time there were still a number of hopeful directions in which an attack might be made on this most difficult of problems.

"The nucleus of a heavy atom," Sir Ernest said, "is undoubtedly a very complicated system, and, in a sense, a world of its own, little, if at all, influenced by the ordinary physical and chemical agencies at our command. When we consider the mass of the nucleus compared with its volume, it seems certain that its density is many billions of times that of our heaviest element. Yet, if we could form a magnified picture of the nucleus, we should expect that it would show discontinuous structure occupied, but not filled, by minute building units, protons and electrons, in ceaseless rapid motion, controlled by their mutual forces."

Doubts Power of Broken-up Atom

Sir Ernest then attacked the popular idea of the immense store of energy to be released if man ever succeeded in breaking up the atom:

> Since it was believed, that radioactive elements were analogous in structure to ordinary inactive elements, the idea naturally arose that atoms of all elements contained a similar concentration or energy which would be available for use if only some simple method could be discovered of promoting and controlling their disintegration. This possibility of obtaining new and cheap sources of energy for practical purposes was naturally an alluring prospect to the lay and scientific mind alike. It is quite true that if we were able to hasten radioactive processes in uranium and thorium, so that the whole cycle of their disintegration could be confined to a few days instead of being spread over thousands of millions of years, these elements would provide very convenient sources of energy on a sufficient scale to be of considerable practical importance.
>
> Unfortunately, although many experiments have been tried, there is no evidence that the rate of disintegration of these elements can be altered in the slightest degree by the most powerful laboratory agencies. With the increase in our knowledge of atomic structure there has been a gradual change of our point of view on this important question, and there is by no means the same certainty today as a decade ago that the atoms of an element contain hidden stores of energy.

The speaker illustrated this by referring—without, however, mentioning his name—to his well-known experiments of trying to drive the electron-forming part of a system of atoms, by "bombarding" it out of position. This he termed "exciting" the atom, and he pointed out that the energy thus released was of an entirely different order of magnitude from the energy that would be released by the disintegration of the nucleus itself. So he went on:

It may be that the elements uranium and thorium represent the sole survivors on Earth today of types of elements that were common in long-distant ages when the atoms now composing the Earth were in course of formation. The fraction of atoms of uranium and thorium formed in that time has survived over a long interval on account of this very slow rate of transformation. It is thus possible to regard these atoms as having not yet completed the cycle of changes which ordinary atoms have long since passed through, and that the atoms are still in an excited state where the nucleus units have not yet arranged themselves in positions of ultimate equilibrium, but still have a surplus of energy which can only be released in the form of characteristic radiation from active matter. On such a view the presence of a store of energy ready for release is not the property of all atoms, but only of a special class of atoms, like radioactive atoms, which have not yet reached the final state for equilibrium.

Sir Ernest concluded his survey with a tribute to the work the British Dominions had done in the scientific field and a plea for adequate support of the universities to train young investigators and for international peace, to permit the cooperation of all countries in the cause of the advance of knowledge.

When Sir Ernest delivered his address he had the unusual experience of being only partly informed as to the number and character of his audience. He was speaking directly in Philharmonic Hall to as many members of the association as could be crowded into the building. Members who were unable both to hear and see the president listened to his address as it was transmitted by telephone and loudspeaker in the lecture hall of the association exhibition. Beyond these tangible gatherings, Sir Ernest had a far wider body of hearers, as his words were broadcast by wireless simultaneously from all the stations of the British Broadcasting Company.

Americans in Attendance

Several well-known Americans and Canadians are attending the meeting, among them Professors F. S. Lee and Michael Pupin of Columbia University, Professors Wilder Bancroft and Ernest Merritt of Cornell, and Professors Adams, Eve, McCallum, Tate and Whitby of McGill.

—September 13, 1923

Discusses Atom from New Point

Professor Niels Bohr of the University of Copenhagen resumed his lecture course at Yale today, giving the fourth address on his theory of the structure of the atom.

Devoting himself today almost entirely to a discussion of the theory of spectral lines which he first proposed 10 years ago and has consistently developed up to the present, Professor Bohr showed how experiments in the study of the series of spectra of the elements had corroborated his theory of atomic constitution, and how spectral phenomena of various complex kinds were explained by the quantum theory of energy.

He cited spectroscopic observations to support the conclusions he has built up from the quantum theory.

After reviewing his lectures on the similar forms of the atom, Professor Bohr said today that from now on he intended to go into the more complex aspects of the reasoning on which his theory had been built up. Today, he said, he would attempt to show the formal development of the quantum theory as applied to the finer points in spectra, such as the fine structure of lines, which, in contrast to the simple lines previously considered, were made up of a number of lines close together.

He pointed out that the first development of the quantum theory was made by [Max] Planck, the German scientist, who used it for a linear oscillator—an electrically charged particle which oscillates or vibrates.

Describing how the Planck formula had been used to explain the spectra of the elements, Professor Bohr said that he had made the so-called integral of action equal to an integral multiple of Planck's constant. He added that the motion of the particles in an atom, despite its minute and simple structure, was very complicated in comparison with the motion of a Planck oscillator.

"This integral," he added, "has the peculiar quality that its value does not change during a slow change in the motion of the system. The invariancy of this integral is the reason for its being used to fire the stationary states of the atom. It also has the important property that if there are two motions which differ only a little from each other, the difference in energy will be equal to the difference in integral action times the frequency of revolution."

Professor Bohr pointed out the relation of the failure of the integral to change during slow motion to the question of the stability of the atom, which he says the quantum theory explains, while the classical laws of mechanics and electrodynamics do not.

He then explained the correspondence principle. He showed how this principle established an intimate connection between the character of the motion of an atom in the stationary states of the atomic system and the possibility of a transition between those two states. The correspondence principle, he said, was believed to offer a basis for a theoretical examination of the process that took place during the formation or organization or reorganization of the atom.

"The correspondence principle," he said, "states that there is a connection between the frequencies of spectral lines and the motion of the stationary states, although it is of a different kind from that given by the classical theory."

Professor Bohr then showed by diagrams that any periodic motion could be represented as the sum of oscillations with different frequencies, which were overtones of a fundamental frequency. He also showed that on the classical theory these frequencies were the frequency of the light emitted, while based on the quantum theory the frequency was determined by the change in energy during the transition between two stationary states.

"The correspondence principle," he added, "states that the occurrence of a transition for which the quantum numbers are charged by a given integer is conditioned by the occurrence of the motion of an oscillation whose frequency is equal to this integer times the fundamental frequency of the motion.

"From this point of view, for example, the green line of the hydrogen spectrum which corresponds to a transition from the fourth to the second stationary state may be considered in a certain sense to be an octave for overtone of the red line, corresponding to a transition from the third to the second state, even though the frequency of the first line is by no means twice as great as that of the latter."

Professor Bohr said that an example of a system which was like the linear oscillator, where the quantum integer could change by only one unit, was to be found in the spectra of diatomic gases, like oxygen.

"The occurrence of very weak overtones for these lines, which are not exactly at twice the frequency, can be beautifully explained on the quantum theory," he added.

The lecturer then explained how the quantum theory accounted for rotational spectra resulting from a superposition of rotations in the ordinary motion of a molecule.

"The theory can also be extended to determine stationary states of systems that are not simply periodic, but belong to a kind of motion known as multiple periodic motion," he said.

—*November 14, 1923*

Atomic Theory Clears Some Cosmic Problems

By W. J. LUYTEN

Three great problems of cosmical interest have recently come to the fore: The creation of chemical elements, the origin of the mysterious super-radiation coming to us from space, and the enormous densities—2,000 times denser than gold—which some stars are known to possess. Now, touched by the magic wand of the modern theory of the constitution of the atom, these three problems have revealed a little of their identity, of their common origin. Years have been spent in building up the magnificent and complicated structure of the atom, and, naturally enough, we are now trying hard to find out what happens when we destroy it.

Nowadays we think that an atom is built up from a very small central particle, the nucleus, around which electrons, the smallest material bodies that carry electricity, are circling at various distances. So small is this central body and so far away are the revolving electrons that the whole structure may well be compared with our planetary system; the nucleus is the sun and the electrons represent the planets.

Although the whole atom is no more than a billionth of an inch in size, it is still so large compared with the actual sizes of the nucleus and the electrons that Sir Oliver Lodge aptly described the whole structure by comparing it to "flies in a cathedral." To make the picture correspond even closer, so far as weight is concerned, we should imagine the fly to be made of platinum and the cathedral built of the thinnest tissue paper. What happens when the tissue-paper cathedral collapses and we are left with no more than the platinum fly forms the basis of some very interesting speculations by Professor Andreas von Antropoff at Bonn.

Whirls to Keep from Falling

It all begins with hydrogen, the lightest and simplest of all gases. In hydrogen gas the nucleus of the atom is composed of just one particle, the proton, around which one single electron runs in a circle. The proton and the electron attract each other very strongly, and the only way the electron can keep from falling into the central nucleus is by whirling around it, the same way the Earth does around the sun.

44

But suppose that some outside force disturbs the motion of the electron, or that another electron, coming from somewhere else, passes close to the proton, an upset may occur. The proton, which is 1,800 times heavier than the electron, may suddenly step in and exert its authority. The electric force between the two particles does not rest until they are united into one body.

Naturally, a complicated atom such as iron, which has a great many electrons whirling about its core, is not easily affected. Such a collection of electrons forms a good protective barrier, hard to penetrate for any outside electron. "Iron for the ironmen" is their Monroe Doctrine, and they resent outside interference. Hydrogen, on the other hand, has only one electron to protect its center, and this electron is easily alienated by heat.

Disappear upon Colliding

When hydrogen gas is heated, the affection between proton and electron is much reduced. The electron may leave entirely, and the now "stripped" atom is wholly unprotected, except by the fact that it is so exceedingly small. A lone hydrogen proton is no more than one ten-quadrillionth of an inch in diameter. An electron, itself no more than one ten-trillionth of an inch, must indeed take good aim if it wants to hit the proton head-on! When such a collision takes place it must be a marvelous sight to watch, for both proton and electron are completely non-elastic. There is no rebound, as in the case of billiard balls.

Once the two parties collide, they hug each other closely and disappear from sight. The combination proton-electron, for which the name "neutron" has been adopted, comes very near to fulfilling Hamlet's query "To be or not to be." It certainly exists, because it has weight, and yet at the same time we might almost say that it does not exist, for we cannot detect it with any means of observation. Chemically it cannot be touched; it is much more inert than even helium; and physically we cannot get at it either, for it is so exceedingly small that it falls and goes through everything material.

A neutron would shoot through a piece of platinum without the slightest difficulty; even this most dense of all metals is no more than a sieve with holes hundreds of thousands of times larger than the neutron. Electrically speaking, a neutron is dead too; the opposing electric influences of proton and electron have killed each other. In fact, were it not that a neutron has weight, we should be inclined to consider it the real "ether."

Penetrates to Earth's Center

The result of these peculiar properties is that as soon as a neutron has been formed it will drop, penetrating everything in its way. It will shoot past the center of the Earth, and swing back and forth like a pendulum, ultimately coming to rest in the center of the Earth, or the sun, or any other star near which it has been formed. It is quite possible, therefore, that at the center of the Earth we might find a collection of neutrons packed exceedingly close together.

Calculations show that the minimum density of this "solid neutronium" reaches the high value of four trillion times that of water. One cubic inch packed solidly with neutrons weighs 60 million tons. A sphere composed entirely of neutrons, with a diameter of no more than half a mile, would weigh as much as the whole Earth.

The densest star we know is the small satellite of Sirius, a star not much larger than the Earth but weighing as much as the sun. The average density of this freak is 50,000 (one ton per cubic inch). Not more than five years ago, when this density was first calculated, no astronomer dared to believe it; everyone was convinced that there was something wrong. Two years ago theory first and observation afterward established beyond doubt the reality of this amazing density. Within a year people may shrug their shoulders at it, for what does a density of "only" a ton per cubic inch amount to when we can juggle with solid neutronium at 60 million tons per cubic inch? The companion to Sirius may be a fairly normal star after all. A small globe of closely packed neutrons fifteen miles in diameter surrounded by a layer of gas 10,000 miles thick would do the trick and give the whole star an average density of 50,000 times that of water.

Explaining Hard X-Rays

How then do we explain the origin of the "hard X-rays" coming to us from all directions in space, on the basis of the "neutron theory"? That, too, is very simple. The modern theories of the atom tell us that, in the case of the hydrogen atom, where we have but one central proton and one electron whirling around it, the electron cannot go wherever it pleases. The radius of the circle it describes can only have a few definite values. But every time such an electron is pushed in from a larger circle into a smaller one a ray of light is emitted. The bigger the jump an electron takes the bluer the light.

Of its own accord, however, an electron will never fall into the proton. But, way up in the atmosphere, where the sun's rays are powerful and play havoc with the hydrogen atoms, electrons may get loose. They may even entirely forsake their proton and begin wandering on their own hook. It sometimes happens that they approach a proton so closely that they fall into its clutches. The joy of the proton in having caught the electron is so great that a ray of light of extreme intensity is emitted, so penetrating that it far surpasses the hardest X-rays we know on Earth. Our eye cannot see them, but our delicate measuring apparatus detects them easily.

Created in Pairs

There is one strange and significant fact about these protons and electrons to which little attention has been paid previously. A proton and an electron are different objects. The electron is comparatively large, weighs little, and has a negative electric charge; the proton is much smaller in size, but 1,800 times as heavy and has a positive electric charge.

The one curious thing is that not only are these electric charges exactly equal, but the total number of electrons and protons in the universe seems also to be equal. This suggests that they always are created in pairs, that perhaps the energy which the stars so wastefully radiate into empty space is caught somewhere and converted into protons and electrons, one proton for each electron.

Have we, perhaps, in this way come one step nearer to the solution of how the universe regenerates itself? The key to the problem of the Infinitely Great lies in the Infinitely Small.

—September 5, 1926

Details Concepts of Quantum Theory

By WALDEMAR KAEMPFFERT

Of 30 addresses delivered today before the various sections of the British Association for the Advancement of Science, one of the most important was that of a young German, Dr. Werner Heisenberg. Fully 200 mathematical physicists listened to his brief exposition of a conception which will make it necessary to modify belief in what we are pleased to call "common sense" and "reality." The layman without a knowledge of higher mathematics, listening to Dr. Heisenberg and those who discussed his conclusions, would have decided that this particular section of the British Association is composed of quiet and polite but determined lunatics, who have created a wholly illusory mathematical world of their own. The conception is that they and their kind alone have a proper view of "reality"; the rest of us live in a dream world fashioned by ill-understood words.

To explain the quantum theory and its modification by Dr. Heisenberg and others is even more difficult than explaining relativity. It is much like trying to tell an Eskimo what the French language is like without talking French. In other words, the theory cannot be expressed pictorially and mere words mean nothing. One is dealing with something that can be expressed only mathematically.

The consequences, however, are startling. Electrons and atoms cease to have any reality as things that can be detected by the senses directly or indirectly. Yet we are convinced the world is composed of them.

Action Supersedes Substance

In the new mathematical universe, events are more important than substances, and energy more important than matter. All mental pictures we have formed of bodies moving through space are thrown into confusion. So simple a conception as a baseball flying from the pitcher to the batter turns out to be obscure, doubtful and even ridiculous.

Planck, the originator of the quantum theory, Heisenberg, Schrödinger and de Broglie have shown that the whole science of mechanics must be rewritten. And when it is rewritten, no one but a mathematician will be able to understand it. The scientific world is faced with an upheaval as great as that brought about by Einstein.

—September 2, 1927

Super X-Rays Reveal the Secret of Creation

By WALDEMAR KAEMPFFERT

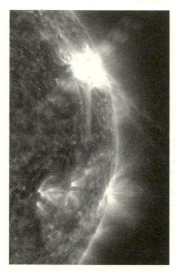

Experiments have been made which, for the first time, confirm the supposition of scientists that somewhere in the universe a building-up process is going on to replace the tearing-down process represented by radioactivity. The announcement of this discovery, with the reservation that it is yet to be more fully proved, was made last Sunday by Dr. Robert A. Millikan, noted American physicist and winner of the Nobel Prize in 1923. The evidence for this new view of the world has been obtained from a study of the cosmic rays, which tell the physicist of the birth of new elements. If the theory is confirmed, Dr. Millikan said, "it will constitute new proof that this is a changing, dynamic and continuously evolving world instead of a static or merely disintegrating one." The article which follows explains the broad meaning of this discovery.

X-rays reveal the process of creation? The notion seems preposterous. What possible connection can X-rays have with Sirius, the sun and the Earth? Yet even before Professor R. A. Millikan made his startling announcement last week, astrophysicists were convinced that the stars are colossal X-ray tubes. Professor Millikan has not only confirmed their conviction with experimental evidence but reinforced the Einstein theory of relativity and the modern conception of the electrical nature of all matter.

There are 3.6 billion stars in the system to which the sun belongs. It stands to reason that they are not all of the same age and the same constitution. So, ever since the beginning of astrophysics, students of the heavens have formulated theories to account for the observed differences in stars and have tried to estimate their ages. Darwin did not see protoplasm evolving from living slimy ooze into such higher organisms as corals, codfish, birds, dogs, apes and men; but he did see that there was a ladder of evolution, on which each living creature found a rung.

It takes eons to make a star, and the human race has not been in existence for more than 1 million years. Hence the astrophysicist must employ Charles Darwin's method—arrange and classify stars into a grand system where each will find its proper chemical and physical place. After that he may assume that evolution shifts stars from rung to rung on the ladder, even though a single shift means millions of years.

Heat from Cooling Star

The great German physicist, Hermann von Helmholtz, explained the generation of heat and light by a star very simply. As it cooled, a star naturally contracted, but this contraction in turn generated heat. Thus heat and light (energy) were literally squeezed out. Thereupon the astrophysicist proceeded to arrange stars according to color and mass.

The contraction theory was accepted until about 20 years ago. In fact, it is still good; for it seems certain that the squeezing process is necessary to create some of the conditions that physicists demand in order to account for stellar energy. But standing by itself the theory proves to be inadequate.

How old is the sun, according to this view? About 20 million years, the mathematicians answer. The geologists and paleontologists protest. The chemist has determined the exact time it takes for such radioactive elements as uranium, radium and thorium to disintegrate spontaneously—a process that occurs under his very eyes. He has found that the minerals in the oldest rocks could not have been formed less than 1.2 billion years ago.

Even an Earth so old is still too young in the face of other evidence. All the lead in the Earth's crust could have been produced from uranium and thorium in not less than 8 billion years. Evidently the energy of the sun and the stars is generated by other forces than contraction alone.

In England, Professors James Jeans, Arthur Eddington, Michelle Milne and Peter Fowler, and, in this country, Professor Henry Norris Russell applied the principles of atomic physics to stars and gave the world views that Professor Millikan has partly substantiated. "We started to explore the inside of a star," says Eddington. "We soon found ourselves exploring the inside of an atom." One reason for this lies in the fact that the stars are curiously alike in chemical composition. Hydrogen, helium, iron, calcium—the elements are the same. All those with which chemists deal in earthly laboratories are found in the great crucibles of the sky. Hence there is a unity of atoms and stars, heaving seas and tossing trees, swimming fish and living men, that makes the whole cosmos kin.

To explain what Millikan has discovered we must explain the modern conception of matter. As everybody knows, the atom is no longer the single, indivisible, hypothetical unit of matter that it once was. It is a complicated system. It has a nucleus, a compact mass of protons and electrons. Around the nucleus revolve electrons. The simplest of all atoms is that of hydrogen. It has but a single electron revolving around the nucleus. Most complicated of all is the uranium atom with its girdle of 92 revolving electrons.

All electrons are alike. Iron, silver, lead, gold, mercury, oxygen differ from one another merely because they differ in the number and arrangement of their electrons. If we could only subtract and add electrons to nuclei at will, we would have the power of transmuting elements. Man's engines are too puny. In the distant stars tremendous forces are at work tearing atoms apart and building them up anew. In terrestrial laboratories all that physical chemists can do is to watch the subtracting process—watch radium and other radioactive elements throw off electrons and reduce themselves to less exciting elements.

Sir Ernest Rutherford has bombarded the nucleus of some atoms with the alpha particles of radium—projectiles that have a speed of 100,000 miles a second—and has at least succeeded in proving that the electron theory is correct. In other words, he knocked hydrogen out of such elements as nitrogen, phosphorus and aluminum and thus proved that the primordial substance of the universe is hydrogen. But no one on Earth has ever succeeded in building hydrogen or any other element into something higher in the atomic scale.

This electron theory, which reveals the very rock-bottom of matter, enables us at last to understand the process whereby light and heat are produced. What happens when the flame of a match or the glow of a filament sends us its rays? Electrons jump from one orbit to another with incredible rapidity. Professor Niels Bohr received the Nobel Prize for developing that view. It is clearly quite incorrect to liken an atom to a miniature solar system, as textbooks do. Imagine the planets leaping from one orbit to another and releasing energy as they do so! There could be no solar system under such conditions.

We never knew how to explain the lines of the spectrum until Bohr developed his theory of the atom and the transfer of energy. The marvelous spectral colors of the rainbow are now accepted as the evidence of tremendous activity in the electrons of the sun's atoms. Similarly the spectrum of any star tells the story of leaping electrons.

But what makes the electrons jump from orbit to orbit? Heat is the most convenient lash. In our insignificant laboratories physicists apply the heat of the electric current to vaporize a metal, such as mercury, or to stir up the electrons in a gas, such as hydrogen, or the neon that now glows ruddily in many a sign over a shop window. Thus the electrons are made to leap from one orbit to another. Bohr showed that when they leap from a high-energy orbit to one of low energy, light is emitted—either visible light or invisible X-rays. He worked out the case of the hydrogen atom very thoroughly and showed that its solitary revolving electron could be made to jump into 30 different orbits. When we think of such a complicated atom as iron, which has 26 electrons, or uranium, which has 92, it is easy to understand why the spectral lines, the visible evidence of these leaps, should become more and more numerous as we proceed up the scale from hydrogen and why, for example, iron should display more than 2,000 lines.

Stripping an Atom

Now the forces that tie an atom together are so tremendous that it is difficult to overcome them and thus tear electrons away. It can be done, however, even on the Earth. When an atom is thus stripped of one or more electrons, it is called an "ion." There are various ways of thus stripping or ionizing an atom. Turning on the X-rays is one; battering an atom with helium particles—Rutherford's method—is another. Both these methods and probably others are responsible for the stripped or ionized condition of atoms that lie deep within the sun, and for the thousands of stars that dot the nocturnal canopy. To such a pitch of perfection has the modern theory of matter been brought that the physicist knows exactly how much energy is required to rob an atom of one or all its electrons. Men like Eddington or Russell will tell you, with the utmost assurance, that at the center of the sun calcium is stripped of all but its two innermost electrons.

But this, in turn, means that the stripping force (heat) must be known. It is. The sun has a surface temperature of about 6,000 degrees Celsius. At the core Eddington and others hold that the temperature must be 40 million degrees. He is not trying to be fantastic—merely to state his mathematical conclusions. Such temperatures actually exist. A temperature of 40 million degrees Celsius means simply that molecules move fast. At ordinary room temperature air molecules rush about with a speed of 500 yards a second; at 40 million degrees Celsius they would dash about at more than 100 miles a second.

The modern physicist is even more certain about pressures. Von Helmholtz was right when he asserted that the sun and stars contract—only he did not happen to hit on the main source of energy. A star contracts because its atoms are pulled together by the attraction of gravitation. The outer layers weigh heavily on the inner. Density in the heavenly bodies is a matter of gravitational pressure. If that pressure is high, atoms are squeezed tightly together; if it is low, they may be separated by miles.

In contraction we have the energy that raises the temperature of the stars until they glow internally with a terrific, inconceivable heat. In this heat we have the force that disrupts atoms into their individual electrons and brings about the transmutation of elements.

The Interior of a Star

Now we are ready to contemplate Eddington's sensational picture or a star's interior:

> Disheveled atoms tear along at 100 miles a second, their normal array of electrons being torn from them in the scrimmage. The lost electrons are speeding 100 times faster to find new resting places. Let us follow the progress of one of them. There is almost a collision as an electron approaches an atomic nucleus, but putting on speed it sweeps around in a sharp curve. Sometimes there is a side-slip at the curve, but the electron goes on with increased or reduced energy. After a thousand narrow shaves, all happening within a thousand-millionth of a second, the hectic career is ended by a worse side-slip than usual. The electron is fairly caught and attached to an atom. But scarcely has it taken up its place when an X-ray bursts into the atom. Sucking up the energy of the ray, the electron darts off again on its next adventure.

Everybody knows that white heat is more intense than red heat. But just what does this mean? Simply that when a metal is white hot it emits waves of light which are shorter than if it were only red hot. Hence the higher the temperature the more dazzling the light and the shorter the waves.

What kind of light would be emitted at 40 million degrees, Eddington's esti-mated temperature of a star's interior? X-rays—light of such short wavelength that we cannot see it, of such penetration that they cannot be produced by our feeble X-ray apparatus. Albert Gockel, Victor Hess, Werner Kolhörster, Millikan are a few of the men who have detected such X-rays in space; but Millikan has done more than any other scientist to measure their intensity and determine their origin.

The higher the measuring instruments are sent, the stronger the rays become. Millikan has sent up instruments in balloons to note automatically how powerful the rays are. He has climbed mountains in Bolivia and California to study them. He is able to assert that they will penetrate several hundred feet of water at high altitudes, which means that they are strong enough to pierce a dozen feet of lead, the very metal that physicians use to protect themselves from the effects of X-rays.

Kolhörster, the German scientist, believes that the supercosmic X-rays of space come from some such nebula as Andromeda. Millikan, whose measurements are the most accurate thus far made, holds that they come not from any particular quarter of the heavens but from every part.

At this point Einstein is invoked by Millikan. According to the older physics, matter is indestructible. Not so, says Einstein. Matter can be converted into energy. And his case is so convincing that every astrophysicist accepts his findings. It is just as legitimate to speak of a pound of heat or light as of a pound of beef in this new Einstein language. Jeans and Eddington calmly tell you that the sun, for example, is radiating away 42 million tons of heat a year.

So, in a very real sense, the stars are throwing themselves away in the form of energy. There is no complete agreement as yet as to the exact nature of the process. Jeans believes that matter is being steadily annihilated and that this is the main source of stellar light and heat—in fact, so good a source that in the past billion years the sun has lost only one fifteen-thousandth of its huge mass. Another source of energy, perhaps a subsidiary source, is found when simple atoms are transmuted into more complex atoms—when, for example, hydrogen is changed into helium. That this latter process actually takes place and that it accounts for his mysterious, highly penetrating cosmic rays, Millikan seems to have established; and therein lies the scientific importance of his recent announcement.

The Birth of X-Rays

But how can the transmutation of matter generate X-rays? It is difficult for us to conceive of any such process, because no chemist has yet succeeded in changing hydrogen into helium. But the electron theory explains the method simply enough.

All matter is fundamentally alike. It all consists of electrons. By adding more protons and electrons to the nucleus of the hydrogen atom, and more revolving electrons, we can theoretically build up any kind of matter we please. So we start with hydrogen and try to build up the next atom in the series, which is that of helium.

Chemists call the mass of the hydrogen atom 1 because it is the lightest of all atoms. In the scale of atomic weights, helium is entered as 4. This means that four hydrogen atoms make one helium atom—theoretically. In actual practice the mass of helium is 3.97. This determination is so precise that only one conclusion can be drawn: when hydrogen is transformed into helium, something must happen to 0.03 parts of the mass. What if matter disappears, what are we to expect instead? Look for energy, says [Albert] Einstein. That 0.03 is converted into radiation—much of it into X-rays of the highly penetrating cosmic type.

On this all astrophysicists are agreed. And now Professor Millikan comes forward with experimental proof that his cosmic rays are born when new matter is created in the stars—when hydrogen is transmuted into helium and heavier atoms. And what is his proof? A spectrograph with lines in it. The lines prove that electrons are jumping from orbit to orbit in an atom quite in accordance with Bohr's theory. What is more, they are in the very positions in which they should be if the Einstein theory is correct—if mass is actually radiated away in the form of energy. In the whole range of physics it would be difficult to find a more satisfying demonstration of the truth of the great principle which Einstein developed mathematically.

Einstein Universe Very Real

Millikan has made one of the striking, fundamental discoveries of our day. Physicists were sure that elements were being transmuted in the stars. He proves their case for them. He furnishes new evidence that the seemingly absurd universe of Einstein—a universe which is finite, in which space is curved, in which light comes by the pound, in which gravitation is regarded as a crumpling up of space—is very real.

Most important of all, the astrophysicist is assured that one source of cosmic energy is the building up of hydrogen into higher elements.

From now on his business is to test and redesign his theories of the constitution of stars.

The first task will probably be that of determining from what type of body the cosmic, super X-rays come. Eddington holds that as matter is destroyed and re-formed within a star the rays reach the outer rind after thousands, even millions of years, but markedly changed. Progressing outwardly, they increase in length—become "softer." In the end they are no longer Millikan's rays, but X-rays of the feebler type we produce in our laboratories to photograph bones and light rays which speed peaceably toward the Earth for centuries, which we can see, and which, therefore, tell us of the star's existence and position. But if these short, hard, piercing rays are thus transformed and tamed, how is it that they are detected in the upper reaches of our atmosphere?

Eddington believes that the main source of Millikan's rays lies in the diffuse nebulae. But this view presents difficulties of which he is fully aware. In a nebula the gases are so rare that the vacuum in an electric incandescent lamp is dense in comparison. In that thin medium, electrons and protons would have to travel for days before they met and grouped themselves into hydrogen, helium and other gases and send out powerful X-rays. This is inconsistent with the doctrine that new elements are built up only when the heat is much more intense than it can possibly be in a nebula, and when atoms and electrons are crowded together. Atoms must be excited if they are to be broken down, re-formed and made to emit X-rays. But, as Eddington says, "The more we face the difficulties of all theories of the release of subatomic energy, the less inclined we are to condemn any evidence as incredible." The main thing is that Millikan has removed the doubt once entertained as to the origin of the cosmic X-rays. They are unquestionably created when new matter is born somewhere in the heavens.

—March 25, 1928

Tests of the Electron Indicate It Is a Wave

By CLINTON J. DAVISSON

Probably no scientific question is the subject of more intensive inquiry today than the nature of the electron. The basic component of all matter, both on the Earth and throughout the universe, its analysis is as important to the astronomer as to the physicist. For long it was regarded as a minute solar system. Dr. Davisson of the Bell Telephone Laboratories caused a stir in scientific circles recently by describing further before the American Philosophical Society his hypothesis that the electron is a complicated vibratory phenomenon. He describes here how he came to this conclusion and outlines its significance.

Scientists frequently make discoveries that lead them to believe they have unraveled the nature of negative electricity, or electrons, tiny specks of electricity so small that even a powerful microscope cannot detect them. The size of these little particles has been pictured by one scientist, who states that if a drop of water were magnified to the proportions of the Earth, each of the electrons composing it would be no larger than a grain of sand.

Electrons are at work for man in thousands of different ways. To a radio receiving set, for instance, they are like blood to the human system. But the more man learns about the magic electron, the more perplexed he becomes. For more than 30 years scientists have been questioning nature about the electron's entity. During all but the last three of these years every one of the answers has permitted them to infer that radiation becomes the source of a new type of radiation which passes through the walls of the tube and is capable of fogging photographic plates even when screens opaque to light are interposed in its path.

It had been discovered that the cathode rays could be bent by magnetic force; the shadows which were cast by objects in their way could be displaced by moving a magnet about in the neighborhood of the tube. We may suppose that the investigator who was first to make this test was not greatly surprised at the result he obtained. The test is one which would occur to anyone familiar with electrical phenomena, and the result is consistent with the acceptable hypothesis that the radiation is composed of electrically charged particles streaming forth from the cathode—negatively charged particles, for the sign of their charge may be deduced from the deflections. They might be negatively charged atoms, but this was not a matter which could be determined by qualitative experiments of this kind.

Observing the paths of the cathode rays, J. J. Thomson in England, and [Emil] Wiechert in Germany, in 1897 and 1898, confirmed the idea that the rays are streams of charged particles. They found that the velocities of the particles varied with the conditions of the experiment, but were always tens of thousands of miles per second.

The Electron as a Particle

During the 25 years which followed upon the work of Thomson and Wiechert there were innumerable experiments of the most diverse kind, and the conception proved adequate for all. Electrons were found to be a universal constituent of matter. They could be abstracted from any kind of matter in a variety of ways; they could be vaporized from matter by heating, a fact which is utilized in the construction of the thermionic tubes in our radio sets; they streamed forth under the action of light and X-rays; they were emitted spontaneously by radioactive materials.

Probably the strongest of the evidence that electrons are corpuscular, or taking the form of minute particles, is afforded by certain experiments of the English physicist C.T.R. Wilson, the "cloud experiments." In a typical experiment of this kind, a beam of X-rays traverses a gas, tearing electrons loose from occasional atoms in their path. Many of these electrons rush off with enormous speeds, tracing long crooked paths through the gas. Each in its turn detaches electrons from many of the atoms which it strikes, leaving behind a train of "ions"—atoms which are positively charged. Wilson's evidence is based upon the fact that upon these ions water vapor is peculiarly likely to condense. He saturates the gas with water, cools it suddenly by letting it expand, and then the trail of each electron is marked by a stream of tiny droplets.

The paths pursued by the individual electrons are individually sharp and clear.

Röntgen discovered X-radiation in 1895. It proved much harder to settle the nature of this radiation than it had seemed to be for cathode rays. Beams of X-rays are not deflected in either electric or magnetic fields. They are not a radiation of electrically charged particles. They might still be a radiation of uncharged particles, or a radiation of waves similar to light, but nature was reluctant to say which. Only after cross-examination which lasted sixteen years was the question put to her in a form to which she would give a clear and unequivocal answer. The successful inquisitor was the German physicist Max von Laue. The answer he received was that X-rays, like light, are a radiation of waves. They are to be compared with the propagation of a disturbance, as when, to use the time-honored simile, the commotion caused by casting a stone into a quiet pool is propagated outward in the form of waves.

The experiment performed by von Laue and his collaborators in 1911 was rather simple; a narrow beam of X-rays was directed through a crystal (of the mineral called zinc-blende) and a photographic plate was set up beyond the crystal to receive whatever rays got through. On developing the plate they found a spot produced by rays which had passed through the zinc-blende undeflected. But in addition—and this was the astounding fact—they saw a multitude of lesser spots sprinkled all around it. There was a beautiful and regular array around the central spot bearing witness that there must have been an array of deviated lesser beams about the undeflected beam. The evidence showed that X-rays are waves.

X-rays and Electrons Both Waves

The question asked of nature by von Laue has been put in as many different forms as experimenters have been able to devise, and always the answer is the same—X-rays are waves.

It occurred to no one in 1911 to repeat von Laue's experiment with cathode rays to find if the pattern of spots produced with X-rays might not be produced also with electrons—and why should it? There was no question about the nature of cathodic radiation. It was definitely corpuscular! The pattern obtained by von Laue was definitely a consequence of the wavelike character of the X-rays. There was no reason to expect that similar experiments with electrons could possibly yield the same or similar results. However, during the last three years, not only the Laue experiment but all of the experiments which demonstrate the wave character of X-rays have been repeated with electrons; and in every instance, wonderful to relate, the result was the same! A beam of electrons projected against a single crystal is broken, like a beam of X-rays, into an array of lesser deviated beams; ring patterns are obtained when electrons are scattered by disorderly masses of little crystals, and so on. And as with X-rays, all the details of these phenomena conform to what one may deduce from the assumption that the incident beam in these experiments is a beam of waves.

A ring pattern resembling the ripples on a pond when a stone strikes its surface was obtained recently by the German physicists Eisenhut and Kaupp, and is for electrons scattered by an aggregate of tiny crystals of silver. Ten or 12 rings appear on the original plate and the diameters of these rings are precisely the ones which would have been observed had the pattern been produced by X-rays of a particular wave length. The reasons for regarding electrons as waves are the same as those for regarding X-rays as waves. If X-rays are waves, so also are electrons.

—June 15, 1930

To Speed Hydrogen to Break Up Atoms

A new apparatus intended to speed up atomic hydrogen particles to 37,000 miles a second in an effort to attain a long-sought goal—the breaking up of the atom—was described to the National Academy of Sciences in Berkeley, California by Professor Ernest O. Lawrence of the University of California.

Professor Lawrence and his colleague, Dr. N. E. Edlefson, have devised an apparatus that promises to give atomic bullets great speed without the use of immense electric fields or super X-ray tubes that have hitherto been considered necessary.

Speeds of some 37,000 miles per second, or energies equivalent to that produced by a million volts, are necessary before the atomic battering can begin. The largest vacuum tubes commonly available are capable of withstanding only about 200,000 volts.

By Professor Lawrence's method, protons, which are hearts of hydrogen atoms, are caught by an electrical impulse of only 10,000 volts, then snatched by a magnetic field, next seized by a current of opposite direction and treated in this way again and again until the protons spiral at ever-increasing speeds and finally are shot out of the apparatus at immense speeds ready to attack other atomic nuclei.

—September 20, 1930

Discovers Neutron, Embryonic Matter

Dr. James Chadwick, working in Cavendish Laboratory at Cambridge, has discovered the neutron—one of the ultimate particles of nature. The discovery, first made public today, was hailed by scientists here as the most important achievement in experimental physics since Lord Rutherford demonstrated the nuclei structure of the atom in 1911.

The particle Dr. Chadwick discovered, according to *The Manchester Guardian's* scientific correspondent, appears to consist of a proton and an electron bound together, and hence without an electric charge. It was therefore named the neutron, representing the first step in the evolution of the elements out of electricity and may well be the material of the famous cosmic rays, which might be streams of neutrons.

Some physicists here are of the opinion that it is one of the fundamental discoveries in physics, ranking with the discovery of the electron, the proton and the X-ray. A question also has been raised as to the practical use of the discovery—whether it will result in entirely new experiments, as the X-ray has done.

The neutron represents the first step in the evolution of matter, the first step in the building up of the common materials of everyday life out of primeval electrons and protons. It is the embryonic form of ordinary matter, growing but not yet born.

Scientists Hail Discovery

"The discovery is of the greatest interest and importance—possibly the greatest since the artificial disintegration of the atom," declared Lord Rutherford, director of the Cavendish Laboratory. "It already has afforded a number of examples of unexpected types of atomic disintegration and offers a promising approach to a number of important problems."

The neutron's principal properties bear out a prediction by Lord Rutherford twelve years ago in a lecture before the Royal Society, when he discussed possible types of matter then unknown.

The discovery recalls Dr. Chadwick's first researches as a student at Manchester University. In 1912, at the age of 21, he discovered that substances struck by the heavy particles ejected from radium emit a penetrating wave radiation. He has now shown that the same kind of particles may also cause neutrons to be emitted.

Later, Dr. Chadwick studied at Berlin and Cambridge and became one of Lord Rutherford's chief lieutenants during a period of radioactivity research at Manchester. Always distinguished by exceptional thoroughness, now he has added special inspiration to his gifts.

Completed in Fortnight

Those who heard him describe the latest research, which was completed within a fortnight, said to be a record for an important discovery in modern physics, were profoundly impressed by his mastery of detail and theory.

During the World War, Dr. Chadwick was interned at Ruhleben, where he improvised a small research laboratory in camp and passed his time experimenting. C. D. Ellis, a fellow prisoner, who had been interested in physics, revived his interest and after the war went to Cambridge, becoming one of the chief figures at the Cavendish Laboratory. Dr. Chadwick and Mr. Ellis collaborated with Lord Rutherford in the recent revision of Lord Rutherford's treatise on radioactive substances. When Lord Rutherford left Manchester for Cambridge, Dr. Chadwick went along as his right-hand man.

—February 28, 1932

Chadwick Calls Neutron "Difficult Catch"

By FERDINAND KUHN JR.

Dr. James Chadwick, 41-year-old Cambridge physicist, explained to this correspondent the experiments which led to the discovery of a new type of ultimate particle, consisting of a proton and an electron bound together without an electric charge and hence called a "neutron."

While British scientists are eagerly discussing his discovery and Sir Arthur Eddington has called it "tremendously important," Dr. Chadwick refused to make any sweeping claims for what he had found. The existence of the neutron is "probable," he declared, but with the true scientist's caution he would not say he was absolutely certain. For the present he is inclined to understate his own case in the interest of scientific accuracy.

So far, Dr. Chadwick said, his experiments have been convincing, but he plans to exhaust every possible means of proof before he is completely satisfied. Experiments in the Cavendish Laboratory will continue for days and possibly weeks, he said, until the chain of evidence becomes unbreakable.

"It is curious that the neutron has been with us almost two years and we never knew it," he explained. "We have suspected its existence more than a year, but it was impossible to get a really conclusive experiment. Perhaps we did not have the proper technical equipment or possibly our first experiments were not correctly performed.

"In any event, the neutron was difficult to catch. Other particles can be seen and their actions watched, but the neutron we could not see and it left no traces of its passage. It was not till we got photographs that we were reasonably certain of the neutron's existence."

The discovery of the neutron was made, he explained, in the process of investigating the atomic nucleus with the help of earlier researches by Irène Joliet-Curie, daughter of the discoverer of radium, and Professor Robert A. Millikan, the California investigator of the so-called cosmic rays. With the technique which has made the Cavendish Laboratory world famous, Dr. Chadwick and his associates proceeded to bombard atoms of metal beryllium with radioactive rays which physicists know as "alpha" particles.

"Myriads of alpha particles bounded off the atom, but one or two penetrated to the inside of the nucleus with curious effect," Dr. Chadwick explained. "Normally the entry of an alpha particle causes the protons to escape from the atom, but in this case no protons escaped. There was simply radiation which was neither waves of light nor protons nor electrons. It arose neither from vibrations nor from positive or negative electric charges. The only explanation was that it was caused by neutrally charged particles which, we decided, must be neutrons."

Many of the neutron's effects in passing through matter resembled those of the quantum of high energy, Dr. Chadwick continued. "But all the evidence so far has been in favor of the neutron. The quantum hypothesis can only be maintained if conservation of energy and momentum is relinquished at some point. Because neutrons are small and have no electric charge, their range in the air is more than one mile and they can pass through many feet of lead. Carrying no electric charge, they are not repelled by charged atoms which they approach. They can find a way between atoms much more easily than protons and electrons.

Dr. Chadwick only smiled, however, when asked if his discovery had practical importance.

"I am afraid neutrons will not be of any use to any one," he said. "It may be, of course, that the so-called cosmic rays consist of neutrons, but I doubt it."

Sir Arthur Eddington is even more dubious than Dr. Chadwick as to the neutron's importance in explaining cosmic rays.

"I see no reason for believing cosmic rays are made up of neutrons," said the noted astronomer-physicist, sitting in his study beneath the dome of Cambridge Observatory. "In fact, that is the aspect of Dr. Chadwick's discovery which interests me the least.

"But I think it is a tremendously important discovery in helping us to understand the structure of nuclei. Hitherto the simplest nucleus we have known has been of six particles in combination. Now we have a nucleus apparently consisting of two particles bound together. In this way Dr. Chadwick's discovery may become important in throwing light on the origin of matter. It will be much easier to investigate the building of a nucleus composed of two particles than one of six."

Sir Arthur did not witness the actual experiments leading to the discovery, but he examined photographs and said he was greatly impressed with them. In one way, he explained, it was an advantage that the neutron could not be seen.

"When alpha rays are photographed," he said, "the plate is all cluttered up with traces of rays which fail to reach their objective inside the atom and usually they hide the most interesting part of the picture. In the case of neutrons, the photograph gives clear evidence of the disrupted atom."

—*February 29, 1932*

Bombardment of Atoms

If we are ever to find out how matter is constituted we must break up the atom. This means that it must be blasted open. Now an atom consists of a highly complicated nucleus of protons and electrons, around which from one to 92 negative electrons are grouped. The electric forces that hold such a structure together are terrific. It takes energy to break through the outer electrons and to reach the inner nucleus.

Lord Rutherford was the first to succeed in this bombardment. He used alpha particles (helium nuclei) ejected by radium so energetically that they have a speed of 12,000 miles a second. It was thus that he managed to break down some elements and to bring about transmutation.

While the energy represented by 12,000 miles a second is high, it is not high enough. Rutherford did little more than chip the nuclei of a score of atoms. Nothing but a complete disruption will satisfy the physicist. So we find experimenters all over the world devising apparatus which would have done Edison credit and which is designed to impart to electrons, protons and alpha particles speeds of 80,000 miles a second and more.

The striking energy of a hammer depends upon its mass and its speed. Electrons have been used in atom bombarding. Their mass, however, is low. A proton is 1,840 times as massive as an electron. Give it speed enough and it ought to plow through the outer electrons of an atom and strike the nucleus. It was thus that John Cockroft and Ernest Walton proceeded in their disruption of lithium and thus that Professor Ernest O. Lawrence of the University of California split atoms.

A Question of Apparatus

To deliver a blow represented by a million volts, it is usually the practice to apply energy of at least a million. There is no insuperable difficulty in thus obtaining accelerating voltages of many millions, but the size of the apparatus and of the laboratory that contains it, not to mention the cost, increases rapidly with the voltage.

Professor Lawrence of the University of California and his colleague Professor Milton S. Livingston adopt a different principle. With a huge electromagnet, weighing tons, they whirl protons (heavy hydrogen nuclei) on the slingshot principle. Faster and faster the protons whirl. Wider and wider are the circles that they describe. When their energy is high enough, they are shot through a slit in the containing vessel at the target of atoms. Thus with only 10,000 volts Lawrence has imparted to protons an energy of 1.5 million volts. He hopes to attain 10 million and even 20 million, of course with correspondingly higher starting voltages. What this means may be imagined when it is considered that a proton with only an energy of 1 million volts can cover about 40 miles while an ordinary rifle bullet is traveling a foot.

Up to the present Lawrence and Livingston have hurled protons at atoms with an energy of 4.8 million volts. Last December they thus disintegrated lithium and boron atoms. About 10 billion protons a second were shot at invisible atomic targets. Hits at the rate of 90 to 118 a minute were made in the case of lithium. Sharpshooting being impossible, it became necessary to discharge billions of projectiles on the chance that once in a while a bull's-eye would be scored.

—*April 9, 1933*

Jekyll-Hyde Mind Attributed to Man

By WILLIAM L. LAURENCE

The Theory of Complementarity, a new theory of knowledge, based on the concepts of modern atomic physics, was presented in Chicago for the first time before an American audience by Professor Niels Bohr of the University of Copenhagen, originator of the Bohr model of the atom and Nobel Prize winner.

The new theory is expected by some of the leading scientists here, attending the summer meeting of the American Association for the Advancement of Science, to take its place alongside the relativity and quantum theories as one of the revolutionary developments of modern scientific thought.

Complementarity is the outgrowth of relativity, quantum mechanics and the Heisenberg principle of uncertainty. The new theory expands the uncertainty principle beyond the realm of atomic physics, where it had been primarily applicable, to include man's entire relation to the world around him and to all processes of knowing and thinking.

Briefly, and in non-technical language, Professor Bohr, after a lifetime of contemplation of both the ponderables and the imponderables of the physical and mental world, has come to discover an inherent essential duality in the nature of things, as they relate to man's ability to know them. The paradox of this duality lies in the fact that the Jekyll-Hyde nature of all things is essentially contradictory, with both aspects being true at different times, but with only one aspect being true at any one given time.

Duality Held Inescapable

In other words, the very process of knowing one aspect of nature makes it impossible for us to know the other aspect. We can know only one side of its nature at any one time. There is a definite discontinuity in all things partaking of existence and of knowledge, so that when one thing is true this very truth perforce makes another thing non-existent as far as any possible knowledge on our part is concerned.

This contradictory duality is inescapable because it lies at the very heart of things.

It is wrong, according to this theory, to say that there are either free will or determinism, causality or chance. Both are essential parts of one and the same reality, the convex and concave sides of the same sphere. Both are true at different times, but it is never possible to know both at one and the same time. When you are inside the sphere, the sphere is concave and it is never possible to experience its convex aspect. In that case it is impossible to have any knowledge of the convex aspect. When one gets outside the sphere, only the convex side becomes the reality.

Dr. Bohr told the story of how he and Einstein had recently joined intellects to find a way somehow to get around the troublesome principle of uncertainty, according to which it is possible to determine either the position or the velocity of an electron but absolutely impossible to determine both the position and the velocity at the same time.

Thwarted by Uncertainty

After wrestling with the problem for many days, the two super-minds believed for a time that they had accomplished their ends. Jubilantly they began working out the details of their solution only to find themselves in another intellectual vicious circle. After believing that they had succeeded in killing the monster of "uncertainty," it was there, Mephistopheles-like in another form, mocking at them.

This finally convinced both Bohr and Einstein that the "uncertainty principle" is not the result of a lack in our knowledge but is an inherent part of the very mechanism of knowledge. Trying to get around "uncertainty" in nature would be, to use a wholly unscientific phrase, like trying to preserve the hole while at the same time wanting to eat the doughnut.

The uncertainty principle, one of the most startling developments in modern physics, was first announced by Professor Werner Heisenberg of Leipzig in 1925, and created as much of a sensation in scientific circles as did the original announcement of the theory of relativity.

Science, depending for its results on accurate measuring instruments and on observation, discovered an inherent weakness in its fundamental structure. It found that in order to observe anything you must have light to see it by. Yet it found to its consternation that light does something to the object observed, changing it somehow, so that actually one could never hope to observe any object at any particular position as it really is before the light had affected it.

Points to Effect of Light

The effect of the light on an observed object does not mean merely the change in the color of the object. Light, it was found by Professor Arthur H. Compton of the University of Chicago, another Nobel Prize winner, acts in the manner of a small bullet when played on an electron, so that it collides with it and changes its momentum and its energy.

By the intricate laws of quantum mechanics, theoretical physics has devised means which make it possible to determine either the position of an electron or its velocity. But the "principle of uncertainty" decrees that one cannot determine both the position and the velocity of an electron at the same time.

In his address Dr. Bohr first pointed out that our recently gained knowledge of atoms and molecules demanded fundamental changes in our general philosophical outlook.

"We have been forced to recognize," Professor Bohr said, "that we must modify not only all our concepts of classical physics but even the ideas we use in everyday life—such as our ideas of space and time.

"Indeed, the features which are so characteristic of atomic phenomena are of such a nature that they cannot be analyzed in ordinary mechanical and electrodynamic concepts, and we have to renounce a description of phenomena based on the concept of cause and effect.

"However, since all measurements must ultimately be interpreted on the basis of classical ideas, we cannot desert or modify these ideas but are compelled to apply them also in our description of atomic phenomena.

"We are thus led to a description of the universe which is 'complementary'; that is, in quantum mechanics the application of any classical concept will invariably exclude the simultaneous use of other classical concepts.

"Logically speaking, it is this peculiar 'complementarity' which leaves room for the atomic phenomena, and the principle of uncertainty."

The apparent contradiction between classical physics and modern atomic physics, according to Professor Bohr, does not mean that both one and the other are wrong or that one must be discarded in favor of the other.

On the contrary, Dr. Bohr holds, this realization of the "complementarity" of our knowledge furnishes us with the long-looked-for bridge with which to link the contradictory concepts of classical and atomic physics.

This "bridge" is our recognition as a fact, or rather as an axiom, that we can apply any classical concept to any modern concept of the quantum theory or of relativity, but that we cannot apply any more than one classical concept at the time.

The difficulty in the apparent inability of modern thought to reconcile the old concepts with the new lies in the fact, Dr. Bohr holds, that we tried to apply all the old concepts to all the new concepts simultaneously. If we do it one concept at a time, the apparent contradictions disappear.

Dr. Compton Presides

Dr. Compton, whose work on the collision of photons, such as light and X-rays with electrons, led to the formulation of the "uncertainty principle," and thus directly to the Bohr theory of complementarity, presided at the meeting, which was held in the Illinois Host Building of the Century of Progress Exposition.

After the meeting Dr. Compton remarked that Dr. Bohr's address was a "brilliant presentation, such as one only of Dr. Bohr's caliber was capable of, which brings into harmony the electron theory, the quantum theory and the theory of relativity."

—June 23, 1933

Fermi Measures Speed of Neutron

Professor Enrico Fermi of Italy reported before the International Conference on Physics in Cambridge, England, that he had measured the time it takes for "an irresistible force" to traverse "an immovable body."

He said that when a neutron, the most penetrating particle so far discovered in nature, was hurled against the nucleus of an atom, the most impenetrable citadel of matter, the neutron penetrated inside the nucleus, but remained there only one thousand million million millionths of a second—this being the time needed for the neutron to traverse a distance of one million millionth of a centimeter.

Dr. Fermi, one of the world's leading physicists, bombarded atoms of uranium, the heaviest of all the elements, with neutrons, and by this process of modern alchemy transformed the uranium into another element. At first it was believed that the new element was a superuranium, heavier than any element produced by nature and occupying No. 93 on the periodic table of elements, on which uranium occupies the last place, or 92. It is now generally believed, however, that the new element created by Dr. Fermi is an isotope of protactinium, Element 91.

So powerful is the neutron, Dr. Fermi reported, that it seems to penetrate the nuclei of the heavy elements with the same ease as it does those of the light elements.

It now seems probable, he said, that the disintegration of elements by the bombardment of neutrons does not depend on their atomic weight, as the heavy elements apparently can be transmuted by the battering of the neutrons as easily as the light.

There apparently exists a difference, however, in the mechanism of the disintegration of the heavy and light elements, respectively, he added.

Tells Neutron's Power

Dr. James Chadwick, discoverer of the neutron, addressing the conference in Cambridge, an ancient city which is now the world's greatest center of experimental physics, showed how far the knowledge of atomic transformations had progressed since the wintry day in 1932 when he detected the new ultimate particle at Cavendish Laboratory and called it the neutron.

By now, according to Dr. Chadwick, the neutron has proved to be the most powerful instrument for breaking up atoms.

"Even slow neutrons are able to penetrate into a heavy nucleus," he said.

"As neutrons are uncharged, they are not repelled by nuclei. Thus we may expect the neutron to prove more effective than the alpha particles as a promoter of nuclear transformations. These general considerations have been supported by experiments."

Both light and heavy elements can be transformed by a stream of neutrons, he continued. He cited cases in which the disintegration of heavy elements by neutrons had led to the production of artificial radioactivity, but he warned that there was no reason to suppose such a phenomenon would always result.

—October 5, 1934

Bohr and Einstein at Odds

By WILLIAM L. LAURENCE

The world of science is watching with the keenest interest the development of a modern controversy which promises to take its place along with the other historic controversies in the realms of philosophy and science. The participants are Albert Einstein and Niels Bohr (show at left), two men who have achieved the rare distinction of having joined the ranks of the immortals during their lifetime. The subject concerns the fundamental nature of physical reality, the problem of problems which has engaged the minds of all great philosophers, poets and scientists throughout the course of human history.

The Einstein-Bohr controversy has just begun this week in the current issue of *Nature,* the British scientific publication, with a preliminary challenge by Professor Bohr to Professor Einstein and with a promise by Professor Bohr that "a fuller development of this argument will be given in an article to be published shortly in the *Physical Review.*" Undoubtedly, Professor Einstein will accept the challenge and very likely Professor Bohr will answer, and there is the possibility that other leading scientists will enter the arena.

Bohr's challenge has no relation to Einstein's theory of relativity, with which he is in full agreement, but concerns itself with a defense of the quantum theory, particularly the famous Principle of Uncertainty, which states that both the position and velocity of a particle cannot be accurately known at the same time. The challenge comes in reply to an attack on the quantum theory in an article by Einstein in a recent issue of the *Physical Review,* written in collaboration with Drs. Boris Podolsky and Nathan Rosen, of the Institute for Advanced Study, Princeton, New Jersey.

Cornerstone of Science

The quantum theory, uncertainty principle and all, constitutes one of the very cornerstones of modern science. All modern knowledge on the structure and constitution of the atom is based on its principles. Einstein himself played an important part in its development. Yet, with the characteristic boldness that stamps his other scientific contributions, Einstein announced his conclusion that the quantum theory is "incomplete" for the reason that it does not provide "a complete description of physical reality," and added his belief that another theory must exist which does provide a complete picture of the universe around us.

Professor Einstein's position, as explained by Dr. Podolsky, may be summarized as follows: Physicists believe that there exist real material things independent of our minds and theories. We construct theories and invent words (such as electron, proton, positron, neutron, etc.) in an attempt to explain to ourselves what we know about our external world and to help us to obtain further knowledge of it.

Before a theory can be considered satisfactory, Einstein and his colleagues hold, it must pass two severe tests. First, the theory must enable us to calculate facts of nature, and these calculations must agree accurately with observations and experiments. Second, we expect a satisfactory theory, as a good image of objective reality, to contain a counterpart for every element of the physical world.

A theory satisfying the first requirement may be called a "correct theory," according to Einstein, while, if it satisfies the second requirement, it may be called a "complete theory." The quantum theory, he maintains, fulfills the first requirement of being "correct," but it does not fulfill the second requirement, and is, therefore, "incomplete."

Theory "Incomplete"

The "incompleteness" of the quantum theory, according to Einstein, lies in the startling results of its uncertainty principle, one of the most revolutionary developments in the realm of thought. On the one hand, it tells us that a collision between a bullet of light, or a light quantum, and a particle of matter, such as an electron, follows the same fundamental principles of the conservation of energy and the conservation of momentum as does a collision between two billiard balls.

But whereas it is possible to predict from the behavior of the billiard balls before the collision how they will move after it, no such prediction can be made, the uncertainty principle tells us, concerning a collision between a bullet of light and an electron. We can measure accurately either the position of the electron or its velocity after collision, but we can never measure both at the same time.

This is not merely a statement of a didactic principle but an experimental fact. No one so far has been able to predict in actual experiment how both the electron and the bullet of light will bounce away after collision.

Since the ability to predict how two bodies will behave under given conditions is inherently linked in our minds with our ideas of cause and effect, the discovery that the behavior of the constituents of the atoms of which the universe is made up is essentially unpredictable led to the conviction among modern scientists that the universe as a whole, instead of being governed by immutable laws of cause and effect, in an endless chain of interactions between physical bodies and energy states, is, after all, governed by mere chance. Indeterminism thus took the place of determinism, and nature ceased to be intrinsically either understandable or subject to immutable law.

It is against this aspect of the quantum theory that Einstein rebels. He refuses to believe that such a picture actually fits reality. Unlike Sir Arthur Eddington, who once said that no facts are true until they have been corroborated by theory, Einstein holds that no theory is true until it embraces all the facts. If, like the quantum theory, it can explain the facts individually, but cannot explain some of them when acting together, something must be lacking in the theory, and another theory must exist that gives a more complete picture of reality.

Suppose we gaze at reality through a distorted mirror. Should we assume that reality itself is distorted? Thus argues Einstein. But, the champions of the quantum theory reply, suppose the only means available to us for gazing at reality is this one and only mirror? How can you tell, under such circumstances, whether or not the mirror is distorted? To which Einstein replies: I have two mirrors, and each gives me a picture irreconcilable with that of the other. There must be a third mirror, in which these contradictions would be reconciled.

Question of Dualism

And right here Bohr, Nobel Prize winner and originator of the Bohr model of the atom, takes issue with Einstein. The trouble with you, he tells Einstein, is that you are a monist who believes that there must exist one unified picture of the whole universe. This is impossible, for the universe itself is made up of an inherent dualism. You are not really looking at the world through two distinct mirrors; what you are actually looking at is a universe with a dual personality, for the universe itself is a cosmic prototype of a Dr. Jekyll and Mr. Hyde.

This is the essence of what Bohr calls "complementarity." Nature has a "dual aspect in which, by a paradox inherent in its very essence, things are fundamentally contradictory; both aspects are true at different times, but only one aspect is true at any one given time. When the one thing is true the very truth of it perforce makes another thing non-existent as far as any possible knowledge on our part is concerned."

It is incorrect, therefore, Bohr tells Einstein, to say that there is either free will or determinism, causality or chance. Both are essential parts of one and the same reality, the convex and concave side of the same sphere. Both are true at different times, but it is not possible to know both at the same time.

The uncertainty principle, Bohr is convinced, is not the result of a lack in our knowledge, as Einstein has grown to believe, but an inherent part of the very mechanism of knowledge.

In this respect we are in a position similar to that of the legendary Lady of Shalott who could look at the reality of "many-tower'd Camelot" only through the magic mirror. The minute she became "half sick of shadows" and "look'd down to Camelot . . . the mirror crack'd from side to side" so that she had no means left of comparing the shadows she had seen throughout her life in the mirror with the corresponding reality that was Camelot outside her window.

The quantum theory, according to Einstein, is like the Lady of Shalott's mirror—it gives a correct reflection so long as you look into it, but when you want to turn from it to compare the reflection with the original, it "cracks from side to side." Such a mirror is undesirable, Einstein holds.

To this Bohr might reply: The fault lies not in the mirror, but in the fact that there is a curse upon the Lady of Shalott. And the Lady of Shalott in this case would symbolize the very essential nature of human knowledge. We may know things only in their contradictory duality and we never can know them otherwise.

—July 28, 1935

Discovery of the Antiproton Ends a Long Search, Confirms Einstein's Equation

By WALDEMAR KAEMPFFERT

Important though it is the announcement from the University of California that, with the aid of its 6.2-billion-volt Bevatron, Drs. Owen Chamberlain, Emilio Segrè, Clyde Wiegand and Thomas Ypsilantis, and Edward J. Lofgren and Herbert Steiner have discovered the antiproton (negative proton) comes as no great surprise. Physicists have been searching for the antiproton for about 30 years. Nobel prize-winner Paul Dirac thought he had discovered it, but his discovery turned out to be the positron or positive electron. It was a mathematical discovery, a prediction that there must be such a thing. Later the positron was discovered by Dr. Carl Anderson at the California Institute of Technology. The function of the positron in the cosmic rays in radioactive disintegration was not apparent at first. So it is with the antiproton. Physicists do not know what to do with this new particle.

The nucleus of an atom is composed of neutrons, which have no electrical charge, as the name implies, and of protons, which are the cores of hydrogen atoms and which are positively charged. What holds the nucleus of an atom together? The answer is what the physicists call "binding energy," a hypothetical force that is far more powerful than gravitation. Physicists would like something that is not so speculative. It would seem that in the antiproton they have exactly what they want. But they know the antiproton will not serve. When two particles of opposite signs are brought together, they annihilate each other. That is what happens when a positive electron (positron) meets an ordinary electron, which is negative. Physicists will not change their conception of the nucleus because of the discovery of the antiproton. They will still have to speculate about "binding energy." But is likely that the antiproton will help them to explain more simply phenomena that are baffling—just what happened when the positron was added to the list of fundamental particles. It may well be that mysteries still associated with the cosmic rays will be more cogently explained than is possible now.

Bevatron in Action

The antiproton was created in the Bevatron, a huge atom-smashing machine for which the Atomic Energy Commission paid $9.5 million to foster fundamental atomic research. Until the Bevatron was built, it was impossible to find out whether there was an anti-proton. The older atom-smashing machines were not powerful enough to batter target atoms with sufficient energy. Probably the antiproton will be discovered in the atmosphere with the aid of cosmic rays, which are even more powerful than those produced by the Bevatron.

What happens in the Bevatron confirms Einstein's famous mass-energy equation which says that energy can be converted into mass and mass into energy. In the Bevatron protons are accelerated; that is, they are given ever higher and higher speed until the maximum is reached. At this critical point they are hurled against a target of copper. When a proton with an energy of 6.2 billion volts smashes into a neutron in the copper target atoms, strange things happen. Out of the collision of the proton projectile and the struck neutron comes a brand new pair of heavy particles, a proton and an antiproton. Part of the bombarding proton's energy is converted into mass in accordance with Einstein's equation. This part is the antiproton. So the antiproton is born out of nothing but energy—the highest that man has so far been able to produce.

Unstable and Floating

This antiproton is annihilated when it encounters another proton or a neutron. When the two collide they decay into mesons, which are short-lived particles that appear when cosmic rays bombard atoms in the air. The mesons, in turn, vanish almost as soon as they are born, which helps to explain why the antiproton is so unstable and fleeting, except in a vacuum.

The discovery of the antiproton was one of the objectives that the University of California physicists had in view when they designed the Bevatron. Long before the machine was built, the physicists knew what the characteristics of the antiproton should be, such as its speed and its effect on the proton or neutron that it struck. It was necessary to look for a particle that had those characteristics. This was done with appropriate instruments.

—October 23, 1955

Discovery of New Particle
Called "Crucial Test" of Theory

By WALTER SULLIVAN

Detection of the new subatomic particle, the omega-minus, was described by its discoverers as a "crucial test" of a theory that could mark a turning point in particle physics.

It may play a role in bringing order out of the chaos of subatomic particles comparable to that played by the periodic table of elements. The latter was devised about 1870 by the great Russian chemist Dmitri Ivanovich Medeleev.

By arranging the elements in parallel columns, according to their atomic weights, Medeleev showed that they fell into groups with common properties. Furthermore, there were obvious gaps in the table and these enabled him not only to predict the discovery of new elements, but their properties as well.

Moreover, the fact that the elements fell into such a striking arrangement showed that there was an inner symmetry in the structures of these atoms—a symmetry that was yet to be discovered.

This is what now has happened on the much more fundamental level of the atomic particles. In Medeleev's case the discovery of his three predicted elements, gallium, scandium and germanium, persuaded the scientific world of the validity of his hypothesis. The omega-minus performs the same role, in that its properties are peculiar and would have been unexpected but for the concept known as "the Eightfold Way."

This theory was proposed independently, early in 1961, by Dr. Murray Gell-Mann at the California Institute of Technology and Dr. Yuval Ne'eman of the Imperial College of Science and Technology in London. Dr. Ne'eman, strange to say, was a newcomer to physics. He was a colonel in the Israeli Army who suddenly decided to turn physicist.

The theory did not come to them out of the blue. Others had recognized various symmetries and relationships between the subatomic particles. Likewise, Medeleev's predecessors had experimented with groupings of elements. In both cases it was dramatic success in prediction that demonstrated the validity of the theory.

As Dr. Maurice Goldhaber, director of the Brookhaven National Laboratory, Upton, Long Island, put it at a news conference, "Most people smiled when they spoke about the Eightfold Way." They do not smile any longer.

It was with the world's most powerful accelerator, or atom smasher, at Brookhaven that the omega-minus was discovered. The experiment is described in the issue of *Physical Review Letters* dated Feb. 24, 1964. The scope and importance of the experiment is illustrated by the fact that the report is signed by 33 participants.

The proliferation of particles produced when the nuclei of atoms are broken apart by high-energy bombardment has been a source of dismay to physicists. The number of such particles, including some that, because of their extremely short lifetimes, are called "resonances," is at least 82. The discovery of the omega-minus, however, has been a cause for rejoicing.

The Eightfold Way was suggested by a form of algebra developed in the last century by the Norwegian Sophus Lie. Part of this algebra deals with eight components and this seemed to Dr. Gell-Mann and Dr. Ne'eman to be applicable to eight "conserved quantities" characteristic of the various particles.

These quantities in general cannot be explained in terms of familiar concepts, in that they involve mathematical relationships. For example, one of them, "isotopic spin," is derived from the electric charges characteristic of a particular group of particles.

Some particles have a positive charge, some a negative one and some have no charge at all—a zero charge. One kind has two positive charges. Isotopic spin is an arbitrary number equal to the number of charge states characteristic of a given particle (plus, minus, etc.) minus one, divided by two.

For example, a group of particles known as pions appear in three charge states: positive, negative and zero. Their isotopic spin is therefore three minus one, divided by two, which is one.

The other quantities of the Eightfold Way are similarly derived. Some of them, as stated in an article by Dr. Gell-Mann and others in the February issue of *Scientific American,* have not yet even been named.

However, it was found that application of the hypothesis to eight particles, some of which had hitherto seemed quite unrelated, showed them to be, in a sense, variations of the same particle, differing only in energy levels (or mass) and electric charge. The eightfold nature of this symmetry seemed further justification for the title of the theory.

Further eight-part groupings were identified. However, there was one misfit group of particles, called by Dr. Gell-Mann and his colleagues the deltas, with a mass of 1,238 million electron volts. It is customary in physics to describe the mass of a particle in terms of its equivalent value in energy, mass and energy being interchangeable in nuclear reactions.

The delta particles, discovered in 1952, appear in four electric states: plus, minus, zero and double-plus. This is the only group of particles with four such states. The eight-part patterns were no more than three units wide and hence could not accommodate four particles. Thus it was proposed that a ten-unit pattern, formed as a triangle or pyramid, was needed to accommodate the deltas.

The four deltas would form the base. And three sigma particles would form the next layer. The discovery of two xi particles with a mass of 1,530 fit the third layer. Hence, at the 1962 high-energy physics conference in Geneva, Dr. Gell-Mann urged that a search be made for the particle needed to crown this pyramid.

He described the characteristics that it must have to fit this slot and named it the omega-minus. Its properties were in part based on the work of Dr. Susumu Okubo of the Universities of Rochester and Tokyo.

Meaning Still Obscure

Its discovery, in the words of Dr. Goldhaber, "forms the capstone in a building which was so far held together only by the bold imagination of Dr. Gell-Mann and Dr. Ne'eman."

Like the periodic table of elements and the complex arrangements of lines in the spectrums of light emitted by atoms, the structures of the Eightfold Way clearly have some inner significance. However, this meaning, in the last-named case, is still obscure.

The Eightfold Way is described by Dr. Gell-Mann in terms of a new set of 10 particle names that many physicists consider more logical than the old system. Grouped together under each of these 10 names are those heavy particles (baryons) with the same isotopic spin and hypercharge.

Hypercharge, like isotopic spin, is one of the quantities that determine the patterns of the Eightfold Way. It is related to the average electric charge of a group of particles and to their "strangeness." The latter is a property of particles so named because it seemed strange when first discovered.

When particles are defined in terms of these quantities, some with the same name have widely varying masses. It is the assumption of Dr. Gell-Mann and others that these differences in weight are actually differing energy levels in the same particle. In high-energy reactions, such as those required to produce these short-lived fragments, energy is converted into matter and vice versa.

Search at Brookhaven

The search for the omega-minus began in late 1963 at several laboratories. At Brookhaven the basic tool was a high-energy beam of protons, or hydrogen nuclei. These smashed into a tungsten target generating a multitude of fragments, including a variety known as negative K mesons.

With a 400-foot array of magnets and electrostatic separators, these were extracted and directed into an 80-inch bubble chamber. This device, filled with liquid hydrogen at its boiling point, is the world's largest of its kind. The mesons occasionally struck hydrogen nuclei in this chamber, and researchers hunted for the telltale decay products of the encounter that theoretically might produce the omega-minus.

The latter would decay, almost immediately, into other particles that could be identified. Thus the omega-minus would be known by its children. On Jan. 31, 1964, one such decay mode was observed and, since then, a second form has been detected.

Dr. Ralph P. Shutt was in overall charge of the experimental team with Dr. Nicholas Samios responsible for conduct of the experiment. Dr. William B. Fowler was in charge of the bubble chamber and Dr. Medford S. Webster headed the group that produced the K mesons.

—*February 23, 1964*

Two Men in Search of the Quark

By LEE EDSON

In the last couple of years an intensive hunt has been going on all over the world for an elusive quarry known as the quark. This is no Alice-in-Wonderland adventure; far from it. The hunters are some of the world's leading physicists. The hunting grounds are almost anywhere from the high atmosphere to the bottom of the sea to the inside of the latest atom smashers. One enthusiastic researcher at the University of Michigan even has been grinding up oysters on the theory that an oyster eats almost anything—so why not a quark?

Despite this painstaking search, the quark so far has remained as hard to track down as Lewis Carroll's Snark. There is an excellent reason. According to modern theoretical physicists, the quark, if it exists at all, is the simplest particle in the universe, out of which almost everything else is made.

Capturing this incredibly wraithlike substance may not help us produce super-bombs—the implications of gaining new elementary facts about nature depend on how man uses them, whether for war or for peace, and are not ascertainable for years, anyway. But to physicists, the excitement in confronting the quark is something more sublime than the discovery of a new application to everyday life. It is the immediate realization that through it we may uncover the missing linkage to our understanding of the structure of matter throughout the universe.

The men largely responsible for sending scientists on this wild quark chase are two California Institute of Technology physicists named Murray Gell-Mann and Richard Feynman. Each has won a fistful of high honors. In 1965, Feynman shared the Nobel Prize for his achievements in explaining some of the abstruse mechanisms in the subatomic world. Many physicists think Gell-Mann is next. One California scientist calls the two men "the hottest properties in theoretical physics today."

What makes them radiate so fiercely on the high-energy physics circuit is a fine blend of showmanship and brilliance. Take the quark hunt, for instance. "Dick and I were batting around some aspects of theoretical physics," Gell-Mann recalls. "We started to get excited about a new theory and threw out words for our ideas. There have been some crazy ones in physics lately. The theory depended on a triplet of particles, with the right characteristics, and we needed a word for it. I started to say 'squeak,' 'squark,' and it came out 'quark.' We loved the word as soon as it was uttered. Much to my surprise, I found the line 'Three quarks for Muster Mark' in James Joyce's *Finnegans Wake*. Nothing could have fitted better."

Of course, both men could have rejected the word for something sane and august and ending with "on" like "electron" or "neutron," or even used "ace, deuce, trey" as later suggested by physicist George Zweig of Caltech, who came to a similar conclusion about the nature of the fundamental particle, but Gell-Mann admits to a certain puckishness in accepting and publicizing the quark. "Maybe it will help C. P. Snow to bridge his two cultures," he says with a smile.

Gell-Mann and Feynman—who claim they work together separately—are nearly unique in modern physics. In a field noted for quiet introversion, they generate a kind of charisma which draws students and faculty members to Caltech to an extent that had not been seen on the campus since the days of Robert Oppenheimer. The lectures of both men play to SRO audiences. When Feynman won the Nobel Prize, one blasphemous sophomore admirer celebrated by slipping his picture into a plaque of the Last Supper in place of Jesus.

Although they have comparable intellectual impact, Feynman and Gell-Mann create their sparks in entirely different ways. Feynman, a lean, intense, dark-haired man of 49 who is beginning to fear that he may soon be regarded as a premature elder statesman, is a natural showman, full of exuberance for his subject. His lectures are couched in pithy, often rough-cut phrases—"I always try to say things differently," he says—and he uses hand gestures and intonations the way Billy Rose used beautiful women on the stage, spectacularly but with grace.

Occasionally, he sets the stage. In one popular lecture on color vision, for instance, Feynman flooded the entire stage with a rainbow of light. "Why use a tiny prism on a small table, which people have to strain to see?" he says. "Nature is too interesting to stick in a corner." But generally he needs no such props. The subject matter to him is so glorious that it is only necessary to see it honestly, as it really is, for everyone to respond to his fervor. "I always come to Feynman's lectures," says an old-time faculty member, "because I am sure there will be at least one good surprise."

Gell-Mann is less flamboyant but equally compelling on the platform. Eleven years Feynman's junior, he is round-faced and bespectacled, and looks somewhat like a jovial neighborhood storekeeper. In the classroom, his lectures are clear, smooth and amusing, and draw upon an extraordinary erudition. But he really excels among small groups. If Feynman can be compared to a star who basks in the warmth of large audiences, Gell-Mann seems to enjoy the give-and-take intimacy of a seminar of bright, articulate graduate students. Like Oppenheimer, he prefers a small but devoted following—and, indeed, he has declared often that a teacher lecturing in front of a class represents a primitive form of education.

Despite this personality difference, the two men work well together—which means, according to a friend, that they argue continuously and loudly but, after a few moments, come to an understanding and move forward rapidly. This rapport was almost interrupted a few years ago when Gell-Mann for several reasons thought seriously of leaving Feynman and switching to Harvard. He decided not to, according to gossip among physicists, when he found that Harvard was willing to meet all his demands except one—changing its name to Gell-Mann Institute.

Since that time, the two men have mellowed, but an undercurrent of rivalry occasionally crops up. "Dick is always calling up to see whether Murray is working," says Mrs. Margaret Gell-Mann, an attractive blonde from Birmingham, England. "If I say he's in the garden, Dick is happy for the rest of the day. But if I tell him Murray is doing physics, then Dick gets nervous and immediately wants to come over."

When Dick married an English girl a few years ago (his third wife), he made a point of calling Murray and saying: "I looked around to see what you and I didn't have in common. I saw that you had an English wife and a brown dog, so I went out and got both."

One quality both men do have in common is an extraordinary capacity to make physics lucid and highly romantic. In part, the romance is made easy by the nature of the world of physicists. In the pecking order of this society, the theoretical physicist is the glamour boy. After him comes the experimentalist, the fellow in the laboratory, and then the engineers and applied physicists who make such things as sonar, rockets and hydrogen bombs.

Right now, the theoretical physicists have captured the front and center of scientific interest because they seem to be on the threshold of answering a long-asked, almost childlike question: What are things really made of? Have we at last come down to the last foundation stone from which we can build anything: a table, a human being, or a universe? Or must we go on looking at smaller and smaller pieces, and going deeper and deeper into a bottomless pit?

To answer these questions, one has to remember that, from the very start of civilization, philosophers have wanted to find a simple idea that would unite everything we experience in the world around us. So there has always been a search for the building block, like the cell or gene in biology. The fifth-century BCE Greek Democritus receives the honor of declaring that the simplest thing out of which everything else is made is an atom (*atomos* means "uncut"), and this idea sufficed for 2,000 years.

In the 19th century, scientists came to realize that the atom was not the ultimate particle after all. Inside the atom was a nucleus, with electrons orbiting around it like the planets around the sun. Then, in the 20th century, scientists began to concentrate on the nucleus, and saw that this was no simple item either. It contained smaller things, such as neutrons and protons, which must be held together by a very strong force, perhaps the strongest in the universe. So the question for years was: What was this "strong interaction," this glue that held the nucleus together?

In 1935, a Japanese physicist named Hideki Yukawa provided the first educated guess. He theorized that there was another particle in the nucleus, which he called a meson. It would act as a carrier of force between the proton and the neutron, so that energy would be exchanged among the particles pretty much as a football is passed between players in a game. Two years later Dr. Carl Anderson of Caltech discovered a particle that he thought might be Yukawa's carrier. Physicists exulted. Things seemed neat again—until it was found that the new particle failed to obey predictions expected from the laws of physics, and indeed created such a mess in attempts to relate theory and experience that Columbia's I. I. Rabi threw up his hands and said: "Who needs it?" It took five years for physicists to realize that the new particle was not the Yukawa meson, but an unrelated particle known as a muon. After the war, as bigger and more powerful atom smashers appeared, a number of mesons were discovered, including the one suggested by Yukawa. This constituted a major advance in understanding the strong interaction—the interaction which is known to us in everyday life only through the power of the A-bomb.

By then the atom smashers were revealing all kinds of new and peculiar particles, whose mass was created out of the energy of motion. Most of them died out in incredibly brief periods—on the order of a billionth of a second—but their trails could be photographed, and their presence raised new questions. How did they disintegrate? Were they complex structures in themselves?

Feynman was one of the Young Turks of physics who in the 1950s addressed themselves to these thorny questions. (He had already earned a formidable reputation in theoretical physics because of his efforts in another area—applying quantum mechanics to electromagnetic radiations. For this work, 16 years later, he was to share the Nobel Prize with Dr. Julian Schwinger of Harvard and Dr. Shinichiro.)

Feynman's interest centered around a phenomenon that had intrigued scientists for years—the emission of fast-moving electrons from radioactive substances. This process, which goes under the name of "beta decay," had introduced physicists to a new force in the nucleus, a force quite different from the "strong interaction" that holds it together. Feynman was fired by the challenge of this insight and the subsequent revolutionary developments.

For one thing, physicists in the 1950s were discovering that the new force—or "weak interaction," as it came to be called—was far more widespread and manifested itself in many more reactions than beta decay. Indeed, it was on a par with the strong interaction, and with the two other universal forces known to science: electromagnetism, which keeps electrons perpetually spinning around a nucleus, and gravity, the weakest force of all. The weak interaction, which is now known to be involved in the decay of many strange particles, is 100,000 times weaker than the electromagnetic force. However, it is exceedingly more powerful than gravity, which is so incredibly feeble compared with the strong interaction that it takes a fraction with 42 zeros to describe its relative strength. This kind of comparison delights Feynman. "Isn't nature wonderful," he says, raising his hands ecstatically, "to make something with 42 zeros!"

Another important development in physics in the 1950s was the overthrow of one of the fundamental laws in nature. It had long been thought that nature operates with a number of conservation laws, ranging from the familiar indestructibility of energy and matter to the conservation of lesser-known properties of the atom, such as those that explain the stability of the proton and the reason why some particles are created in atom smashers only in multiples. These laws were regarded as unchanging and universal. One of the most important of them was the one known as "the law of the conservation of parity." It said that if an object had a mirror image, that image would obey the laws of physics just like the real object. To satisfy this law, particles in the subatomic world could exist in one of two ways. The particle and its mirror image could be completely identical—like the word MOM, which reads the same in the real world and in the looking-glass world. Or there could be two particles, a "left-handed" one and a "right-handed" one—one the mirror image of the other, like the words MAY and YAM.

Since the strong interactions obey this rule of parity, it was assumed that the weak interactions should also. But a striking thing was discovered in atom-smashing experiments involving a weak interaction. A particle was found that did not have an image fitting the pattern of MOM.

Were there actually two particles, with reversed symmetry, like MAY and YAM? But that suggestion did not jibe with further experiments, which continued to point to the existence of just one particle. In an inspired moment, Feynman and another physicist, Martin Block, offered the view that the law of parity might have failed for this particular weak interaction.

It was a prophetic suggestion. In an epic-making paper, two Chinese-born scientists, Chen Ning Yang and Tsung-Dao Lee, then of the Institute for Advanced Study at Princeton and Columbia University, suggested that perhaps all weak interactions violate the law of parity. They proposed experiments, which were carried out and proved them right. Their intuition won them the Nobel Prize in 1957. More important, they turned nuclear physics topsy-turvy.

Freed from the confines of parity, Feynman and Gell-Mann (along with E.C.G. Sudarshan and Robert Marshak) tackled the problem of finding a way to describe the law of the weak interactions. In 1957, they developed a theory that shows how this force depends on various properties of the particles, such as their directions of spin, thus providing what is now generally regarded as a major con-tribution to our understanding of the nucleus.

Feynman says that the discovery of this new law was the most exciting thing in his life, more exciting than his earlier work, which led to his Nobel Prize. "I won the prize for shoving a great problem under the carpet," he says, "but in this case there was a moment when I knew how nature worked. It had elegance and beauty. The goddamn thing was gleaming." So shiny was the new law that when several eminent physicists conducted experiments that seemed to vitiate it, Feynman insisted the experiments must be wrong. And so indeed they proved to be.

Gell-Man experienced the thrill of another major and beautiful discovery a few years later. By then the number of particles emerging from the nucleus was increasing fantastically. Almost 100 had been counted by 1962. They had been classified into two main groups known as leptons (weakly interacting particles), or weaklies, as Feynman calls them, and hadrons, or stronglies. Examples of the first group are positive and negative electrons, muons and neutrinos; the second group includes neutrons, protons and pions. (Just to confuse matters, each of these particles has an antiparticle carrying an opposite charge which annihilates the particle on contact.)

A subdivision of the hadrons is a group of particles which are known as "strange particles" because, instead of dying out as they should, they live to a relatively ripe old age. To round out this picture, there is also the photon, a particle that carries the electromagnetic force—and, supposedly, there is a graviton for gravity, though this particle is yet to be found.

To bring some order into this nuclear grab bag, Gell-Mann introduced two new concepts. The first, which had developed as early as 1952, was a quality which he called "strangeness." As with the quark, it had a literary counterpart, this time in Sir Francis Bacon's line: "There is no excellent beauty that hath not some strangeness in the proportion."

Each particle could be assigned a degree of strangeness, depending on the number of steps in its disintegration, and thus it could be distinguished from its neighbors—just as a neutron can be distinguished from a proton by its different electric charge. (Gell-Mann did not know until later that a Japanese scientist named Kazuhiko Nishijima working independently in Tokyo came to the same conclusion at about the same time.) With the differences known, Gell-Mann set about seeing in what ways the particles were similar and whether they could all be slipped into a neat, organizational chart, more or less as Mendeleev had done with his periodic table of the chemical elements in the 19th century.

Gell-Mann recalls that he and Feynman tried one pattern after another without luck. Finally, one scheme seemed to work. The particles seemed to fit into families of eight or 10 members, with similar characteristics of strangeness, electric charge and other properties, such as mass and spin. Even while Gell-Mann was doing this, Yuval Ne'eman, an Israeli military attaché in London, studying for his doctorate in physics while trying to buy guns for his country, came upon the same scheme.

However, one family of particles in the table had only nine members instead of ten. It seemed incomplete. Gell-Mann thereupon predicted that, if a new member were found, it would have certain properties as indicated by its relatives in the same family of particles. Experimental physicists took up the hunt.

A couple of years later, a team of 33 scientists at Brookhaven National Laboratory hit pay dirt. They had bombarded nuclei and peered at more than

100,000 photographs of the interactions. In one of those pictures they saw the track of the missing particle, which was called omega-minus. It had a life expectancy of a 10-billionth of a second, and it had the basic properties predicted by Gell-Mann. The existence of the omega-minus was shortly verified by scientists at the University of Maryland and thereafter independently at CERN, the great nuclear establishment in Switzerland.

Gell-Mann, with his flair for literary analogy, called his chart the Eightfold Way, as in the Buddhist dictum: "This is the noble truth that leads to cessation of pain. This is the noble Eightfold Way—right views, right intentions, right speech, right action . . ." More prosaic physicists call it the SU-3 theory because it is a symmetrical structure based on a triplet of fundamental particles.

The Eightfold Way hit physics like a bombshell. Gell-Mann occasionally marvels at the apparent simplicity of nature that is revealed in the scheme. "Why should an aesthetic criterion be so successful so often?" he asks. "Is it just that it satisfies physicists?" Then, echoing Feynman, he declares: "I think there is only one answer—nature is inherently beautiful."

A number of physicists have tried to go beyond the Eightfold Way and explain why nature operates with so many particles in this rather neat and poetic style. At Harvard, Dr. Julian Schwinger claims to have developed a simple mathematical theory to explain it all. Others, following Gell-Mann, have constructed a "quark model" in which neutrons and protons behave as if they were made of quarks—and, more than that, as if they were made of three quarks, each of which may come in three forms. The reason why a number of scientists are seriously hunting real quarks (though Gell-Mann himself is not sure that they exist) is that theoretically they cannot decay into something else and thus a stable quark must be somewhere around, and may indeed have lasted since the birth of the universe.

Not all physicists agree, of course. At the University of California at Berkeley, Dr. Geoffrey Chew has taken a radical approach to the entire problem. He has developed a theory, known whimsically as the "bootstrap theory," which says that there really is no rock-bottom fundamental particle like the quark or anything else; indeed, that the strong particles are made of one another, pulling themselves into existence, so to speak, by their own bootstraps. "At first glance, this theory and our quark model may seem contradictory," Gell-Mann says, "but they may actually be quite compatible— and both may even be right—especially if the quark should turn out to be, as is likely, a useful mathematical figment rather than a concrete building block of matter."

Although Gell-Mann and Feynman were both born in New York City of middle-class families, they came to theoretical physics by different routes. Feynman traces almost everything in his makeup to the influence of his father.

"When I was a boy," he says, "Dad and I took long walks in the woods and he showed me things I would never have noticed by myself. He told me about the world and how it looked many years ago. He would say, 'See this leaf? It has a brown line; part of it is thin and part thick. Why?' And when I tried to answer, my father would make me look at the leaf and see whether I was right and then he would point out that the line was made by an insect that devotes its entire life to that project. 'And for what purpose? So that it can leave eggs which turn into new insects.'

"My father taught me continuity and harmony in the world. He didn't know anything exactly, whether the insect had eight legs or a hundred legs, but he understood everything. And I was interested because there was always this kick at the end—a revelation of how wonderful nature really is."

Dick's love of science flowered rapidly, and occasionally it got an unexpected boost. In Far Rockaway High School, which he found pretty dull, a teacher disciplined him for talking too much in class and not paying attention. He was sent to the back of the room and given a book. "After you read this," the teacher said, "you can open your mouth again."

"So I learned calculus," says Feynman. He went on to MIT, where he graduated with honors in 1939, and did his graduate work at Princeton. In World War II, he ended up as a group leader at Los Alamos working on theoretical aspects of the material in the A-bomb. He was present at the first test explosion at Alamogordo in July 1945.

As befits one who challenges the fundamental laws of nature, Feynman is an inherent iconoclast. He loves to play the bongo drums, and he is probably the first physics professor in history—certainly the first Nobel Prize winner—to have a picture of himself on the drums included in the introduction to a three-volume college textbook of physics.

Gell-Mann's extracurricular tastes run in more erudite directions. With his wife, who studied archeology at Cambridge, he often enjoys digging for artifacts in such places as Greece and Palestine. He is also a devotee of linguistics; he has studied many languages, including some odd dialects of Africa and the Middle East. "I like diversity," he says, "and I like the natural history behind diversity. Why are there so many different tongues, so many different birds, even so many different human neuroses? It's interesting to find the pattern behind them."

Gell-Mann, the son of a language teacher and a prodigy who entered Yale at 15, says he came into physics almost inadvertently. "I had to fill out an application form," he recalls, "where I had to list my future occupation. I started to put down archeologist but my father said I'd never make a living in archeology, and suggested engineering instead. I couldn't stand engineering, so I put down the closest thing—physics."

He went on to get his doctorate at MIT, and then to Princeton, where he worked with Oppenheimer. The Gell-Mann and Feynman merger occurred in 1954, during a visit by Gell-Mann to Pasadena. Some words were passed between them, along with howls of laughter over jokes they both found funny, and the next day, Gell-Mann recalls, he was being interviewed by the dean of Caltech. He joined the staff as an associate professor in 1955 and became a full professor the following year.

Gell-Mann has been a missionary for the development of the world's largest atom smasher, the proposed 200-billion-electron-volt machine, scheduled to be built at Weston, Illinois. Along with other top physicists, he believes the $200 million machine is vital to maintain United States supremacy in particle physics.

"I think particle physics is where atomic physics was in the early years of the century," Gell-Mann says. "We're getting an outline of an underlying structure, but there is still no complete theory of either strong or weak interactions which enables us to understand what is really happening at the bottom of everything.

"Recently an experiment performed by Princeton physicists has shown the violation of another law of symmetry that was thought to be valid [like parity 10 years ago]. Some theorists went so far as to speculate that a fifth natural force was involved, but that doesn't seem to be true. How is the violation occurring? Nobody seems to know at the moment, but I think we are on the verge of an important discovery."

Feynman says much the same, but in terms of a metaphor—playing chess with a Martian. "If you don't know the rules," he says, "and you see only parts of the board, how do you know how to play? If you know all the rules, can you tell what's in the Martian's mind when he moves the pieces in a certain way?

"The biggest mystery of physics is where the laws are known, but we don't know exactly what's going on. We don't know the strategy in the middle game. We know castling, or how the different pieces move, and we know a little bit about the endgame, but nothing in the middle.

"We get reports from the experimentalists, the watchers of the chess game, and we try to analyze the information. We may even suggest a new experiment. But we're still waiting and hoping for the big strategy. Then maybe we'll really understand how wonderful is nature."

—October 8, 1967

Einstein: Relativity in the Kitchen

By WALTER SULLIVAN

His public image is that of the great theorist, the dreamer—but Albert Einstein had a keen interest in the world's practical problems, as well: the invention of novel cameras and refrigerators, for example.

This little-known aspect of his career has come forth as a sequel to the recent announcement, by Princeton University Press and the Einstein estate, of plans to publish his papers in a series of some 20 volumes. From a variety of sources has come information on a number of patents taken out by Einstein and his coworkers.

When he was doing his first great theoretical work on relativity, Einstein earned his living as a technical specialist in the Swiss Patent Office in Berne. It was not a job that he relished, but it gave him an early feel for the practical, as well as the abstract, uses of science.

During the 1920s while he was living in Berlin, Einstein undertook, with Leo Szilard, to invent a new kind of refrigerator. Szilard, who was himself to become a leading physicist, was then a student in his 20s who became a close associate of Einstein.

Last week Dr. Gertrud Weiss Szilard, his widow, told of efforts by the two men to patent something that would produce enough income so they could pursue their research without financial worries.

One of the problems, in those days, was to produce a refrigerator quiet enough to make a kitchen livable. In the early 1920s, two engineering students in Sweden—Carl Munters and Baltzar von Platen—had invented a refrigerator that required no moving parts. It was known as the Electrolux absorption system. While it was considerably less efficient than refrigerators in which the refrigerant is circulated by pump, it was completely silent.

Einstein and Szilard then patented what they considered to be an improved version of this system. But it failed to make them rich. For just at that moment, others had found a means to make refrigerators which ran quietly. The Einstein-Szilard patent was never put to use.

The two scientists also patented a pump for circulating liquid metals, using electromagnetic forces. At the time its applications were limited, but today its successors are widely used in nuclear power plants.

Szilard's most famous patent was one that he filed in 1934 concerning the laws governing a chain reaction—which is the process in which neutrons released by the splitting of one atom go on to split another in an endless succession. Realizing that this could be used to create a bomb of frightful destructiveness, he assigned the patent to the British Admiralty, thereby (he hoped) ensuring its secrecy. Szilard, by then, had fled Germany to live in Britain.

Although Einstein and Szilard took out a number of patents in Germany and Britain, as well as one in the United States, none proved lucrative. Some of these patents will be published, Mrs. Szilard said, in the first volume of her husband's papers, to be published by the Massachusetts Institute of Technology Press later this year.

Dr. Thomas Bucky of Westport, Connecticut, told of a collaboration between his father, Dr. Gustav Bucky, and Einstein that suffered a similar fate. The two men had known each other in Berlin, he said, but it was after Einstein settled in Princeton, following the Nazi takeover in Germany, that the collaboration began.

It included development of a special camera for medical use and another, which was patented, in which a photocell, or "electric eye," controlled the lens aperture. This forerunner of the modern camera, with automatic control by built-in light meter, was not, however, adopted by any camera maker.

—April 9, 1972

Signs of Quark Discovery Cited in Stanford Report

By WALTER SULLIVAN

After years of searching, scientists reported in Washington the first evidence that quarks, those hypothetical building blocks of all heavier particles of matter, may have been discovered.

The evidence comes from the first measurements of partial charges on particles that make up atoms. According to theory, the elusive quarks would have either one-third or two-thirds of an electron charge.

The annual meeting of the American Physical Society was told that in two out of eight critical measurements at Stanford University, electric charges roughly equal to one-third that of the electron or proton were observed. If confirmed, this would be a major discovery in fundamental physics.

In reporting the results, Dr. William M. Fairbank, professor of physics at Stanford, termed them "tentative" and said a long series of follow-up tests was planned.

"It is an inference to think that this is the same as finding the quark," he added, but did not specifically make such a claim.

Dr. Fairbank added at a news conference, however, that if atoms with nuclei containing quarks (and therefore fractional electric charges) could be isolated, it would make possible an entirely new chemistry. The positions of the electrons that determine the chemical characteristics of ordinary atoms would be changed. Such atoms, unlike those that form the known elements, could not be made electrically neutral.

Otherwise, all electric charges in nature are integral multiples of the basic charge (a single negative charge for the electron and a single positive one for the proton).

The experiments reported today were a modern adaptation of those used by Robert A. Millikan, working at the University of Chicago from 1909 to 1913, to demonstrate that electric charge normally is not fractional. The Stanford experiments have been under way for almost 12 years. Seven years ago the experimenters thought they might have detected a fractional charge, but they were not sure.

Quarks, as independently envisioned in 1964 by Murray Gell-Mann and

George Zweig, would carry electric charges of one-third or two-thirds. A proton, according to this hypothesis, would combine two quarks with a charge of plus two-thirds and one with a charge of minus one-third, for a net charge of one. The neutron would combine a quark with a charge of plus two-thirds and two quarks with a charge of minus one-third, resulting in zero net charge.

In Millikan's classic experiments, droplets of oil so tiny that their fall was substantially slowed by air friction were dropped through an electric field. If the droplets carried an electric charge, they could be slowed further, stopped or even made to rise by varying the nature of the field. Their response indicated the extent of their electrical charge.

The method was so sensitive that charges equal to a small fraction of that on the electron or proton could be recorded. In thousands of tests, however, no fractional charges were recorded—with a single exception that Millikan was unable to repeat. The exceptional charge that he observed, like that of the Stanford group, was roughly one-third that of the electron.

The new experiments were initiated in 1965 as a doctoral project by Arthur F. Hebard, who was then a student of Dr. Fairbank and is now at the Bell Telephone Laboratories. The work has been continued by George S. LaRue, a doctoral candidate at Stanford. Instead of oil droplets, he is using niobium balls that are about one-quarter the size of the tip of a ballpoint pen.

Because the balls are 10 million times heavier than Millikan's oil droplets, they are suspended magnetically and then subjected to varying electric fields. To achieve such suspension, the balls are made superconducting, that is, deprived of all resistance to electricity, by first being baked at 3,300 degrees Fahrenheit and then chilled almost to absolute zero (the absence of all heat).

It is estimated that each ball is formed of about 50 billion billion protons and roughly a matching number of electrons. An excess of one or the other would give it a plus or minus charge.

In an interview, when the findings had become widely known but were embargoed for publication, Dr. Fairbank said that a typical initial charge might indicate an excess of 100,000 electrons. This could gradually be reduced as the ball floated in the magnetic field by shooting positrons at it, derived from radioactive material. A positron, the antimatter counterpart of an electron, carries a single positive charge.

In this way the charge on the ball could be gradually brought down to zero—or to a fraction of one charge, if there is such a thing among the billions of normal atoms.

Dr. Fairbank likened it to balancing books where billions of dollars, but no smaller denominations, are supposedly involved and ending up with 25 cents left over. In one case, the residue was roughly plus one-third, and in another minus one-third. Mr. LaRue said one of the balls retained its fractional charge through two further repetitions of the test. The other failed such a repetition, Dr. Fairbank said, possibly because the fractional charge "rubbed off" in handling.

The experiment was based on the assumption that, although quarks are so tightly bound inside atomic particles that their attraction apparently is impossible, some may never have found a "mate."

In preparing the balls for the Stanford experiment, three of them were heated on tungsten and five on a niobium surface. Both of those showing a fractional charge had been heated on tungsten, leading Dr. Fairbank to suspect that tungsten may have been the source of the particles.

The next step, he said, would be an attempt to separate such particles electrically "and observe exactly what they are."

—April 27, 1977

Detection of the Elusive "Gluon" Exciting Scientists

By MALCOLM W. BROWNE

A thrill of excitement has reverberated among scientists around the world since the disclosure that a hypothetically vital component of all matter has apparently been found to exist in reality.

The discovery has powerfully buttressed a theory that promises profound insight into the nature of all things, scientists agree. From the fundamental particles of nature, everything in the universe is derived, and understanding of these particles not only forms the basis of contemporary astronomy and physics but also is vital to progress in chemistry, biology and ultimately the behavior of human beings.

The general results of the dramatic new experiments were outlined in a symposium at Fermilab, the Fermi National Accelerator Laboratory at Batavia, Illinois. In the past several days the scientists involved have described their landmark experiment in greater detail, and several were interviewed.

Some physicists who did not participate in the work feel it should be crowned with a Nobel Prize. Experimenters themselves said, however, that the research had involved so many scientists over such a long time that bestowing such an honor would be virtually impossible.

For the first time in the history of international team research projects on nuclear particles, a major contribution came from China, 27 of whose scientists worked on the key experiment. In all, about 300 physicists from many countries pooled their research in four related experiments, all of them using a powerful new machine in West Germany for the achievement. They believe their work has made a significant step toward establishing the actual existence of a theoretical packet of energy called a gluon.

Team Leader from MIT

If a fundamental theory of matter called quantum chromodynamics, or QCD, is correct, the gluon must exist, and if the scientists had failed to find it in their new experiment, much of the theoretical work in physics in the past decade would have been in serious doubt.

The leader of the 56-member team that produced the most dramatic results was Dr. Samuel C. C. Ting of the Massachusetts Institute of Technology, who shared the 1976 Nobel Prize in Physics with Burton Richter for their discovery of a nuclear particle called the J/Psi particle.

"Our latest experiment has not absolutely clinched the QCD theory," Dr. Ting said in a telephone interview from Europe, "but if we had not seen this evidence, the theory would have been in trouble. As it is, every experiment of the last 10 years has supported QCD, and we've now reached the point that we believe it is probably correct."

An understanding of quantum chromodynamics requires a look at the history of man's study of the atom.

Growth of Theories

Until about the end of World War II, physicists believed the atom could be described by a rather simple model. Each atom was believed to contain a nucleus made up of presumably indivisible heavy particles—positively charged protons and particles of equal size but without electric charge, known as neutrons. Around this nucleus whirled a layered cloud of negatively charged light particles called electrons, the number of which, in a simple, uncombined element, matched the number of protons in the nucleus.

The discovery that an atomic nucleus is made up of protons and neutrons was made by banging nuclei apart, using fast-flying particles or atoms as hammers. The impacts cast off debris—fragments of the original nuclei—whose flights could be studied through the tracks they left on photographic plates, through microscopic streaks as they passed through bubble chambers and through other devices.

To increase the precision of their experiments, scientists over the years have needed ever more powerful tools to cause harder collisions and break fragments and particles into ever smaller bits. As technology improved, the accelerators they used grew increasingly powerful, but instead of confirming traditional theories, surprises began to occur with disquieting frequency. The protons, neutrons and other heavy particles once thought to be the ultimate building blocks of nature were themselves breaking up or transforming into many puzzling particles never previously suspected.

Mysteries were proliferating faster than answers. A longstanding joke among physicists has it that nuclear particles are like onions—you strip off one layer and there's always another one underneath.

"Who knows, there may really be no end to it," Dr. Leon M. Lederman, director of Fermilab, has remarked.

Physicists have so far catalogued more than 100 different kinds of nuclear particles. The confusion has been further complicated by awareness that four fundamental forces governing all of nature are very imperfectly understood.

The existence of gravity and the electromagnetic force have both been long known, but two other forces were needed to understand how an atomic nucleus behaves.

One of these, dubbed the "strong force," was defined as the force that keeps nuclear particles from merely drifting apart, since gravity is far too weak to hold them together. Another force, the "weak force," would account for the radioactive decay of certain atoms.

In the 1960s, science found a possible way out of the confusion through a theory, developed by Murray Gell-Mann of the California Institute of Technology, that postulated the existence of a fundamental particle named the "quark."

Dr. Gell-Mann found a logical way of grouping the proliferation of particles into species and subspecies and postulated that the massive particles, such as protons and neutrons, actually consist of triplets or pairs of different kinds of quarks bound together.

If this were correct, all matter could be broadly grouped into two major classes—the light, dimensionless particles called leptons (electrons, for example), and the more massive quarks, which are believed to occupy definite amounts of space.

But reality could not be explained by the existence of just one kind of quark.

Experiments have led to the grouping of different quarks in terms of at least five "flavors" and three "colors." Such names are merely conveniences, and have no more meaning than numbers.

Origin of Chromodynamics

The differentiation by "color" refers to the fact that quarks seem to behave as if they carried three different kinds of charge—not electrical charge, but something analogous to it. The study of the "color" of both matter and antimatter suggested the name *chromodynamics,* which has nothing to do with real, visible color.

Complicating the problem of sorting out particles was the discovery that for every kind of matter there is a corresponding form of antimatter. Each of the major particles making up antimatter have the opposite charge of its corresponding form of matter. When matter and antimatter are combined, they annihilate each other, to be instantly reborn as new forms of matter and energy.

The mechanisms by which these strange changes occur must be driven by forces of various kinds, but a complete understanding of the nature of all forces has defied science.

It is relatively easy to predict how matter will respond when subjected to the familiar gravitational and electromagnetic forces, but how are these or any other forces carried from one thing to another to cause their observed effects?

In the case of electromagnetic force, the problem was solved earlier in this century. The carrier of electromagnetic force was shown to be the photon, a massless packet of vibrating energy that makes up such forms of radiation as visible light, X-rays, radio waves and so on. But what about gravity and the newly discovered strong and weak forces?

The way gravity is conveyed has proved by far the most difficult to explain, partly because gravity is much weaker than other forces, and must be studied on a cosmic scale, rather than an atomic scale.

Research has moved faster in the investigation of the weak force, and in the past few years its distinctive carrier has been fairly well established as the elusive and barely detectable neutrino. Neutrinos have sometimes been called the "ghost particle" because of their ability to pass through any matter almost unimpeded.

That left the strong force as the best remaining target for research having a realistic hope of success. The trouble was that to find the carrier of the strong force would involve smashing particles together at energies beyond the scope of the most powerful existing accelerators.

German Accelerator Used

Finally, an accelerator equal to the task was built by West Germany at the Deutsche Elektronen-Synchrotron (DESY) complex at Hamburg. The accelerator, called PETRA, was brought up to its full power in recent months, and scientists from the world's leading physics laboratories began organizing teams to work there.

Their target was the particle presumed by the chromodynamic theory to exist as a carrier of the strong force, which holds atomic nuclei together. They had whimsically named this particle the "gluon" because of its supposed ability to hold together the groups of quarks making up stable particles, which in turn are bound together in the atomic nucleus.

There was no hope of directly detecting, or "seeing," the gluon, according to theory, because it could exist as a separate entity for only an instant of time too short to measure. Actually, no one has yet "seen" a single, uncombined quark, and scientists are undecided as to whether such a "free" quark could exist.

But despite the impossibility of "seeing" a gluon, it can leave an indirectly detectable signature. Once formed, the gluon would be expected to transform itself into a cascade of pairs of quarks, converting its energy into matter to create the quarks. These quark pairs would combine to form relatively stable particles, whose behavior would reveal how they had come into being. The daughter particles born from this train of transformations, called pi mesons, live for only about one 100-millionth of a second, but that is long enough to tell the tale.

For the gluon research, the PETRA accelerator was gradually throttled up to near its maximum energy, and when the machine had reached a power of about 30 billion electron volts, the experiments began.

In one direction, electrons were whipped around at increasing speed, while their antimatter equivalent, positrons, were accelerated in the opposite direction. The huge energy of the machine was pumped into these tiny particles, to be combined in their occasional head-on collisions.

Chromodynamics had predicted that the pi mesons streaming out of these collisions as the result of the birth and transformation of gluons should group themselves in three jets moving at slightly different angles. If only two jets of pi mesons were detected coming from each collision, the entire chromodynamic theory would be in serious doubt.

When the experiment was actually performed the third jet of pi mesons "clearly and unambiguously" appeared in the sensitive recording equipment of the teams led by Dr. Ting and others. The gluon had been found.

"Of course, we'll never actually see the gluon itself," Fermilab's Dr. Lederman said. "The nature of such particles can only be learned by inference, like the nature of a ball or boomerang that you cannot see but which is being thrown between two boats. You study the angles and speeds at which the boats recoil after each throw, and from that you deduce the properties of the object itself and how the boats interact with it and each other."

—*September 2, 1979*

New Quarks Stir Debate
on Basic Laws of Nature

By WALTER SULLIVAN

The discovery of new families of atomic particles and the development of new theories to explain them have plunged the scientific community into a ferment of speculation over the nature of matter and the laws that have made the universe what it is.

The findings suggest that there may be as many as 18 quarks, those supposedly "basic" constituents of matter, and some physicists believe the discoveries bring them closer than ever to understanding nature at its most fundamental level. This belief has prompted a revival of talk about a "grand synthesis" that would at last provide a logical explanation for seemingly disparate phenomena.

Such a single theoretical framework would, for example, explain atomic energy, radioactivity, the swing of a compass needle and the flash of an electric spark.

Yet the proliferation of particles has led other scientists to fear that the more they learn, the more elusive final answers will become.

At a meeting of the American Physical Society in the New York Hilton, it was evident that many scientists were most excited about the suggestion of a particle, called by some researchers the "baseball," that would be roughly a billion billion times more massive than the proton and neutron, which form the atomic nucleus.

The "baseball" would be the key to unifying three basic forces in nature—the electromagnetic force that drives electrical appliances, the "weak" force responsible for some forms of radioactivity and the "strong" force binding the nuclear particles together.

The massive particle represents the energy level at which all those forces would become equal and integrated into a single theoretical structure. Optimism that some sort of unification may be possible has soared following the confirmation of several predictions implied by the theory that the electromagnetic and radioactive forces are the same phenomenon.

The theory was advanced independently in 1967 by Dr. Steven Weinberg, then at the Massachusetts Institute of Technology and now at Harvard, and Dr. Abdus Salam, head of the International Center for Theoretical Physics in Trieste, Italy.

Dr. Weinberg and others are now working toward a unification of the powerful nuclear force with the already unified weak and electromagnetic forces. All three would become equal at the extremely high energies that would be released when the proposed supermassive particle decays.

The concept of such a particle is also being advanced to explain one of the great paradoxes: why the universe is apparently made almost entirely of matter rather than a mixture of matter and antimatter.

In laboratory experiments, energy readily converts itself into matter, but it always forms equal numbers of matter and antimatter particles. Such symmetrical behavior seems to affect much of what happens on the atomic level.

Particles of matter and antimatter are identical except that they are opposite in such features as electric charge or spin. For example, the proton, with a positive charge, is matched by the antiproton, with a negative charge. The electron, which is negative, has a twin, the positron, which is positive. Yet the world—and probably the universe—is predominantly made of matter.

Theorists have long wondered how this imbalance came about if, as is widely believed, the universe was born in a fireball of energy. One explanation would be that the proposed supermassive particle—which could have existed only under extreme conditions like those that prevailed right after the primordial fireball—decayed in a "lopsided" way, slightly more often into particles of matter than of antimatter.

More May Be Found

The particles recently discovered—or deduced—have been produced in accelerators, machines that cause collisions by accelerating beams of particles to extremely high energy. Scientists say they expect to find more particles when they are able to use more powerful machines now nearing completion.

When Drs. Murray Gell-Mann and George Zweig independently proposed the quark concept in 1963, they envisioned only three such basic components of matter. All the heavier particles, including protons and neutrons, they speculated, are formed of such hidden constituents. Given that background, the fact that there now seem to be 18 quarks is disturbing to some theorists.

"Is this going on forever?" asked Dr. Martinus Veltman of the Institute for Theoretical Physics in Utrecht, the Netherlands, as he discussed quarks at an international meeting on astrophysics in Munich.

Unlike other atomic particles, all of which have positive or negative electric charges equal to the charges on the proton or electron, quarks carry fractional charges equal to one-third or two-thirds of those on protons and electrons. Two or three quarks combine within a particle to produce an integral charge.

The proton, for instance, is formed of two quarks, each with a positive charge of two-thirds and one quark with a negative charge of one-third. The cumulative result is a positive charge of 1.

Negative Charge Recorded

Last month. Dr. William M. Fairbank and his colleagues at Stanford University reported that on two occasions they had recorded a negative electric charge of one-third after purging a sphere of all other charges. They reported a single such event two years ago, after many years of experiments.

Whether these observations represent the detection of quarks remains problematical. Some scientists have theorized that a few free quarks may remain from the formation of the universe. All attempts to blast them out of protons and other particles have failed.

The quarks are assumed to be bound together by a "glue" that, unlike normal forces, gets stronger at increasing distance—as does a rubber band. Hence quarks can never be torn apart. The force is embodied in particles called gluons.

The theoretical proliferation of quarks has come about because more and more of them seemed necessary to explain newly discovered classes of short-lived particles.

Inevitably, parallels have been drawn between this approach and the efforts of astronomers in ancient Greece to explain apparent motions of the celestial bodies not in terms of the Earth's rotation and orbital flight around the sun but in terms of transparent rotating spheres within rotating spheres. More and more hypothetical spheres had to be added as more precise observations became available.

But, some scientists argue, these theories represented progress because, as with the quarks, they symbolized a recognition that some orderly, underlying system was at work.

The particle properties that suggest the existence of additional quarks are "conserved" in all transformations. It has long been known, for example, that if the electric charges on particles entering a reaction add up to a positive charge of 1, those emerging from the reaction must do so, too.

The original quarks were designated the "up" quark, with a positive charge of two-thirds, the "down" quark, with a negative charge of one-third and the "strange" quark, which also has a negative charge of one-third. The last quark was so named because scientists were puzzled by the slow rate at which it decays. This property of "strangeness" seems indestructible since it, too, is "conserved" in a reaction.

The "Charmed" Quark

In 1964, Sheldon Glashow of Harvard and James Bjorken of Stanford suggested that there was another property that involved slow decay but was different from "strangeness." That property, which was named "charm" and which is conserved in a reaction, would have to be represented by a fourth quark.

Beginning a decade later with detection of the J/Psi particle, which mates a charmed quark with a charmed antiquark (a particle with the same mass as the quark but the opposite electrical charge), an entire new family of particles was found. They are more than four times more massive than the "up" and "down" quarks.

Late in 1977, Dr. Leon Lederman of Columbia and his colleagues at the Fermi National Accelerator Laboratory found the upsilon particle, the first of yet another family. Its properties imply the existence of a fifth ("bottom") quark, three times heavier than the charmed variety. The upsilon, formed of a bottom quark and a bottom antiquark, is 10 times more massive than the proton.

The upsilon's discovery left scientists with what seemed to be an incomplete "table of organization"—two categories of quarks with three members each and one category with only two members. This led to the suspicion that there is a sixth quark that may be as much as 13 times the proton mass. It would be the "top" quark, and researchers hope that it can be observed by a new machine in Hamburg that collides high-energy beams of electrons and positrons.

Dr. Samuel Ting of MIT, who shared a Nobel Prize for discovery of the J/Psi particle, says that he and other scientists using the Hamburg machine have been working toward collision energies of 40 billion electron volts, more than they expect is needed to detect the top quark.

Converting Energy to Matter

In both the Hamburg machine and a similar facility that is to begin operating this fall at the Stanford Linear Accelerator in California, electrons and positrons collide head-on at very high energy, annihilating one another and forming a "fireball" of energy that materializes into new particles.

At a given energy level, the particle-antiparticle pairs produced in greatest profusion are those with a combined mass (in terms of equivalent energy) that matches the energy of the fireball.

Not only do there appear to be six types of quark, but each apparently comes in three "colors," making a total of 18. The term *color* describes a property that many theorists contend must exist because of a principle that says no identical particle like quarks can remain side by side. Since they seem to do so, it is assumed that they differ in what scientists have labeled "color."

The "colored" quarks always combine in a particle so their colors neutralize one another. Dr. Gerard 't Hooft, a leading theorist from the Netherlands, compares the effect to combining the colors red, green and blue to produce white.

Evidence of Unity Is Found

In the worldwide effort to find the unity that is believed to lie within nature, a recent experiment has provided the strongest evidence so far that the electromagnetic and weak forces are one. In reactions normally controlled by electromagnetism, the experiment showed that there is sometimes an asymmetry, presumably caused by an effect of the weak force.

Such asymmetrical behavior was first observed in 1957, when it was shown that electrons emitted in radioactive decay come out spinning in a left-handed manner slightly more often than in a right-handed one. The discovery that nature is not entirely even-handed was revolutionary, but at the time it applied only to weak processes such as radioactivity.

The new experiment, involving what are normally electromagnetic interactions, was carried out last year at the Stanford Linear Accelerator by 20 physicists from five institutions led by Dr. Richard E. Taylor of Stanford. Beams of polarized electrons—those spinning in a single direction—were fired at a tank of liquid deuterium or hydrogen, and the researchers watched how the electrons scattered off nuclear particles in the tank.

The scattering was usually controlled by the electromagnetic force, but it was found that, because of an effect of the weak force, left-polarized electrons scattered slightly more readily than right-spinning ones.

Dr. Weinberg and others are now discussing the possibility that a similar asymmetry may have controlled the decay of supermassive particles early in the life of the universe, tilting the scales in favor of matter over antimatter.

As explained recently by Drs. Sam B. Treiman and Frank Wilczek of Princeton University, who participated in development of the hypothesis, it is possible to estimate how many particles of matter and antimatter annihilated one another early in the history of the universe. Each such annihilation should have produced one light wave, now stretched by the expansion of the universe into a radio wave.

Radio Waves Observed

It is these waves that are observed as the radio glow that permeates the universe— often described as the residue of the "big bang" in which the universe was born. The ratio of surviving particles to these annihilation relics works out to about one per billion.

To account for the predominance of matter, therefore, it is only necessary to tilt the early decay processes sufficiently to produce one excess particle of matter for every billion particles of antimatter.

The supermassive particle idea arises from evidence that, in reactions at increasingly high energies, the strong, or nuclear, force becomes weaker relative to the weak, or radioactive, force.

At an energy perhaps a billion billion times that required to produce a proton, it would appear that there should be democracy among the forces other than gravity.

A theoretical by-product of this hypothesis is a suggestion that until recently would have been a shocking idea—that the proton, the building block of our world, is unstable. Its decay rate would be so slow that its average life would be a million billion billion billion years—far longer than the assumed age of the universe.

While proton decays would be extremely rare, a large volume of fluid contains so many protons that occasional decays may be observable. Two groups in the United States and one in Europe are reported to be planning experiments to test that hypothesis. If proton decay is observed, what are now highly speculative ideas will become credible.

—February 13, 1979

Microscopes Peer Ever Deeper into Small World

By WALTER SULLIVAN

Since the 17th century, when Anton van Leeuwenhoek's microscope first discovered the teeming world of bacteria, protozoa and other microorganisms, every advance in the ability to see objects on smaller and smaller scales has led to fundamental increases in the understanding of living and physical processes.

Today, scientists are testing and using new techniques that go far beyond those of conventional optical or electron microscopes, enabling them to see the most intimate structures of living and nonliving systems. In ways never before possible, scientists can magnify and examine the surfaces, internal structures and even the atomic compositions of specimens, including living cells. The applications in medicine, electronics and other fields are virtually unbounded.

While the original microscopes used light waves, these powerful new devices use a variety of technologies, including beams of sound waves, X-rays, polarized electrons or the nuclei of various atoms. The imaging often depends on advanced computers and other complex processing methods.

Despite the multitude of approaches, however, there is "virtually no overlap" among them, said Dr. George A. Hazelrigg of the National Science Foundation, which is financing some of the projects.

Although most of the advances depend on illumination of the specimens with high-energy particles, there is much excitement over new approaches that use visible light and therefore do not damage living specimens. Electron microscopes, on the other hand, "fry" the subject with electron bombardment.

One visible light method, developed by Alan Boyde at University College in London, produces three-dimensional pictures of highly magnified subjects for the first time. A series of sharply focused images is obtained at successive depths in the specimen. As described in the Dec. 13, 1985 issue of the journal *Science*, the images are then stacked for a three-dimensional effect.

At Cornell University, Michael S. Isaacson and his colleagues have devised a way to produce images showing details far smaller than the wavelength of light used to scan the material. The trick is to produce an extremely narrow beam of light by passing it through a hole whose diameter is only one-tenth to one-twentieth the light's wavelength. To avoid dispersion of the beam once through the hole, the specimen must be extremely close. The technique, near-field scanning optical microscopy, is still in its infancy.

Most of the new approaches, however, depend on illumination with radiation whose wavelengths are far shorter than those of visible light. Just as one cannot dissect an insect with ax and shovel, it is difficult to study something smaller than the wavelengths used to illuminate it.

This led to the invention of the electron microscope in 1932, making use of two critical discoveries. One was that electrons move through space in a wavelike manner, their wavelengths dependent on their energy. High-energy electrons have wavelengths far shorter than those of light. The other crucial discovery was that a magnetic field can focus electrons as a glass lens focuses light.

In the transmission electron microscope, the first type developed, a beam of electrons is fired through a thin slice of specimen, just as light passes through a specimen in a conventional microscope.

Scattering of Electrons

Material in the specimen scatters the electrons, producing an image that can then be enlarged by the "lens" of a magnetic field, and recorded. An image must be obtained, however, before the electrons alter the specimen. The results are two-dimensional, but slight lens adjustments, as in a light microscope, can focus different layers in the specimen.

The scanning electron microscope, which came into widespread use in the 1960s, creates a three-dimensional image, not from electrons fired through the specimen but from secondary electrons released from the specimen's surface by the electron bombardment. This produces an image in the same manner that light scatters from a surface.

In microscopes that depend on light or conventional electron transmission, the entire specimen is illuminated. But a scanning electron microscope's beam scans a raster of successive lines similar to the way an image is "painted" on a television screen. The results have included greatly magnified, three-dimensional pictures of insect heads, human blood cells and other specimens, although enlargement cannot be as great as it is in transmission microscopes.

A new approach, combining features of both earlier types, is the scanning transmission electron microscope, or STEM. The most ambitious version is being built at the University of Chicago under the direction of Dr. Albert V. Crewe, a pioneer in microscope design. It is a refinement of a device he completed in 1966 and subsequently upgraded with which he was first able to produce images showing individual atoms.

Seeing Chemical Properties

The new version is designed to reveal chemical properties as well as to distinguish objects only 0.6 angstroms apart. The angstrom equals one 10-billionth of a meter. Magnification by light microscopes is limited because wavelengths of visible light are measured in thousands of angstroms. The wavelength of an electron accelerated to 100,000 volts is only 0.04 angstroms. Atoms in a crystal are two to five angstroms wide.

At present the most powerful such device is at the Lawrence Berkeley Laboratory of the University of California. Its "resolution"—defining the smallest scale at which it can observe—is 1.6 angstroms. The existing Chicago machine now achieves 2.4-angstrom resolution. It is being used to examine the atomic structure of such substances as the hemoglobin in earthworm blood.

The multimillion-dollar project to achieve 0.6 angstrom resolution depends on high technology. A deep vacuum must be maintained to prevent electrons from being scattered by molecules of air, and the electron beam must be kept within a narrow energy range to prevent the blur of multiple images caused when electrons of different energies are bent to different degrees by the magnetic lens.

The greatest challenge, however, is to compensate for the spherical aberration that has blocked progress toward greater magnifications. Such aberration occurs because electrons bent by the spherical configuration of a magnetic lens focus on a line, rather than a point.

The distortion becomes more serious at the very short wavelengths needed for great magnification. Dr. Crewe has devised a system of sextupol, or six-pole, magnets to correct for this effect. To be free of magnetic asymmetries the sextupols must be fashioned to extreme precision from iron melted in a vacuum, to draw off impurities, and hammered in a special forge, to eliminate large crystals. It is hoped that the microscope will be ready for testing next summer.

Other new approaches are those using X-rays of relatively long wavelength to show very tiny structures without destroying them. This technique was used in 1983 to obtain the first X-ray image of a living cell. The resolution was 75 angstroms, almost enough to show individual molecules.

Early this year Ralph Feder of IBM and his colleagues presented in *Science* a series of "flash" X-ray images showing living blood platelets reaching out with "pseudopods" that bind them to other platelets. This is the process that, in healthy individuals, leads to blood coagulation.

The specimens had been placed on top of X-ray-sensitive material and exposed to a flash of X-rays, producing a shadow image on the X-ray-sensitive material that was then viewed with a scanning electron microscope. The images, being produced by the specimens' response to X-rays, provided information not obtainable with visible light or electron beams. Thus, the IBM group reported, direct, high-resolution viewing of many cellular processes should now be possible.

The availability of high-intensity X-ray sources, such as the new National Synchrotron Light Source at Brookhaven National Laboratory on Long Island, is making possible microscopy in which X-rays are focused by devices called zone-plates. X-rays cannot be bent by ordinary mirrors or focused by lenses.

But they can be bent, or diffracted, toward a focal point by a grating or pattern of concentric, circular grooves. In this way Janos Kirz and his colleagues at the State University of New York at Stony Brook have mapped the calcium content of human skull tissue, scanning it at two X-ray wavelengths, only one of which is strongly absorbed by calcium.

A device called the scanning tunneling microscope, in whose development IBM is playing a major role, looks at the bumps and valleys of surfaces down to the scale of individual atoms. A needlelike electrode scans across an electrically charged specimen, and electrons that escape, or tunnel, out of each spot on the specimen can be used to map its surface atom by atom.

This technique, which can achieve a resolution of about 2 angstroms, will enable researchers to study corrosion and other metal surface reactions on the smallest scale. It will reveal the structure of coatings deposited to fulfill a multitude of roles in electronics and other fields.

The acoustic microscope, under development at Stanford University and elsewhere, offers special advantages in that its images reflect the mechanical qualities of the specimen: density, elasticity and viscosity. These, according to Calvin Quate of Stanford, are properties "far more vital to the function of living tissue" than are the properties revealed by light waves.

The challenge has been to generate wavelengths of sound as short as those of light or other radiation used in microscopes. The frequency used for tuning by musicians, the A below middle C, is 440 hertz. As reported by Dr. Quate in the August issue of *Physics Today,* frequencies of 3 million hertz have been produced in water, and 8 billion hertz in helium, where the wavelength is only 300 angstroms.

The vibrations are electrically generated in a sapphire crystal whose concave face focuses the sound onto the specimen. Sound reflected back onto the crystal is then converted into electrical energy by the sapphire. "We can see fibers inside various composite materials," Dr. Quate wrote, adding that "we can see features of biological cells that are difficult to image in the optical microscope" and the structure and adhesive properties of metal films and other coatings.

Another approach is the scanning ion microprobe, developed by Riccardo Levi-Setti at the University of Chicago in conjunction with Hughes Research Laboratories. Instead of electrons, it fires a beam of ions, or atoms that have shed some of their electrons, at the specimen. The ions are usually those of the metal gallium.

This not only maps elements of the specimen with a resolution of about 400 angstroms, but also indicates which of their isotopes are present. It has been aimed at tooth and bone specimens and should help reveal how they change from infancy to old age.

The scanning ion microscope has also been used to examine linen fibers from the Shroud of Turin, the cloth that is said to have covered the body of Christ.

—*December 10, 1985*

Reagan to Press for $6 Million Atom Smasher

By BEN A. FRANKLIN

The administration has announced that it will ask Congress for funds to start planning and building a giant $6 billion atom smasher.

The device, which is a superconducting supercollider in a 52-mile oval tunnel, would dwarf existing machines used to probe the secrets of matter and energy.

The project to build the world's largest research machine, in which subatomic particles moving at high speed would collide and burst, is as scientifically significant as America's 1969 manned landing on the moon, Secretary of Energy John S. Herrington said.

The device would accelerate atomic particles to an energy level 20 times greater than possible in existing laboratories. Mr. Herrington said that on completion in 1996, the supercollider would "bring answers to unsolved questions that have fascinated the world since the earliest times, such as what are the fundamental building blocks of matter."

The need for the device has been debated for years. Some scientists say it is crucial if the United States is to stay on the frontiers of particle physics. By enabling scientists to experiment at higher energies and smaller scales than before, proponents say, the supercollider may provide new insights into the elementary forces and particles of the universe.

Opponents of the project contend that it is too costly and unlikely to produce commensurate results.

At least 20 states have sought to be the home of the supercollider, but Mr. Herrington said a site decision was still yet to be made.

The announcement by Mr. Herrington at a news conference followed a showdown at meetings at the White House. Mr. Herrington was reported to have persuaded President Reagan to support the huge project over the objections of high administration officials concerned about its effect on the budget and the opposition of other Cabinet members who one administration official described as feeling that "the country didn't need it."

But one administration official said that after "a lot of missionary work in recent months by Secretary Herrington, the decision was not close." The fact that the president did not mention the project in his State of the Union Message on Tuesday was described as "just a scheduling thing—it didn't get through the Cabinet and the Domestic Policy Council until yesterday."

Foreign governments, and perhaps the state in which the project is eventually set up, will be asked to share in the costs and will share in the scientific and economic benefits, the secretary said.

Construction of the aboveground research facilities and the vast oval underground tunnel would create 4,500 jobs, the Energy Department said. A staff of scientists and technicians would total at least 2,500, with 500 others to have visitor status.

How Collider Would Work

The accelerator will send two beams of protons speeding through the tunnel in opposite directions. At several junctures the beams will cross and collide, creating a flash of energy out of which subatomic particles will burst. The belief is that such collisions, taking place at energy levels higher than ever achieved before, will disclose the existence of a host of new particles and forces.

The speeding protons will be contained in their path by powerful electromagnets cooled by liquid helium to temperatures so low that their coils lose all resistance to electricity; that is, they will be superconductive.

Among the states vying for the project are California, Colorado, Illinois, New York and Texas. Alvin W. Trivelpiece, the department's director of energy research, said some states were offering $10 million to $15 million in inducements.

Mr. Herrington said the "fair and open" site-selection process would take months or years. Asserting firmly that "there is no frontrunner" among the states, he promised to say more about site selection at another briefing on Feb. 10, 1987.

The project would be a costly venture into basic research, yielding possibly major scientific insights into the creation of matter—of the universe and the world—but with few firmly predictable practical applications.

"No Military Applications"

Mr. Herrington stressed that there would be "no military applications," adding: "The nature of basic research is that you don't know, going into it, what you will find. But the American track record in this has been good."

At the secretary's hurried presentation, called suddenly after Senator Phil Gramm, Republican of Texas, said Thursday night that the president would approve the project, Mr. Herrington seemed to give as much emphasis to the psychological and national morale implications as to the scientific.

Reflecting the president's emphasis on American "competitiveness" in the State of the Union Message, Mr. Herrington said Mr. Reagan had made "a watershed" decision for the United States.

"This is a watershed for America's scientific and technological leadership and another clear sign that President Reagan is committed to keeping this nation on the cutting edge of world leadership and competitiveness," Mr. Herrington said, reading a statement.

Calling the decision "of tremendous scientific significance and historical consequence," the secretary said: "It is a tremendous leap forward for America and for science and technology. Once again, this nation has said there are no dreams too large, no innovation unimaginable and no frontiers beyond our reach. By virtue of this decision we are embarking on an adventure of unlimited opportunity, tremendous promise and a new scientific world to be won."

Major Atom Smashers

The world's largest existing atom smashers, circular devices four miles in circumference, are at the Fermi National Laboratory in Batavia, Illinois, and at CERN, the European Laboratory for Particle Physics in Geneva. Larger devices are on the horizon. Soviet physicists are building a 13-mile atom smasher, and in Western Europe, physicists are constructing a 16-mile accelerator.

Throughout the development of the supercollider plan, the project has had passionate supporters and bitter critics.

One advocate restrained his glee today. Dr. Stanley G. Wojcicki, of the Lawrence Berkeley Laboratory of the University of California, said no champagne corks were popping because the team's efforts, both technical and political, would have to be redoubled.

"Everybody is obviously very, very pleased," he said. "But it's just the first step. The next is to convince the Congress and the American public that this is a scientific project that justifies the expenditure."

Critics have contended that the project would sap federal funds from less glamorous, but equally important, areas of scientific research and that there are no guarantees that the facility will yield more discoveries than current or planned facilities.

"Approaching a Budgetary Limit"

The supercollider "may be close to the end of the line of large science projects," said John E. Pike, associate director of the Federation of American Scientists, a private group in Washington.

"The assumption has always been that we're going to be able to understand everything," he added, but with this project "we're approaching a budgetary limit."

"These devices are becoming so expensive, and what they're trying to find is so obscure, that we may be at the point where scientists can no longer justify the cost," he said.

Dr. Arno Penzias, a Nobel laureate in physics at AT&T Bell Laboratories, has criticized the supercollider as a threat to the rest of physics research in the United States, much of it based in small laboratories at universities.

"For scientists the question to be answered is, what contribution of resources should the rest of science be asked to make to permit high-energy physics to build and operate the superconducting supercollider?" he has written. "The supercollider's capital cost will clearly squeeze capital expenditures for the other sciences."

Asked by reporters to be more specific in describing the collider's scientific importance, Mr. Herrington said he would do that at his Feb. 10 briefing.

The secretary said the $60 million required for an immediate start on design and site selection could come from funds appropriated for other projects. He predicted bipartisan support for it in Congress. Preliminary studies begun three years ago have cost about $20 million a year.

—*January 31, 1987*

The Supercollider's Demise Disrupts Many Lives and Rattles a Profession

By MALCOLM W. BROWNE

Still stunned by the congressional decision last month to terminate the $11 billion Superconducting Supercollider project, hundreds of scientists are casting about for new positions in high-energy physics, in hopes not only of pursuing their profession but also of continuing to support their families.

But their prospects seem bleak.

Many of the 200 or so physicists dispossessed by the demise of the big Texas accelerator hope to find jobs at CERN, Europe's premier accelerator laboratory, on the border of France and Switzerland near Geneva. CERN now seems destined to be the world's center for high-energy physics research well into the 21st century.

For the last decade, CERN has been the leading competitor of America's major accelerator laboratories, and if the Europeans succeed in building their Large Hadron Collider at CERN, which is scheduled for completion in 2002, CERN will move into a preeminent position in this field of research.

The Hadron collider, like the supercollider, will smash beams of protons together in the hope of revealing the mysterious mechanism that gives mass to all matter. But the Hadron collider will be far less powerful than the supercollider, and many physicists give it only about one chance in three of finding this scientific quarry, known as the Higgs mechanism. Physicists say they believe that the supercollider's energy would certainly have shed light on the Higgs mechanism, and that if a special particle associated with this mechanism exists, it would have been found.

Problems of Its Own

CERN itself has financial problems because of the European recession, and unless the United States Congress chooses to grant American scientists substantial dowries to take to the European laboratory, employment prospects there seem poor.

Dr. Christopher Llewellyn Smith, the British physicist who will become CERN's director on Jan. 1, 1994 said in an interview, "We must try to find a way to make it possible for many of our stranded American colleagues to come to CERN." But, he added, "this will take difficult negotiations, because it's quite obvious that there would have to be financial contributions from the United States" to pay for the added CERN staff and the experiments that they would undertake.

The demise of the supercollider project, Dr. Llewellyn Smith said, "will be a catastrophe if it turns out to mean that the United States is leaving the field of high-energy physics, which that country pioneered."

Blessing or Curse?

Dr. Stanley G. Wojcicki, who is a professor of physics at Stanford University and the chairman of the High-Energy Physics Advisory Panel—a group of physicists who advise the Department of Energy and other federal agencies—says that there is little left of the federal funding pie to divide up. "Congress has spoken rather firmly that this country is not willing to embark on any large scientific projects," he said.

Such views are by no means those of all scientists, many of whom regard the passing of the supercollider as a blessing.

Delighted by the recent action of Congress, Dr. Rustum Roy, a professor of materials science at Pennsylvania State University, said in an interview that "this comeuppance for high-energy physics was long overdue."

Discouraging Scientists

Dr. Roy expressed the view of many scientists that "there is an acute oversupply of scientists in the United States," and that it is the responsibility of the educational system to try to reduce it.

But to Dr. Roy Schwitters, who was replaced as director of the supercollider, the United States is about to lose "the best and brightest" of its high-energy physicists.

"But for most of the thousand or so scientific and engineering staff members of the SSC, the future is terrifying," he said, referring to the supercollider. "Many have sunk their life savings into houses near Waxahachie, and now they face jobless futures in a collapsing real estate market."

Among the victims of the supercollider demise, he said, are 180 foreign scientists, many of them Russians, who are working in Texas on visitor visas. They will presumably have to leave the United States soon. "We talk a lot about finding alternative work for ex-Soviet nuclear weapons scientists, to keep them from working for terrorist states," Dr. Schwitters said, "and that's exactly what we were doing."

A Costly End

Dr. Schwitters believes that Congress is unlikely to provide the financial contribution that the European research consortium will demand in exchange for hiring American physicists. "Congress won't spend the money in Texas, so why would they spend it in Switzerland?" he asked.

The cost overruns that plagued the supercollider in life seem likely to affect even its funeral. Congress voted to provide $640 million to close the project down, but it appears that this will not be nearly enough. Environmental regulations, for instance, will require the filling-in of all the tunnels and shafts already completed for the supercollider, which was to have been 54 miles in circumference. This will be costly. Dr. Wojcicki said he had seen estimates as high as $1.5 billion.

Other problems that will have to be resolved include the disposition of the land that was seized from private owners to make way for an accelerator that occupies most of Ellis County, Texas, as well as the hundreds of houses that were moved off the site and onto tracts of storage land. The large factories built at the supercollider site to mass-produce huge superconducting magnets, as well as a stockpile of completed magnets, will have to be dismantled and, presumably, sold for scrap.

Dr. Schwitters, however, said he hoped that the uncompleted site and its materials could be kept intact for scientists in the future who might be able to resume the supercollider's work.

Others Could Benefit

Perhaps the one bright spot that accelerator scientists see in the supercollider demise is the possibility that Congress will spend some of the savings on other physics projects.

Leading the list of potential applicants is Fermilab in Illinois, which is still the highest-energy accelerator in the world. Scientists believe that Fermilab's Tevatron accelerator has a reasonable chance of finding the "top" quark—the only one of the six hypothesized quark types that has not yet been discovered. Fermilab's decade-long quest for the elusive quark has so far failed to find conclusive evidence of its existence.

Finding the top quark would neatly fill out the "Standard Model," the generally accepted theory of how the basic building blocks of matter fit together. With luck, Fermilab might also find evidence of "supersymmetry," a theoretical mechanism that might relate the force of gravity with the three other known forces of nature: the electromagnetic force, and the strong and weak nuclear forces.

Quarks, Matter and Antimatter

But the Tevatron's two trillion electron volts of colliding energy have so far failed to reveal the top quark or supersymmetry, and physicists believe the accelerator's colliding particle beams may not contain enough protons to yield a good statistical chance of success. For this, the machine needs a new "main injector," and the hope is that additional money will be found to hasten completion of the main injector.

Another promising accelerator project, the Stanford Linear Accelerator Center's "Asymmetric B-factory," will get $237 million in federal money, President Clinton announced on Oct. 4. One of the objects of this project is to discover why the newborn universe seems to have contained slightly more matter than antimatter. This assumption explains why, after the primordial matter and antimatter annihilated each other, a small residue of matter was left over, from which the present-day universe formed.

—November 14, 1993

Europe Is Ready to Pick Up
the Pieces in Particle Research

By BARRY JAMES

The decision of the U.S. House of Representatives to halt funding of the Superconducting Supercollider could leave Europe the leader in high-energy particle research at a time when scientists appear to be on the verge of unlocking important secrets about the universe.

Many prominent figures in particle research were meeting at the Renaissance château in Blois, France this week, together with astrophysicists, to examine new evidence that the infinitely small and the infinitely large are aspects of the same reality.

The vote in the House surprised U.S. delegates at the conference, who assumed that the $8.5 billion SSC project was going ahead.

Particle research, scientists said, could be the key element in identifying the mysterious "dark matter" that makes up 90 percent of the universe. This would help explain how the universe was formed and expanded.

Neil Calder, a spokesman for the European Center for Nuclear Research, or CERN, said the U.S. decision was "astonishing, absolutely shattering."

Mr. Calder said that more than 1,000 CERN physicists were working on a feasibility study for building a similar facility, a hadron collider. Hadrons are heavy particles comprising protons and neutrons.

Because the collider would be built in the 27-kilometer (16.5-mile) tunnel housing CERN's existing electron positron collider, it would cost about 2 billion Swiss francs ($1.41 billion), far less than the U.S. project's estimate.

CERN member nations are scheduled to decide at the end of the year whether to go ahead with the hadron collider on the basis of the feasibility study.

Although the European project is less ambitious, Mr. Calder said research indicated it would be able to operate at the energy range at which scientists expect to be able to do important research into the nature of dark matter.

The research consists of accelerating streams of particles in different directions around a circular tunnel at nearly the speed of light and smashing them into one another. By studying the shards of such collisions, physicists can gain knowledge of the universe as it existed shortly after the theoretical "big bang" at the beginning of time and space, when matter consisted of almost infinite energy.

NASA's Cosmic Background Examiner found the lingering echoes of such a cosmic event 15 billion years ago. George F. Smoot, a principal member of the COBE team, who was attending the Blois conference, said that researchers were close to understanding the mechanism of the big bang, which presents new challenges for scientists and for philosophers.

"It raises the question of what caused it, what made the design," he said.

The dark matter that settled around the universe as it expanded has been detected with the help of Einstein's theory of general relativity, which postulated that light bends around a mass.

By observing the behavior of light as it passes distant galaxies and quasars, scientists are able to measure their mass.

But although scientists know that matter is there, most of it is hidden. They do not know whether it is the same as the matter that makes up the visible universe or whether it consists of exotic particles unknown to science.

"If the Large Hadron Collider shows us that we are not even made of the same stuff as the vast majority of the matter of the universe, this will have philosophical and cultural repercussions probably as important as the findings of Galileo," Mr. Calder said.

"This is probably the single most important question in physics today," said James Cronin of the University of Chicago, a Nobel laureate responsible for the asymmetry principle that explains why there is more matter than antimatter, and why the universe therefore exists.

Sheldon Glashow of Harvard University, another Nobel laureate at Blois, said science was raising in dramatic new ways "the big questions we all had when we were children; How did the universe begin? What is it made of?"

Congressional critics of the SSC questioned the wisdom of pouring vast amounts of money into pure research that is unlikely to ease the pressing problems of the world.

Mr. Calder said such doubts had not surfaced in Europe. For one, he said, CERN had reversed a transatlantic brain drain. The number of U.S. physicists working at CERN is far larger than the number of European physicists working in the United States. CERN is the second-largest physics research center grouping of Americans after the Fermi laboratory in Batavia, Illinois, he said.

Mr. Calder also provided an economic justification for continuing with expensive particle research. "Seventy-five percent of research scientists go into industry," he said. "If CERN did not exist, there would be a great hole in the number of people exposed to pure research at the absolute limits of technology."

—*June 20, 1992*

Top Quark, Last Piece in Puzzle of Matter, Appears to Be in Place

By WILLIAM J. BROAD

The quest begun by philosophers in ancient Greece to understand the nature of matter may have ended in Batavia, Illinois, with the discovery of evidence for the top quark, the last of 12 subatomic building blocks now believed to constitute all of the material world.

An international team of 439 scientists working at the Fermi National Accelerator Laboratory will announce the finding, bringing nearly two decades of searching to a dramatic conclusion.

The Fermilab discovery, if confirmed, would be a major milestone for modern physics because it would complete the experimental proof of the grand theoretical edifice known as the Standard Model, which defines the modern understanding of the atom and its structure. The finding is likely to produce waves of intellectual satisfaction for physicists around the world and to give American physics a significant boost.

The discovery in all likelihood will never make a difference to everyday life, but it is a high intellectual achievement because the Standard Model, which it appears to validate, is central to understanding the nature of time, matter and the universe.

"The exciting thing is that this is the final piece of matter as we know it, as predicted by cosmology and the Standard Model of particle physics," Dr. David N. Schramm, a theoretical physicist at the University of Chicago, said in an interview. "It's the final piece of that puzzle."

Dr. Hans A. Bethe, a Nobel laureate in physics at Cornell University, said the finding was "a very big deal" that "makes the whole picture of subnuclear particles much more believable and better established."

"We've needed the top quark," he said. "It figures in all our calculations for further processes, and none of them would be right if it weren't there."

If the top quark could not be found, the Standard Model of theoretical physicists would collapse, touching off an intellectual crisis that would force scientists to rethink three decades of work in which governments around the globe had invested many billions of dollars.

All matter is made of atoms, but nearly a century ago physicists discovered that atoms, long considered to be the smallest units of matter, were themselves composed of smaller, subatomic particles like protons and neutrons. But these particles later showed signs of being made of yet smaller building blocks.

The field was plunged into confusion for many years until a grand unifying theory pioneered by Dr. Murray Gell-Mann, a physicist at the California Institute of Technology, sought to explain the structure of particles like protons and neutrons in terms of new units that he whimsically named quarks.

His theory called for the existence of six different kinds of quarks, named up and down, charm and strange, top and bottom. The quark family parallels a six-member family of lighter particles, known as leptons, that includes the electron.

Various combinations of these 12 particles are thought to make up everything in the material world. In addition to matter, the universe contains potent forces like electromagnetism and gravity, and perhaps many other exotic particles as yet to be discovered.

Five of the six quarks were eventually found, but the sixth remained painfully absent. For nearly two decades rival teams of scientists around the world have sought the top quark by performing ever-more-costly experiments on increasingly large machines that accelerate tiny particles almost to the speed of light and then smash them together in a burst of energy. The resulting fireball can yield clues to nature's most elementary building blocks.

The team at Fermilab, which includes scientists from the United States, Italy, Japan, Canada and Taiwan, cautioned that the evidence they had gathered over the past year and a half for the top quark would be convincing to many scientists but not definitive. They said further work would be needed to firmly establish the top quark and its attributes.

"Some people will say, 'Hey, nice piece of physics but you need more data to make sure,'" said Dr. Melvyn J. Shochet, a physicist at the University of Chicago who worked on the Fermilab experiment and is a spokesman for the discovery team. "To that I can only agree."

"We don't have a discovery," said a senior Fermilab official, who spoke on the condition of anonymity. "We have evidence. It's good evidence. It's tightening up to where the top quark lives. The next step is to get more events."

5,000-Ton Detector

The experiment was run on Fermilab's Tevatron, a four-mile, circular accelerator in an underground tunnel that hurls counterrotating beams of protons and anti-protons at each other with a combined energy of 1.8 trillion electron volts. It is currently the highest-energy accelerator in the world. The detector that gathered the evidence is the size of a large house and weighs 5,000 tons. A 150-page manuscript describing the work was mailed to *Physical Review*, the world's preeminent journal of physics.

Dr. Shochet, the team spokesman, said the mass of the top quark, its most important attribute, was calculated to be 174 billion electron volts, with an uncertainty range of plus or minus 17 billion electron volts.

"That's quite heavy," he said. "It's almost as heavy as an entire gold atom. It's by far heavier than any other elementary particle that's been observed, which is why it's taken so long to find."

As Fermilab, which is run by the federal Department of Energy, reports the finding, simultaneous announcements are being made in Rome, Tokyo, Ottawa and Taipei.

Predictions about Quarks

Dr. Gell-Mann took the word *quark* from a line in *Finnegans Wake* by James Joyce: "Three quarks for Muster Mark." So too, Dr. Gell-Mann predicted that quarks in normal matter came in groups of three. Protons would be made of two up quarks and one down quark; neutrons of two down quarks and one up quark. Dr. Gell-Mann's ideas were radical and strongly resisted, partly because the fractional charges of his quarks seemed implausible. But his theories explained much, and were soon partly confirmed by particle discoveries. In 1969 he won the Nobel Prize in physics.

Low-mass quarks, the up and down, are the only ones thought to ordinarily exist in this world. Physicists believe that the higher-mass ones, charm and strange, top and bottom, were present naturally only for a tiny fraction of a second at the beginning of time during the big bang—the primordial explosion thought to have given rise to the universe. Top quarks, having the highest mass of all, are believed to have vanished from the universe after existing for less than a billionth of a second.

Thus, a time machine is needed to see most quarks. Particle accelerators slam together tiny bits of matter to create intense fireballs almost as hot as those that existed at the beginning of time, creating streams of nature's most rudimentary particles.

In 1977, when the bottom quark was discovered at Fermilab in a particle accelerator, physicists calculated that its top-quark companion would have a mass of 13.5 billion electron volts, making it an easy target for any number of accelerators then planned around the world.

In July 1984, a European team of 151 scientists headed by Dr. Carlo Rubbia announced that it had confirmed the existence of the top quark, calling it a major breakthrough. That fall, Dr. Rubbia won the Nobel Prize in physics for other discoveries. But it turned out that his top-quark claim was premature. The particle was far heavier, and more difficult to detect, than had generally been anticipated.

A Kind of Alchemy

Physicists at Fermilab have been hunting the top quark for nearly two decades, looking at increasingly high energies. The process, they say, is like slamming together two tennis balls and trying to find a bowling ball in the rubble—a hint of the top quark's huge mass. The tennis balls can create things heavier than themselves because of their high energies, a kind of alchemy first suggested by Einstein in his famous law of equivalence between matter and energy.

The rub is the rarity of collisions that make top quarks. Dr. Shochet, the experiment team's spokesman, said many billions of proton-antiproton collisions were needed to produce just one top quark and that even then, subtle clues to its existence might be lost amid a clutter of spurious signals. The quarks themselves exist for only a fraction of a second, and cannot be detected directly. Their presence is inferred from ghostly showers of particles produced as they perish.

Dr. Shochet said the team's evidence gathered over a year and a half amounted to 15 clues from 12 collisions. Those results, he added, were about twice as high as expected from false positives in the background noise. He said really nailing down the top quark would require a mass of evidence three or four times above background levels.

Dr. Claudio Campagnari, a team physicist, said in a Fermilab brochure: "Rather than one 'Eureka!' event, top discovery will come by accumulating a lot of different evidence, bit by bit. You could compare discovering top with what happens in a courtroom in a case where there's no smoking gun and you must convince the jury by the accumulated weight of circumstantial evidence."

A separate team of 420 scientists at Fermilab is now using a different detector in an effort to confirm the first team's findings during the Tevatron's current 18-month run. Its work, and that of the original team, should be eased somewhat by recent accelerator improvements that will increase the number of collisions.

Fermilab is also completing a $230 million upgrade of the Tevatron that should sharply increase the collision rate, perhaps producing hundreds or thousands of top-quark candidates. It should be completed by 1998 or 1999.

After that, the only other accelerator powerful enough to join the hunt would be one under consideration at CERN, Europe's premier accelerator laboratory, on the border of France and Switzerland. Known as the Large Hadron Collider, it might be completed by the year 2005.

If the top quark has indeed been discovered at Fermilab, particle physicists will turn their attention to other enigmas, such as why all matter has mass. In the United States, such questions were to be addressed by the Superconducting Supercollider, which was to have measured 54 miles around and cost up to $11 billion. In October 1993, Congress canceled the half-built machine in Waxahachie, Texas, calling it an inordinate drain on the federal budget.

American inventors are now trying to create small, innovative accelerators in lieu of the big machine.

"Any new particle that's found" in the years ahead, said Dr. Schramm of the University of Chicago, "is going to be exotic in a much greater way than any quark."

—April 26, 1994

Physicists Manage to Create
the First Antimatter Atoms

By MALCOLM W. BROWNE

Physicists at the European Laboratory for Particle Physics have announced that they had created, for 40 billionths of a second, the first complete atoms of antimatter ever made by human beings or seen in nature.

In an antiatom, the antimatter equivalent of an ordinary atom, the electrical charges of all the component particles are reversed; while an ordinary atom has a positively charged nucleus with one or more negatively charged electrons orbiting it, the antimatter atom has a negatively charged nucleus with positively charged orbiting electrons. An ordinary atomic nucleus contains positively charged protons, while its antimatter counterpart contains negatively charged antiprotons.

Unless an antimatter atom is kept from coming into contact with an ordinary atom, the two atoms annihilate each other in a violent flash of energy—a fact that may explain the apparent absence of antimatter in the natural universe. Antiprotons are routinely made in physics laboratories, as are antielectrons, which are also called positrons. But no one had heretofore succeeded in nudging a positron into orbit around an antiproton, making an atom of antimatter. The announcement by the European laboratory near Geneva, known by its former acronym, CERN, establishes that this bizarre kind of atom can actually exist.

Physicists hope one day to make comparative measurements of the properties of atoms and antiatoms in terms of their gravitational attraction, their interactions with light and other features. Subtle differences between atoms and their antimatter counterparts may shed light on the origin and evolution of the universe and help solve the puzzle as to why we are made of matter instead of antimatter.

Although most physicists discount the idea that antihydrogen might one day be developed as a very high potency fuel for interstellar rockets or superbombs, some scientists have not abandoned the dream of exploiting antimatter as a propellant. When combined with ordinary matter, it annihilates, converting mass to energy far more efficiently than does a nuclear bomb.

Dr. Walter Oelert of the Juelich Institute for Nuclear Physics Research in Germany and his German and Italian colleagues reported that they created the 11 atoms of antihydrogen during a three-week experiment at CERN last September, but withheld the news until they and independent experts had thoroughly checked their results, which will be published in a forthcoming issue of the journal *Physical Review B.*

"We're absolutely sure now," he said in an interview, "and the experiment shows without doubt that antihydrogen can exist. No one really doubted it, but it's nice to have the experimental proof."

The antihydrogen atoms created in the experiment were moving at nearly the speed of light and survived only some 40 billionths of a second before colliding with atoms of ordinary matter and annihilating themselves. But from the pattern and types of debris created by these collisions, the scientists were able to establish the identity of the projectiles as antihydrogen atoms.

Dr. John Eades, the British coordinator of experiments at CERN, said that the real challenge had been in producing enough of the right kind of collisions between ordinary particles to create a few antihydrogen atoms. To do this, antiprotons from one of CERN's accelerators were boosted to very high energy and hurled into a target of xenon atoms, each atom containing a nucleus with 54 protons and about 77 neutrons.

Some of the antiprotons survived and passed through, while others collided with xenon nuclei, converting part of their collision energy into the creation of antielectrons. In a few very rare cases, the speeds and directions of the newly born antielectrons and the surviving antiprotons coincided enough that the antielectrons were captured into orbits around the antiprotons, thus forming antihydrogen atoms. These atoms, like ordinary hydrogen atoms, are electrically neutral; the charges of their components cancel out.

But the neutrality of antihydrogen, like that of ordinary hydrogen, renders it impossible to contain or manipulate using magnetic fields. Moreover, an antiatom cannot be contained in an ordinary vessel, since the slightest contact with the container's wall causes it to annihilate. Consequently, other groups are developing enormously sophisticated methods, including interacting lasers, to manipulate and secure antiparticles inside vacuum chambers.

Dr. Oelert acknowledges that the antihydrogen atoms his group made cannot be used as the basis of far-reaching experiments in the fundamental interactions of physics, including such cosmological questions as to why the universe seems to consist entirely of matter rather than antimatter.

"We just wanted to have fun and see if you could make antihydrogen," he said. "It maybe is not such a great scientific achievement in itself, because our antihydrogen atoms are moving around much too fast to study in detail before they annihilate."

Meanwhile, physicists at Harvard University, Pennsylvania State University, Los Alamos National Laboratory and other groups are attempting to create, capture and control antihydrogen in a more useful form. Their efforts focus on "cooling" antiprotons and antielectrons using lasers and other tools, and assembling atoms not by violent collisions but by manipulation. Such atoms, when they are eventually made, will be held almost motionless in their chambers, isolated from contact with ordinary matter while their gravitational, spectral, charge-conjugation, parity and other characteristics are measured.

"Everyone makes shrewd guesses about the probable behavior of antihydrogen and other antiatoms," Dr. Eades said, "and we don't really expect a big surprise—that an atom of antimatter would fall up instead of down, for instance. But there may be subtle differences of great importance. For instance, an atom and an antiatom might fall at slightly different speeds toward a gravitating object like the sun. Our orbit around the sun is elliptical, so the sun's gravitational pull on the Earth varies slightly over a year, and by observing its effect on an antihydrogen atom, we might learn interesting things."

In principle, scientists believe that atoms larger than antihydrogen—the simplest possible atomic form of antimatter—might be created. But each increase in the size and complexity of an atom complicates the assembly problem. Antihelium, the next most complicated atom, would have a nucleus of two antiprotons and two antineutrons, with two orbiting antielectrons.

"We're especially interested in hydrogen and antihydrogen," Dr. Eades said, "not only because of their structural simplicity, but because 90 percent of the mass of the universe is hydrogen. Even slight differences in the properties of hydrogen and antihydrogen could help explain why the universe, as we know it, consists entirely of matter rather than antimatter."

Dr. Gerald Gabrielse of Harvard, whose research group is working to slow down particles of antimatter contained in special traps, commented that the CERN synthesis of antihydrogen "is an important experiment demonstrating that it can be done. The payoff will be down the road, when one is eventually able to study the properties of these atoms."

—January 5, 1996

Stuck in Traffic? Consult a Physicist

By MALCOLM W. BROWNE

 Hateful though they are to most people, traffic jams have fascinated some of the finest scientific minds of the 20th century, who see in them similarities to the freezing of water, the triggering of avalanches, the formation of galaxies and the advent of life itself.

Although physicists have made little progress in dissolving these clots in the arteries of automotive civilization, the enormous power of new computers has opened new possibilities that may one day reduce traffic congestion, improve the reliability of weather forecasting and solve other problems that so far have proved intractable.

The latest volley comes from Germany, where Dr. Boris S. Kerner, a research physicist at Daimler-Benz in Stuttgart, and Dr. Hubert Rehborn, a traffic consultant living in Aachen, have developed a theory that road traffic is subject to the kind of "phase transitions" that water undergoes when it abruptly changes to steam or ice.

Their paper, published last week in the journal *Physical Review Letters*, reports a series of measurements they have made since 1991 on a heavily traveled stretch of Autobahn highway near Frankfurt. They found that traffic along this German equivalent of a freeway flowed in three sharply differing modes: free flow, in which fast vehicles can change lanes and pass; "synchronized" flow, in which high-traffic density prevents lane changes and passing; and jams, in which vehicles come to at least a momentary stop.

Although other physicists in the last 40 years have theorized about phase transitions in traffic, most have conceived of these transitions as a type known as "second order," meaning that they occur gradually in response to gradual changes in average vehicle speed and traffic volume. But Dr. Kerner and Dr. Rehborn found, from data recorded by sensors under the roadway, that the phase transitions between free flow, synchronized flow and traffic jams were "first order" transitions. These occur abruptly and spontaneously, sometimes without any change in traffic volume or speed to trigger them.

Moreover, the two physicists found, when something interferes with freely flowing traffic, like an influx of slowly moving cars from a ramp, the resulting phase transition in the main stream of cars acquires a life of its own. A sluggish mass of synchronously moving vehicles propagates its infuriating mode of movement miles upstream from the obstruction and may persist for two or more hours, even after the obstruction is eliminated. In this, the clog seems to resemble a block of ice floating in water just above the freezing point: it neither grows nor rapidly melts.

Other investigators have shown that when traffic on a highway reaches about 85 percent of the highway's capacity, traffic becomes unstable; it may flow normally for a time, but it may congeal abruptly and without warning.

The German scientists concluded that the tendency of freely flowing traffic to "self-organize" spontaneously into synchronous flow or jams might reflect the influence of chaos, a condition in which tiny perturbations of a system's initial state can set off big changes in the system's evolution.

Chaotic sensitivity to initial conditions affects many phenomena, like weather or the frictional drag created by water flowing around a ship, and the effects of chaos are notoriously difficult to predict.

But if it was possible to create a vast network of sensors and traffic guidance systems under computer control, Dr. Kerner said in an interview, it should be possible to take a substantial bite out of congestion. A mechanism to delay a phase transition until after rush hour, Dr. Kerner said, would be a big improvement.

"Even though it's not possible to do this today, it's in our future," he said. Such a system would need to forecast accurately the occurrence of phase transitions that reduce traffic speed to a crawl.

"Traffic phase transitions are like cancer," he said. "Once a transition occurs, it may be too late to fix. The answer is timely and accurate forecasting of an approaching phase transition."

Dr. Kerner offered no specific suggestions for blocking incipient phase transitions. In general, he said, a neural network of sensors, a powerful computer traffic analysis system and responsive highway controls of some kind would be needed.

Before much can be done about jams in a practical way, physicists believe that they need to understand the fine details of the physics underlying traffic dynamics. To that end, the federal government is conducting studies in the Dallas–Fort Worth area and in Portland, Oregon.

With financing from the Department of Transportation, physicists at Los Alamos National Laboratory in New Mexico plan to spend $25 million over seven years to study traffic in the two urban areas. Their aim is to use this information to create computer simulations that mimic the behavior of real traffic.

"The beauty of this thing," said Dr. Richard J. Beckman, a project leader at Los Alamos, "is that you can try out a policy decision before you actually implement it. For example, you can ask the simulation to show what would happen if you add three new lanes to a freeway."

The mathematical model of traffic in the Dallas–Fort Worth area is nearing completion, he said, and a much more detailed study of Portland traffic is about to begin. In the Portland study, each one of the city's 1.5 million residents will be simulated mathematically in terms of his or her expected activity. That will include information about travel between various points, the likelihood of travel by car, bus, train or bicycle, and other related activity.

"Then you put the whole population together in a microsimulation, run it for 24 hours and see what happens," Dr. Beckman said.

The simulation can even include driver behavior that deviates from "rational norms," including aggressive, weaving drivers who lean on their horns. Major traffic disruptions like three-car collisions can also be inserted in the model.

Traffic theorists agree that major congestion reduction will require much more than mathematical simulation and that politics will always play a defining role in efforts to improve traffic flow.

"But we can provide policy makers with useful advice," Dr. Beckman said. "For instance, our model can show what would be likely to happen to traffic in a large city if the price of gasoline were to jump to $10 a gallon."

Among the most influential traffic theorists since the 1960s is Dr. Ilya Prigogine, a Belgian chemist who was awarded the 1977 Nobel Prize in chemistry. Dr. Prigogine, director of the Prigogine Center for Statistical Mechanics, Thermodynamics and Complex Systems in Austin, Texas, received the Nobel Prize for his theory of "dissipative structures." That theory showed how an "open" system like a living being could spontaneously increase in complexity, provided that it was "far from equilibrium" and could import energy from the outside and export entropy, the physicist's word for disorder.

Among many other things, that theory may help explain how the complex chemistry of life arose from simple precursors, in apparent defiance of the Second Law of Thermodynamics, which, in simplest terms, decrees that everything in a closed system, be it a steam engine or a chemical reaction, runs downhill.

Dr. Prigogine was retained by General Motors in the 1970s to explore possible applications of his theory to traffic flow in cities. He collaborated with the late Dr. Robert Herman, also an adviser to General Motors, who gained fame as a physicist by predicting in the 1940s that a microwave echo of the big bang would someday be found. (The echo was found in 1978 by Dr. Arno Penzias and Dr. Robert Wilson of Bell Laboratories. The two scientists shared the 1978 Nobel Prize in physics for their discovery.)

Dr. Prigogine and Dr. Herman developed a theory of traffic flow dependent on some of the differential equations used to describe fluids in motion. The theory improved understanding of traffic flow, but scientists realized that a really useful traffic theory would have to take all the fine details into account, not just the broad trends that can be described by equations.

Dr. Steen Rasmussen, a physicist at the Los Alamos Laboratory, praised Dr. Prigogine. "His way of looking at nature and sociotechnical systems certainly shaped our thinking," Dr. Rasmussen said in an interview. "But he lacked the tremendously powerful computers we have today, and he was brought up to think about science in a different way. In thinking about traffic, for example, the approach of Dr. Prigogine's generation was to assume that traffic flow is a continuum in which the role of each single vehicle is insignificant. By contrast, we include every vehicle and person in our models."

If enormous numbers of vehicles are included, the results of such simulations closely resemble the solutions yielded by calculus equations, he said, but much greater precision and reliability result from detailed simulations.

Even today, planners in New York City are using a detailed simulation of expected traffic patterns near Chelsea Piers, a riverside development project. The program allows engineers to visualize the results of even small perturbations, like illegally parked cars.

Most of the simulations physicists use today depend on variants of mathematical objects called cellular automata, which can be represented on a computer screen as squares or hexagonal cells that change with respect to their neighboring cells according to some set of simple rules. Motivated by these rules alone, groups of cellular automata organize themselves and evolve complex patterns, sometimes resembling the spread of forest fires, the creation and evolution of new species, the birth of galaxies and the appearance of traffic jams, among many other things.

A decade ago, a Danish physicist, Dr. Per Bak, proposed a theory of "self-organized criticality," based partly on attacking complex problems in terms of cellular automata. His theory sought to explain many kinds of abrupt change in nature: the transition of stable sand piles into avalanches, the collapse of silos and the origin of complex structures, including the molecules of life.

Astonishing advances in the power of computers in the last five years have greatly changed the way physicists attack complex problems like traffic, Dr. Kerner said. Increasingly, scientists analyze problems "bottom up," programming computers to tackle mountains of detail to yield the insights once deemed accessible only by applying some powerful mathematical theory.

For this reason, scientists are hopeful that giant computational problems, including traffic as well as climate and weather forecasting, are close to solutions. Dr. Kai Nagel, a leading builder of mathematical models at Los Alamos, said that complexity was one of the similarities between traffic and weather.

"Another," Dr. Nagel said, "is that you can't conveniently do experiments on either one." But with good computer simulations, he said, "you can get very good estimates."

Hopeful though that may sound, physicists agree that real progress in traffic control will take time.

Samuel I. Schwartz, the former deputy commissioner of transportation in New York City who coined the word *gridlock,* acknowledged that physics-based attacks on traffic problems "are fun and may prove to be useful."

"But it will be 100 years before computers can cope with traffic without the guidance of human minds," Mr. Schwartz said.

—*November 25, 1997*

Mass Found in Elusive Particle; Universe May Never Be the Same

By MALCOLM W. BROWNE

In what colleagues hailed as a historic landmark, 120 physicists from 23 research institutions in Japan and the United States have announced that they had found the existence of mass in a notoriously elusive subatomic particle called the neutrino.

The neutrino, a particle that carries no electric charge, is so light that it was assumed for many years to have no mass at all. After today's announcement, cosmologists will have to confront the possibility that a significant part of the mass of the universe might be in the form of neutrinos. The discovery will also compel scientists to revise a highly successful theory of the composition of matter, the Standard Model.

Word of the discovery had drawn some 300 physicists to Takayama, Japan to discuss neutrino research. Among other things, the finding of neutrino mass might affect theories about the formation and evolution of galaxies and the ultimate fate of the universe. If neutrinos have sufficient mass, their presence throughout the universe would increase the overall mass of the universe, possibly slowing its present expansion.

Others said the newly detected but as-yet-unmeasured mass of the neutrino must be too small to cause cosmological effects. But whatever the case, there was general agreement here that the discovery will have far-reaching consequences for the investigation of the nature of matter.

Speaking for the collaboration of scientists who discovered the existence of neutrino mass using a huge underground detector called Super-Kamiokande, Dr. Takaaki Kajita of the Institute for Cosmic Ray Research of Tokyo University said that all explanations for the data collected by the detector except the existence of neutrino mass had been essentially ruled out.

After Dr. Kajita's remarks, the powerful evidence he presented elicted pro-longed applause from an audience of physicists from dozens of countries who packed the conference hall in Takayama.

Dr. Yoji Totsuka, leader of the coalition and director of the Kamioka Observatory where the underground detector is situated, 30 miles north of Takayama in the Japanese Alps, acknowledged that his group's announcement was "very strong," but said, "We have investigated all other possible causes of the effects we have measured and only neutrino mass remains."

Dr. John N. Bahcall, a leading neutrino expert and astrophysical theorist at the Institute for Advanced Study in Princeton, New Jersey, said in an interview that there had been many claims in recent years of the discovery of neutrino mass by other groups. "But this one is by far the most convincing," he said. "Besides the strong evidence they have found, this team has a magnificent track record of discoveries."

But because the elusive particles cannot be seen, the evidence that they have mass is indirect.

Transformation Is Evidence of Mass

Neutrinos come in three types or "flavors." The data gathered by the Super-Kamiokande team during the two years the detector has operated indicate that at least one of these three "flavors" can "oscillate" into one of the other flavors as it travels along at nearly the speed of light. According to the theories of quantum mechanics, any particle capable of transforming itself in this way must have mass.

Study of the neutrino particle has been glacially slow since its existence was hypothesized in 1930 by the Austrian physicist Wolfgang Pauli as a way to explain the mysterious loss of energy in certain nuclear reactions. The particle was finally discovered in 1956 by two physicists at the Los Alamos National Laboratory, Dr. Frederick Reines (who was awarded a Nobel Prize for the discovery) and the late Dr. Clyde Cowan.

But understanding of the particle since then has been acquired painfully slowly, because neutrinos have no electric charge and rarely interact with any kind of matter. A neutrino so rarely collides with an atom of ordinary matter that a typical neutrino can easily penetrate a one-light-year thickness of lead—some six trillion miles—without hindrance.

As the writer John Updike put it in a poem he wrote in 1960:

Neutrinos, they are very small.
They have no charge and have no mass
And do not interact at all.
The earth is just a silly ball
To them, through which they simply pass
Like dust maids down a drafty hall.

But once in a great while, a neutrino does hit an atom and the resulting blast of nuclear debris supplies clues about the neutrino itself. The debris generally includes many particles that can race through water, mineral oil or even ice, sending out shock waves of blue light. This light, called Cherenkov radiation, can be detected by sensitive light sensors and measured.

During the past few decades, scientists have learned that matter is made up of three distinct flavors or types. This means that there are three flavors of neutrinos—the electron neutrino associated with the electron, the muon neutrino, associated with the muon particle, which is a kind of fat electron, and the tau neutrino, associated with the tau particle, an even fatter relative of the electron. The role of the muon and tau particles and their associated neutrinos in the universe has mystified physicists. "Who ordered that?" the Columbia University physicist Isidor Rabi is said to have remarked when the muon was found.

The Super-Kamiokande detector was built two years ago as a joint Japanese-American experiment. It is essentially a water tank the size of a large cathedral installed in a deep zinc mine one mile inside a mountain 30 miles north of Takayama. When neutrinos slice through the tank, one of them occasionally makes its presence known by colliding with an atom, which sends blue light through the water to an array of detectors.

The enormous volume of water in the detector increased the likelihood of neutrino impacts to the point at which the discovery of neutrino mass became possible.

The Super-Kamiokande collaboration is studying several neutrino phenomena simultaneously, but the one that led to this announcement was based on "atmospheric" neutrinos created when highly energetic cosmic ray particles from deep space slam into the Earth's upper atmosphere.

Finding a Reason for a Puzzling Shortage

Physicists knew that different flavors of neutrinos constantly arrive from the upper atmosphere and they have calculated that the ratio between muon neutrinos and other flavors must have a certain value. But over the years detectors found only about half the muon neutrino predicted by theory.

The apparent shortage of muon neutrinos was explained by the recent observations that led to the announcements in Takayama. The physicists found that when neutrinos come from the sky directly over the Super-Kamiokande detector—a relatively short distance—the proportion of muon neutrinos among them was higher than among the neutrinos coming up from beneath the detector after having passed through the Earth.

The scientists reasoned that by traveling through the entire Earth these neutrinos had had time to oscillate, probably many times, between muon neutrinos and some other type, especially the tau neutrino, and this accounts for the deficit seen in muon neutrinos. (The tau neutrino has not yet been directly detected, but it must exist to make observations consistent.)

A related problem has to do with neutrinos produced by the fusion process in the sun. This process, which merges the nuclei of hydrogen atoms to form helium nuclei and energy, produces neutrinos. Astrophysicists believe they understand the mechanism in complete detail.

The trouble is that all the best detectors ever built find far fewer neutrinos than should be present according to contemporary understanding of the fusion reaction.

Scientists believe the anomaly can be explained by the oscillation of detectable solar neutrinos into types that cannot be detected by existing instruments. But no one has proved this explanation.

Worldwide Efforts to Unlock Secrets

Members of the Super-Kamiokande collaboration have not limited their investigations to huge underground detectors.

The leader of the collaboration's University of Hawaii Group, for example, Dr. John G. Learned, has also worked on an underwater detection system in the Pacific Ocean off the Hawaiian coast (which ran out of money before completion) and a project at the South Pole where a neutrino detector had been buried under thousands of feet of ice.

Another approach to penetrating the neutrino secrets involves the use of particle accelerators capable of producing intense beams of neutrinos. In two experiments currently being prepared, one in Japan and the other at Fermi National Accelerator Laboratory in Illinois, beams of neutrinos will be directed through the Earth toward detectors several hundred miles away. The goal will be to observe changes the neutrinos undergo in transit, both in numbers and types. Physicists expect the experiment to confirm the existence of neutrino oscillations like those seen in the Super-Kamiokande detector.

Although the neutrinos are now known to have some mass, most physicists agree that the mass must be very small. The Super-Kamiokande experiments suggest that the difference between the masses of muon neutrinos and other types of neutrinos is only about 0.07 electron volts (a measure of particle mass). This does not yield a value of the masses themselves, only of the difference between those of muon neutrinos and other types.

Although the mass of the neutrino of any flavor must be small, Dr. Totsuka said, it may be several electron volts, and if so, the overall gravitational effect on the universe would perhaps be significant. It has been estimated that at any given moment, every teaspoon worth of volume of space throughout the universe contains an average of 300 neutrinos, so their aggregate number is staggering.

(The electron volt is used by scientists as a unit of particle mass. One electron volt is the energy, or mass equivalent, that an electron acquires by passing through an electric potential of one volt. By this standard a neutrino is believed to have a mass only about five-hundred-thousandth as much as that of an electron, which itself is a light particle.)

In the last 68 years, a legion of distinguished physicists has devoted inquiries and careers to the puzzling neutrino, which was given its name by the great Italian-American scientist Enrico Fermi. Fermi quickly came to believe in the particle's existence, even though it was not proved in his lifetime, and named it "neutrino," which means "little neutral one" in Italian.

Representatives of dozens of neutrino experiments meet once every two years to exchange ideas at conferences like the one under way in Takayama. Present are representatives of teams that have installed neutrino detectors on the bottom of Lake Baikal in Siberia, under the Aegean Sea off the Greek coast, inside the Gran Sasso tunnel under the Alps, and in many other places.

Lively debate has characterized the discussions in Takayama. For example, Dr. Bahcall, who had high praise for the Super-Kamiokande experiment, challenged assertions by the detector team that neutrinos might have sufficient mass to slow the expansion of the universe. But there was agreement that progress in understanding neutrinos has accelerated tremendously in the last few years.

Another detector built deep within a mine is nearing completion at Sudbury, Ontario. When scientists finish filling it with heavy water, water that includes a heavy isotope of hydrogen as part of its molecule, the Sudbury detector will be uniquely capable of distinguishing between electron neutrinos—one of the three types—and the other two flavors. This is expected to cap the investigation of neutrino oscillations for which Super-Kamiokande has now furnished the "smoking gun."

—*June 5, 1998*

Almost in Awe,
Physicists Ponder "Ultimate" Theory

By GEORGE JOHNSON

When many science fans last tuned in, physicists in search of the universe's deepest secrets were pinning their hopes on a dazzlingly beautiful and bewildering invention called superstring theory. The ultimate explanation of nature would finally come together, it was promised, if one thought of the hundreds of subatomic particles as musical notes produced by incredibly tiny strings vibrating in a realm made of 10 dimensions. Never mind that the universe seems to be made of only four dimensions (counting time). The other six were said to be conveniently curled up into vanishingly tiny balls, out of sight and mind.

Since its quiet beginnings in the 1960s, the idea has flared and faded through a first "superstring revolution" in the mid-1980s and a second revolution a decade later. In recent months, a new wave of discoveries is fomenting what some physicists are excitedly hailing as a third revolution, leading toward a day when all the laws of creation may fit into a single, elegant frame.

But the subject of the latest enthusiasm is not called superstring theory anymore. Along the way, the name has changed to M theory, with the M standing for "magic," "mystery," "mother" (as in mother of all theories) or, more prosaically, "meta," "matrix" or "membrane." For the wiggling superstrings, which were at least vaguely possible to visualize, have been joined (and possibly supplanted) by even more abstract entities: membranes, or "branes," which come in as many as nine dimensions.

All of creation, according to some recent speculation, may be concocted from these barely imaginable objects: God's Tinkertoys. If this realization is correct, then physics may be closer than ever to writing down the elusive theory of quantum gravity, a feat that would unify quantum mechanics and general relativity—the two hitherto irreconcilable pillars of modern physics—and explain all the forces of nature in the same terms.

"People are going to look back on this as one of the most important periods in 20th-century physics, as significant as the development of quantum mechanics and relativity," predicted Dr. John Schwarz, a physicist at the California Institute of Technology and an early pioneer of string theory.

Enthusiastic pronouncements from the string theorists themselves are nothing new. But the excitement is also leaking into other domains, like cosmology.

Dr. Andrew Strominger, a Harvard University physicist who recently used M theory to leap across disciplines and solve a problem involving black holes, said: "We were once considered semi-crackpots working on some bizarre idea. While that may still be true, at least we're no longer perceived that way."

One evening in the summer of 1998, at the annual superstring fiesta, Strings '98, in Santa Barbara, California, some 200 physicists heralded the latest developments by dancing the Macarena, or, rather, a new version called the Maldacena, named after a young Argentine theorist, Dr. Juan Maldacena (pronounced mal-dah-SAY-nah) of Harvard University, whose new theory is the source of the latest excitement:

> *You start with the brane*
> *and the brane is B.P.S.*
> *Then you go near the brane*
> *and the space is A.D.S.*
> *Who knows what it means*
> *I don't, I confess*
> *Ehhhh! Maldacena!*

As Dr. Jeffrey Harvey, a University of Chicago theorist, rapped out the lyrics, one esoteric verse piling on top of another, the physicists worked their way through the 14 steps of the dance. (See page 151.) The high spirits continued later in the summer as theorists flocked to the Aspen Center for Physics in Colorado for a workshop on M theory and black holes.

What the message came down to was this: Physicists have a very successful framework called quantum field theory that describes three of the four forces. The strong force holds the atomic nucleus together; the weak force governs radioactive processes, and electromagnetism combines electrical and magnetic effects. All three can be portrayed as fields transmitted by particles called quanta. For electromagnetism the carriers are photons, for the strong force gluons, and for the weak force W and Z particles. But no one has been able to fit gravity into the picture. It is assumed that the force must be carried by particles called gravitons, but getting these to obey the laws of quantum field theory has proved impossible.

Gravity can be described, however, by superstring theory, or M theory, using the completely different vocabulary of strings and branes. Dr. Maldacena's conjecture, elaborated in an explosion of more than a hundred recent papers by superstring theorists, suggests the possibility of a deep, hidden connection between quantum field theory and string theory, these two seemingly immiscible worldviews.

"This is a very dramatic claim," said Dr. Nathan Seiberg, a theorist at the Institute for Advanced Study in Princeton, New Jersey. In addition to bringing gravity and the other forces closer together, the tentative links Dr. Maldacena has found may provide a powerful new calculational tool for solving difficult problems in particle physics.

The Origin:
New Vision for a Quandary

String theory first arose in the late 1960s and early '70s as an ill-fated attempt to understand the strong force. Parsing the world in terms of particles and fields had already led to a spectacularly successful theory of electromagnetism, and the weak nuclear force was on the verge of succumbing to a similar explanation.

But the strong force seemed, at the time, stubbornly resistant. Some physicists were taking this as a sign that field theory needed to be scrapped and replaced with a whole new vision. What emerged was the curious possibility that particles were really different notes produced by vibrating strings.

The potential payoffs seemed immense. One big problem of dealing with the infinitesimally small particles of quantum field theory was that they caused mathematical absurdities to pop up in the equations, the equivalent of trying to divide a number by zero. The result was infinite terms that rendered the calculations nonsensical. The problem crippled attempts to explain the strong force. If sizeless particles were replaced by little strings, the mavericks proposed, maybe the cancerous infinities would go away.

But there were many problems to overcome. If one could believe the equations, the strings would have to be vibrating in a space of 25 dimensions (with a 26th representing time). Where were the extra 22? The equations also kept coughing up a weird massless particle whose spin (a quantum mechanical analogue of rotation) was 2. The only such particle anyone knew about was the purely hypothetical graviton. If physicists ever succeeded in devising a quantum field theory of gravity, the graviton would be the carrier. But what was it doing in a theory of the strong force?

In any event, physicists soon succeeded in explaining the strong force with a field theory called quantum chromodynamics, or QCD, and most theorists turned their backs on strings. According to QCD, elemental building blocks called quarks come in three "colors" (somewhat analogous to electrical charge). The quarks are bound together by gluons, the carriers of the strong force, to form protons, neutrons and their subatomic kin. By the end of the '70s, QCD had been incorporated into the Standard Model, an amalgam of quantum field theories describing the strong force, the weak force and electromagnetism. The most daring theorists were trying for a "grand unified theory" in which all three forces were shown to be manifestations of a single superforce. But gravity remained far out of the game.

The Revolution:
Tying in Gravity with Superstrings

But not everyone gave up on strings. In the mid-1970s, two physicists, Dr. Schwarz and Dr. Jöel Scherk, tried to turn one of string theory's flaws into a virtue: maybe the persistent appearance of the graviton in the equations was no accident. Maybe what they were looking at was not a model of the strong force, but a model of gravity—a new way to formulate Einstein's general theory of relativity. And if gravity could be described by string theory, then maybe the other forces could also be reformulated this way. All would then be unified into the same package.

Around this time, string theorists found they could pare the 26 space-time dimensions required by the original theory to a mere 10. Along the way, the theory came to be called *superstrings* when it was endowed with a hypothetical quality called supersymmetry, in which the force-carrying particles like gluons and the matter-making particles like quarks are closely knit together.

Ten dimensions were still a lot to swallow. And, in an embarrassment of riches, it appeared that one could potentially construct an infinite number of different 10-dimensional string theories. How would physicists ever know which one described this universe? A breakthrough came in the mid-1980s when, in the first revolution, it was shown that of all possible string theories, only five were mathematically sturdy; the rest would come crashing down because of various inconsistencies. But this was still four theories too many. Even worse, there were still tens of thousands of different ways to roll up the six extraneous dimensions to get the theories to describe a four-dimensional world.

A small band of die-hards remained optimistic. Dr. Edward Witten, a rising young star of string theory, romantically described it as "a piece of 21st-century physics that fell by chance into the 20th century."

They kept quietly toiling away until the second revolution of the mid-1990s. The many ways to hide the extra dimensions were found to be closely related. And the five 10-dimensional theories turned out to be just different views of a single underlying 11-dimensional theory. All could be connected by "dualities," mathematical lenses through which the seemingly different turn out to be the same.

"It reminds me of the story of the blind people examining the elephant," said Dr. Seiberg, the theorist at the Institute for Advanced Study. "We used to look at different pieces and did not see the big picture."

In funneling the plenitude of theories into one, physicists realized that their equations spoke of a world made not just from strings but also from membranous things called p-branes, with the p standing for the number of dimensions. What is normally thought of as a membrane is a two-dimensional surface, like a bedsheet, stretching across a three-dimensional space. This is now called a 2-brane. A point is a 0-brane, and a line is a 1-brane. Extending the idea in the other direction, one can have 3-branes, 4-branes, 5-branes, all the way up to 9-branes: nine-dimensional surfaces flapping inside a 10-dimensional world.

Especially important to M theory is a special type called a D-brane, named for the 19th-century mathematician Peter Dirichlet. In 1995, Dr. Joseph Polchinsky of the University of California at Santa Barbara showed that D-branes, which also come in as many as nine dimensions, described surfaces on which strings can end. But these surfaces are more than mere boundaries: D-branes are now seen as entities at least as fundamental as strings. According to a controversial version of M theory called Matrix theory, D-branes may be the fundamental objects from which strings and everything else is made.

Before long, physicists like Dr. Strominger of Harvard were finding that some of the puzzles about black holes could be better understood if they were thought of as being made from D-branes. In fact, D-branes themselves could be conceived of as extremely tiny black holes. A string came to an end because the rest of it was sucked down one of these infinitesimal wells. A closed string, shaped like a loop, became an open, two-ended string when a chunk of it was bitten off by a D-brane.

And D-branes are an essential part of the choreography of the mathematical dance called the Maldacena, in which string theory and field theory pirouette on the same floor. Dr. Maldacena used D-branes to construct a quantum field theory similar to QCD, in the ordinary four dimensions. He also used D-branes to build a 10-dimensional string theory (with 5 of the dimensions curled up and hidden away). By their nature, string theories include gravity. Thus the excitement when

Dr. Maldacena showed that the two theories were intimately related. The unification of all four forces may now be a step closer to realization.

The Universe:
Linking Inner Space to Outer Space

But the finding is still just a conjecture, lingering in the netherworld between hunches and fully developed theories. To get his model to work, Dr. Maldacena had to pull some clever theoretical tricks. In QCD, quarks come in three "colors." Drawing on an idea from the Dutch physicist Gerard 't Hooft, he simplified the calculations by using a toy theory with many more colors.

And so far, the connection Dr. Maldacena found only works in something called anti-de Sitter space (after the Dutch astronomer Willem de Sitter). An anti-de Sitter universe would be "curved" in such a manner that the expansion from the big bang would gradually decelerate and collapse into a big crunch. Recent evidence hints that in our own universe the expansion may be eternally accelerating. But that is far from certain.

Taking into account these qualifications, the bottom line of the Maldacena conjecture is this: The curvature of the space-time continuum described by the string theory is equivalent to the number of colors in the field theory; more colors mean less curvature. An unexpected bridge may have been found between two different theoretical worlds.

Physicists are now trying to extend the work so it applies to more realistic situations. Dr. Strominger, for one, is betting that the relationship will be found to hold across the board, showing that "string theory and quantum field theory are just two sides of the same coin."

"We're not at the bottom of it yet," he said. "We're all in a good mood because we think there is a lot more to be learned."

Dr. Maldacena's work also supports a hot new theory that the universe is holographic. In laser holography, a three-dimensional object is projected onto a two-dimensional plane, retaining the richness of the original image. In the Maldacena model, the four-dimensional field theory can be thought of as a holographic projection of the five-dimensional string theory (remember that the other five dimensions are rolled up and tucked away). In a holographic universe, the information about everything in a volume of space would be displayed somehow on its surface. The bizarre implications of this notion are only beginning to unfold.

The Meaning:
Just Fancy Math or Real Science?

Dr. Maldacena concedes that his conjecture is burdened with the criticism that applies to all of M theory: that it cannot yet be tested by experiment.

"Up to now, all that has been done is mostly from the conceptual point of view," he said. "There are no experimental predictions so far, but hopefully there will be in the future. We don't know whether that will happen soon or not. How far you can extend a new method is a question that is always hard to answer."

Some physicists still maintain that, for all the conceptual revolutions in string theory, there is little to show for it but a lot of beautiful mathematics.

"No observable physical phenomena have been explained," said Dr. 't Hooft. "So it is tempting to be sarcastic about these developments."

And even M theory's enthusiasts are baffled by what it all really means.

"Before the second superstring revolution, life was simple," said Dr. Steven Giddings, a theorist at the University of California at Santa Barbara. "We believed that everything in the universe—quarks, photons, gravitons, electrons and the rest—were all made out of strings. The recent upheaval has shattered that view, and we've yet to find a convincing logical structure to replace it."

Dr. Giddings continued: "We no longer know what the fundamental constituents of the theory are. Strings and D-branes appear equally fundamental, and it's not clear whether either one of them is made out of the other. Perhaps they're all made from something even more fundamental. It's like climbing a mountain to reach the top and discovering that it's just a foothill to a more distant range. We've made an enormous amount of progress in the past few years, but now realize the greater depth of our ignorance."

He was leaving that day, for a peak in the Sierras, to do something he considered easy: climbing a thousand-foot wall of vertical ice.

—September 22, 1998

New Dimension in Dance: Thinking Man's Macarena

By GEORGE JOHNSON

At the Strings '98 conference in the summer of 1998 in Santa Barbara, California, physicists were so excited about a recent paper by Dr. Juan Maldacena, a Harvard theorist, that they danced in celebration. The lyrics, written by Dr. Jeffrey Harvey, a physicist at the University of Chicago, are a takeoff on the 14-step version of the popular Spanish dance called the Macarena.

You start with the brane [1]
and the brane is BPS [2]
Then you go near the brane
and the space is ADS [3]
Who knows what it means
I don't, I confess
Ehhhh! Maldacena!

Super Yang-Mills [4]
With very large N [5]
Gravity on a sphere
flux without end.
Who says they're the same
Holographic [6] he contends
Ehhhh! Maldacena!

Black holes used to be
a great mystery
Now we use D-brane
to compute D-entropy [7]
And when D-brane is hot
D-free energy [8]
Ehhhh! Maldacena!

151

M-theory is finished [9]
Juan has great repute
The black hole we have mastered
QCD we can compute [10]
Too bad the glueball spectrum
is still in some dispute [11]
Ehhhh! Maldacena!

1. One of the latest crazes in superstring theory: a membranelike object that can come in up to nine dimensions.

2. Bogomolny-Prasad-Sommerfield (named after three physicists): a specific type of supersymmetric brane (see note 4) important to Dr. Maldacena's conjecture.

3. So far the conjecture only works in a special, saddle-shaped universe called anti–de Sitter space.

4. Yang-Mills is the type of field underlying quantum chromodynamics (or QCD), the theory of the strong force. Dr. Maldacena simplified his calculations by attributing a quality known as supersymmetry to the field.

5. N is the number of "colors" in the field theory. In QCD, quarks come in red, green and blue.

6. Maldacena's four-dimensional field theory can be thought of as a holographic projection of a higher-dimensional string theory.

7. Recently, physicists have used things called Dirichlet-branes (see note 1) to verify a prediction made by Dr. Stephen W. Hawking and Dr. Jacob Bekenstein: that a black hole's entropy (a measure of disorder) is proportional to the area of its horizon (the surrounding region from which nothing can escape).

8. One can also use D-branes to compute a thermodynamic quantity called free energy.

9. A bit of sarcasm. No one, including Dr. Maldacena, believes M theory is anywhere near completion.

10. On a practical level, Dr. Maldacena's method may be used to simplify difficult calculations in quantum chromodynamics (see note 4).

11. Glueballs are particles made entirely from gluons, the carriers of the strong force. The glueball spectrum is the range of masses and spins (and other quantities) that these particles can assume. Whether calculations of the spectrum are reliable is still uncertain. If not, physicists may be dancing another tune at Strings '99 in Potsdam, Germany.

—September 22, 1998

In Quantum Feat, Atom Is Seen in Two Places at Once

By GEORGE JOHNSON

Appalled by the weird implications of quantum mechanics, the rules that explain the workings of the tiny particles that make up the universe, Albert Einstein used to stroll the streets of Princeton wondering why the moon wasn't smeared all over the sky.

After all, the moon is made of these particles, and quantum theory holds that until it is observed, a particle doesn't have a definite position. It remains suspended in a mathematical limbo: a state of pure potentiality consisting of all the possible positions it could conceivably occupy.

Do you really believe, Einstein once asked a younger colleague, that the moon exists only when you look at it?

Einstein had come face-to-face with the fundamental paradox of quantum mechanics: why the laws that apply so precisely to the subatomic realm do not appear to carry over into the domain of everyday things.

"According to quantum mechanics, a bottle of Coke should be able to exist in a superposition of two locations, both here and there," said Dr. Wojciech H. Zurek, a theorist at Los Alamos National Laboratories in New Mexico. "We do not see such states. Ever. Bottles are either here or there. What eradicates quantum weirdness?" And could the solidity of reality really be dependent on the presence of observers?

Some physicists believe they are coming closer to an answer with a phenomenon called decoherence, in which the particles themselves constantly "observe" one another, eliminating the quantum fuzziness and yielding the familiar world of solid objects. In recent weeks this theoretical notion received what might be its strongest support yet. Experimenters at the National Institute of Standards and Technology in Boulder, Colorado, observed decoherence in action. They used laser beams to gently manipulate an atom, essentially putting it in two places at once. Then they measured the breakdown of this perplexing state of existence more methodically than ever before.

Dr. Seth Lloyd, an associate professor in the department of mechanical engineering at the Massachusetts Institute of Technology and an expert in quantum theory, called the work an "experimental triumph."

"Certainly no experiments have been done previously that so carefully and thoroughly examine the decoherence process," he said. "The results confirm to a high degree of precision the theoretical predictions of the last 20 years."

Though other groups have measured decoherence in the past, the Boulder experiment is the first to systematically observe how quickly quantum ambiguity is resolved when a particle is exposed to different kinds of environments.

"This is indeed a significant piece of work," said Dr. Gerard J. Milburn, head of the physics department at the University of Queensland in Australia. Dr. Milburn's calculations about how quickly quantum fuzziness gives way to tangible objects were borne out by the demonstration. "This is the most definitive experimental validation of these predictions to date," he said.

In pure isolation, sealed off from the influence of its surroundings, a particle is represented by a mathematical device called a quantum wave function: All of the particle's possible states (its position or momentum, for example) cling together in a condition known as "quantum superposition." In traditional interpretations of quantum mechanics, it is the act of observation that causes this wave function to "collapse," forcing the particle to choose one state or another.

Physicists have tried to discourage mystics by emphasizing that the observer need not be a conscious being: An electronic detector or a photographic plate will do. It is the collision with the rock-solid world that resolves the particle's ambiguous existence.

But many physicists find this explanation dissatisfying. They would like quantum mechanics to be a completely self-contained theory, with no need to invoke any kind of outside measurer.

After all, if quantum mechanics is taken to its logical extreme, the universe itself can be described by a wave function, all its possible histories hovering together in superposition. By definition, there can be nothing outside the universe, no external observer or measurer to conjure up this particular universe from the plentitude of possibilities.

Maybe all it takes to collapse the wave function, Dr. Zurek and his colleagues propose, is for a particle to undergo some kind of tiny disturbance, to come into contact with other particles. The delicately balanced superposition in which all the possibilities stick together, or "cohere," would come unglued. It would "decohere." Then the particle could assume a particular position.

Or, as Dr. Zurek has described it, "the watchful eye of the environment"—the particles and waves that pervade creation—is constantly making measurements, banishing quantum ambiguity and conjuring up hard-edged reality, the familiar world dominated by the commonsensical laws of classical physics.

There would be no need for a curious observer or even a measuring instrument to solidify Einstein's moon or to put Dr. Zurek's Coke bottle on one side of the table or another. The decoherence caused by the jiggling of an object's own atoms and the particles around it would be enough.

To dramatize the problems of applying quantum theory to the classical world, the Austrian physicist Erwin Schrödinger devised his famous thought experiment in which the fate of a cat is tied to that of a single subatomic particle. In one version, a photon (a particle of light) is fired at a half-silvered mirror, giving it a 50-50 chance of reflecting back or sailing through. If the photon passes through the mirror, it strikes a photoelectric detector, activating a circuit that breaks a vial of poison and kills the cat. If the photon is reflected away from the detector, the cat is spared.

Schrödinger argued that until the box was opened and the outcome of the experiment was registered, the photon would linger in a superposition of the two possible paths it could take, leaving the cat in the uncomfortable position of being simultaneously dead and alive.

Decoherence suggests why this is not worth worrying about. Each atom in the cat is tied into a complex environment of other atoms that constantly interact, spiriting away the quantum effects. More specifically, the theory predicts that the speed at which quantum superpositions collapse, giving rise to what theorists call "classicality," depends on how far apart the alternate possibilities are in something called Hilbert space. Put more colloquially, it depends on how different they are.

A Coke bottle on one side of a table is far removed from a Coke bottle on the other side, and a live cat is very different from a dead one. Hence, the super-positions—the bottle both here and there, the cat both dead and alive—almost immediately go away.

To observe decoherence in action, the researchers in Boulder trained their sights on the "mesoscopic" realm—between the submicroscopic world, where particles can hover indefinitely in superposition, and the macroscopic world of objects, where superpositions disappear too soon to be noticed.

In the experiment, reported Jan. 20, 2000 in *Nature*, Dr. David Wineland and his colleagues trapped a charged atom inside an electromagnetic field. Using laser pulses, the researchers coaxed it into a "cat state" in which its outer electron simultaneously had two opposite "spins."

It was as though a tiny sphere were rotating clockwise and counterclockwise at the same time. Laser beams were then used to nudge apart the two states, separating them by about 100 billionths of a meter.

The experimenters had created weird superpositions like this before. This time, though, they went on to see what would happen if the "cat state" was exposed to various disturbances in the form of electrical fields. As predicted by decoherence theory, this interaction with an environment rapidly forced the superposition to come undone and the atom took its place in the world. How fast this happened depended, as theory predicted, on how far the two superpositions were separated.

"We hope these kinds of experiments may shed some light on the inconsistencies between what quantum mechanics predicts and our everyday experience," Dr. Wineland said. He also noted that a deeper understanding of decoherence could help scientists build experimental quantum computers.

In these theoretical devices, all the calculations needed to solve a problem would be performed simultaneously in quantum superposition. The result could be extremely powerful machines that solve problems now considered impenetrable. But first scientists would have to learn how to control decoherence, keeping the superpositions from collapsing before a calculation was done.

Decoherence has also been measured in other laboratories under entirely different conditions. In 1996, Dr. Serge Haroche and his colleagues at the École Normale Supérieure in Paris created cat states by putting an electromagnetic field into a superposition in which its waves were simultaneously in different phases— reaching their crests and troughs at different times. Then they upset the quantum balancing act by sending an interloping atom—a "quantum mouse"—through the field. They too found that the larger the separation between the alternate quantum states, the faster decoherence came into play.

"In all of the cases studied, decoherence behaves as predicted by theory," Dr. Zurek said. "What is best, they confirm the prejudice of theorists (like yours truly) that quantum theory can explain this emergence of classicality without any modifications (which have been invoked by other, more desperate but no less reputable theorists)."

The University of Oxford physicist Roger Penrose, for example, has proposed that the mystery of how classicality arises cannot be completely understood without radically overhauling quantum theory and uniting it with Einstein's general theory of relativity—one of the biggest challenges facing physics.

Even if decoherence succeeds in solving the problem of Einstein's moon, a central mystery will remain: the theory may explain why people do not see the weird quantum state in which the Coke bottle is on both sides of the table. But nothing explains why, when the superposition collapses, the bottle ends up on, say, the left side rather than the right.

As some physicists see it, decoherence must cause the universe to somehow split in two, spawning this world, where the bottle is on one side, and another, parallel world where it is on the other. According to this "many worlds" interpretation, all the different ways history might have unfolded coexist in superposition.

Some physicists, like Dr. David Deutsch at Oxford, insist that these parallel universes are as solid and real as our own. Others, like Dr. Lloyd, believe they should merely be thought of as abstract possibilities—things that did not occur.

"The criterion for things being real ought to be that we're able to get information about them," Dr. Lloyd said. "The alternatives in the other worlds are inaccessible and therefore unreal. I really got up this morning and fried an egg for my daughter and myself. There is another world where my daughter and I had cereal.

"The cereal world is in the wave function of the universe, but it's not real in the sense that any information I'm going to get will falsify the hypothesis. All the information says we had eggs. Look at my cholesterol level!"

Is there another universe where he is picking a milk-sogged Cheerio from the end of his spoon and wondering how anyone could possibly think he had eaten eggs? As far as Dr. Lloyd is concerned, it does not really matter.

—*February 22, 2000*

Art + Physics = Beautiful Music

By JAMES GLANZ

Can the sense of acoustic intimacy created by a fine concert hall be measured in how many milliseconds it takes sound waves to ricochet from the walls and balconies and reach a listener in the seats? Can a hall's aural warmth be calculated from how efficiently bass notes rebound from the same surfaces? Can the prized quality called resonance be estimated from the rate at which the entire hall fades to silence after a blast of electronic sound?

More to the point, can an architect rely on studies of these quantities, using computer calculations and measurements in scale models, to ensure that a structurally innovative, visually inspiring design for a new concert hall will be an acoustical triumph rather than a disaster?

For years, the answer to all these questions seemed to be no—the field of concert hall acoustics has had only spotty success. But now an unusually intense collaboration between architects and acousticians has put the science of acoustics to the test, with two major new successes in Tokyo.

The halls in question are the 1,632-seat concert hall of the multipurpose complex called Tokyo Opera City, and the 1,810-seat opera house of the adjacent New National Theater. Both have architecturally daring designs, yet both have been praised by musicians who have performed in them.

"This hall simply has some of the best acoustics in which I have ever had the privilege to play," the cellist Yo-Yo Ma wrote in a commentary on the concert hall that appeared recently in a technical journal. He said its visual and acoustic aspects combined in a rare synthesis—"a miracle," he called it.

Miracle or no, this is no small feat. "Going to the moon is much simpler as a physics problem," said William J. Cavanaugh, an acoustician at Cavanaugh Tocci Associates who consults on both the construction and restoration of concert halls. In a moon shot, he said, "you've got one source, you've got one trajectory that will get you there, and you've got one 'listener,' or destination."

But in a concert hall, the trajectories of the sound waves begin at any number of places on the stage, bounce in complicated ways from every cornice and pillar, and reach their ultimate destinations in hundreds of occupied seats.

The research for the new halls, whose principal architect was Takahiko Yanagisawa, president of TAK Architects in Tokyo, may be the most extensive use yet of acoustical measurements and calculations in efforts to design concert halls that are not simply copies of great halls of the past.

"If you make a copy of the old, great halls, you'll have a great hall," said Dr. Leo L. Beranek, an architectural acoustician in Cambridge, Massachusetts, who was the principal acoustical consultant for the projects in Tokyo. But the Tokyo concert halls, he said, "are different in appearance and they have the sound of great halls."

The research is described in three papers, published earlier this year in *The Journal of the Acoustical Society of America,* by Dr. Beranek and Dr. Takayuki Hidaka, chief researcher of the Takenaka R & D Institute, which conducted acoustical measurements and built models. The papers describe how, as the designs took shape, scientists analyzed and worked to maintain acoustical variables like reverberation time, spaciousness and intimacy, each with a precise mathematical definition and musical meaning.

Without those studies, "you're gambling" on the acoustics, Dr. Beranek said.

Still, the success of two halls in Tokyo is unlikely to persuade all critics that the science of concert hall acoustics has finally arrived. There are still many acousticians who maintain that a knowledge of the technical issues, while helpful, is less important than experience and a repertoire of acoustically successful designs that one can fall back on in a pinch.

"If you know your craft and you know your art," said Russell Johnson of Artec Consultants, "the math today may not help you very much. And if you believe some math that's wrong, you can get into trouble very quickly."

That cautionary note was echoed by Dr. Cyril M. Harris, a professor emeritus of architecture and electrical engineering at Columbia University. The purely technical approach "works much of the time, but sometimes it doesn't, and you don't know the reason why," he said. "So you get trapped. And you get real disasters."

Yet both Dr. Harris and Mr. Johnson acknowledged that recent advances in acoustics could yield crucial clues to designers.

The history of concert hall acoustics is nothing if not contentious. No single approach, whether based on science, experience, art or pure intuition, has been without its heralded successes and unexplained failures.

In a way, in fact, musical styles and performance venues have engaged in what a biologist might call coevolution, developing in ways that were inextricably dependent on each other.

It is no coincidence that the unhurried, vowel-rich Gregorian chants sound best in medieval cathedrals, whose "reverberation time"—the time it takes a burst of sound to fade away—is 5 to 10 seconds. Later in musical history, the polyphonic, highly articulated Baroque compositions of Bach, Handel and Vivaldi benefited from being played in relatively small rooms with hard, reflecting walls, in which the reverberation times might be less than 1.5 seconds.

Those close walls also added a sense of acoustical intimacy. That is, the delay between the arrival of sound directly from the instruments or voices, and sound reflected off the walls and other surfaces is slight. The smaller the delay, the greater the sense of intimacy.

Later still, as the Classical style of Haydn and Mozart gave way to the Romantics, large concert halls with correspondingly longer reverberation times were built to accommodate both the music and its growing audience.

Beethoven's later symphonies were composed "almost as though he anticipated the large, reverberant halls that would be built in the next half-century," Dr. Beranek wrote in his book *Concert and Opera Halls: How They Sound* (Acoustical Society of America, 1996).

In another biological analogue, poor concert halls often did not survive the wrecker's ball. In a kind of natural selection, the best halls of the 19th century were more likely to survive. The three halls most often cited as models of sonorous pleasure are the Grosser Musikvereinssaal in Vienna (built in 1870); the Concertgebouw in Amsterdam (1888); and Symphony Hall in Boston (1900).

Each of those halls is roughly shoebox-shaped, leading to quick reflections from the fairly close side walls and balconies. Each has a reverberation time of about two seconds. And each displays numerous irregularities, like coffered ceilings and rows of buxom statues on its interior walls. The diffusion of sound created by reflections from those objects, acousticians agree, prevents a nasty acoustic "hardness" or "glare" that smooth surfaces can generate. (This diffusion is one characteristic that still awaits a precise mathematical definition, being determined for now by visual inspection of the interior.)

Symphony Hall rates a special status in the field, since the person generally regarded as the first to apply science to architectural acoustics, the Harvard physicist Wallace Clement Sabine, consulted in its design. Sabine had just discovered a crucial formula that relates a hall's reverberation time to a hall's volume and the amount of sound-absorbing material, like people and curtains, inside it.

"He really, literally, put the first numbers to this whole question," Mr. Cavanaugh said. But reverberation time alone could not divide the good halls from the bad ones. "Anybody that was working on it knew there was a lot more to it than one number."

Dr. J. Christopher Jaffe of the Norwalk, Connecticut, firm Jaffe Holden Scarbrough Acoustics, who also directs a program on sonics in architecture at Rensselaer Polytechnic Institute in Troy, New York, said that a more complete translation of the warm, rich sound of the great 19th-century halls into scientific terms owed much to the development of fresh acoustical measures, or "metrics," by Dr. Beranek.

"Leo got the Rosetta Stone to get that traditional sound developed into reflecting patterns," Dr. Jaffe said. "That gave the architects some great freedom."

That freedom, he said, is most useful for designs, like those of the new halls in Tokyo, that do not carefully copy the tried-and-true, shoebox-shaped halls.

Like intimacy and reverberation time, the additional metrics have deceptively simple names like *spaciousness, bass ratio, acoustical texture* and *clarity*. But each has a precise mathematical meaning that seeks to isolate a specific aspect of acoustical quality in a hall. In the studies leading up to the design of the new Tokyo halls, said Dr. Hidaka of the Takenaka R & D Institute, measurements of those and other metrics were made in 20 opera houses and 25 symphony halls in 14 different countries. (While New York's Carnegie Hall is considered among the world's best, it was not included in the study because there was no opportunity to make acoustic measurements there, Dr. Beranek said.)

The idea, Dr. Hidaka said, was to get a quantitative measure of what made the good halls good and the bad ones bad. For the studies, his team generally produced a burst of sound from a 12-sided speaker on the stage—actually a dodecahedron with a small speaker on each face. Each burst and its acoustic aftermath was recorded on tiny microphones placed in the ears of dummies, and in some cases the ears of real people, scattered around the seats.

The team worked out the value of the various metrics for each hall by analyzing detailed forms of the sound waves picked up by the microphones. Intimacy, for example, was defined as the time delay between the direct arrival of the sound from the stage and that of the very first reflections, which have presumably bounced off protruding side balconies.

Bass ratio gauges how efficiently low notes, compared with middle notes, carom from the walls and other surfaces; a high bass ratio gives a hall what musicians call warmth. Spaciousness is an estimate of what fraction of all the sound bathing a listener has been reflected laterally, from interior surfaces, as opposed to having arrived straight from the stage.

Dr. Beranek, Dr. Hidaka and their collaborators then compared those measurements with an acoustic ranking of the halls based on a survey of conductors and music critics. They found that the most beloved concert halls had reverberation times near two seconds, intimacy times of not much more than 20 milliseconds and relatively high bass ratios and spaciousness factors. Other metrics also took on fairly consistent values in the best halls.

Because of the need for greater clarity in understanding voices, the optimum reverberation times for opera houses turned out to be shorter, around 1.5 seconds.

Then the acousticians turned to large computers that had been programmed to simulate the acoustics in the basic architectural designs of Mr. Yanagisawa.

The team eventually built a 10:1 scale model of the proposed designs and made just the same measurements, using tiny speakers, microphones and one-inch "heads" of dummy audience members, all scaled down in proportion to the model. Even the wavelengths of the sound in the model measurements were scaled down.

This work led to numerous adjustments in the original designs, including changes in the height of the ceiling near the stage in the concert hall, giving some of the balcony fronts a rakish, forward slant and adding a special sound-diffusing material to the pyramidal ceiling.

The reflected wave patterns in the finished Tokyo Opera City concert hall, wrote the team in one of its papers, "appear to be closest to those for Boston Symphony Hall." Dr. Hidaka said acoustic data for the opera house resembled those of the famed Vienna Staatsoper.

For all the apparent success of the Tokyo projects, Mr. Yanagisawa emphasized that acoustical studies were far from the whole story. He believes that his general immersion in the music of the great halls of the world played as large a role in his understanding of good acoustics as the results of the technical work did.

"I believe an excellent hall can only be realized by a design that assimilates nuances beyond description by scientific data," Mr. Yanagisawa said. "The final work is a world of sense created therefrom."

Vindication

As a chapter in the fractious history of architectural acoustics, the success of the new Tokyo concert halls can be seen as a vindication for Dr. Leo L. Beranek, who received years of negative publicity after the 1962 opening of New York's Philharmonic Hall, for which he was principal architectural consultant. The bad reviews garnered by the hall's acoustics cast a shadow over the scientific approach to acoustics that he was then developing.

But in the long run, the message from that failure has been decidedly mixed. The science was far less developed in those days, and the architects did not follow many of Dr. Beranek's suggestions, partly out of financial concerns.

Even after Dr. Cyril M. Harris, now a professor emeritus of architecture and electrical engineering at Columbia University, led a complete reconstruction in the 1970s—when it became Avery Fisher Hall—and Russell Johnson, an acoustician at Artec Consultants, made further adjustments to the stage in the 1990s, a kind assessment would say that the hall is still not considered among the world's best acoustically.

"At that point, in a sense, the architects had much more say about things," said Dr. J. Christopher Jaffe, an acoustician. There used to be a feeling, he said, that in acoustics, everybody thinks differently, so none of the recommendations need to be taken seriously. But that attitude has changed, Dr. Jaffe said.

—April 18, 2000

No Hope of Silencing the Phantom Crinklers of the Opera

By JAMES GLANZ

A team of physicists has made a giant leap in understanding a prime scourge of audiences for opera, theater and concerts—the crinkling and crackling of candy wrappers.

But the physicists' work will be scant help to audiences maddened by the noise. The team showed what plenty of dismayed concertgoers have already intuited: The noise is inevitable no matter how the candy is opened.

The research, which was presented in Atlanta at a meeting of the Acoustical Society of America, is the most detailed examination ever of just how and why the mood-destroying noise is generated in the materials that wrap candy.

The researchers, Dr. Eric Kramer, a physicist at Simon's Rock College in Massachusetts, and Dr. Alexander Lobkovsky of the National Institute of Standards and Technology in Gaithersburg, Maryland, asked why even the most careful, painfully deliberate extraction of a piece of hard candy from its wrapper does not seem to reduce the amount of sound produced as a desperate audience member tries to relieve a dry throat, invariably just as Tosca reaches the sublime pathos of "Vissi d'Arte."

By recording crinkled wrappers as they were slowly stretched out in an otherwise silent chamber, and then digitizing and analyzing the sound emissions on computers, the team found that the noise was not continuous, but consisted of individual bursts or pops just thousandths of a second long. Their loudness, the researchers found, has nothing to do with how fast wrapping is undone. Rather, the pops occur at random volumes as tiny, individual creases in the paper suddenly rearrange themselves while the candy is unwrapped.

The physics of wrappers turned out to be surprisingly complex, said Dr. Kramer, who found parallels in the different shapes that large protein molecules can assume in the human body and the properties of magnetic materials. But because the series of minute shape changes from wrapped to unwrapped is unavoidable, the work's implications were tragic news for every alas-poor-Yorick moment of great drama.

"You can follow that sequence as slowly as you want, but each step in the sequence is accompanied by a click," Dr. Kramer said. "And there's nothing you can do about it."

On Broadway, where announced reminders to turn off cell phones and beepers are accompanied by the suggestion that hard candy be unwrapped before the curtain rises, the results may be about as welcome as a bad review. But this parsing of the intricate behavior of such a familiar object will find a ready audience among scientists who study neither atomic nor cosmic realms, but rather the middle ground between the very small and the very large.

"One, I think, should not confuse familiarity with comprehension," said Dr. L. Mahadevan, a mechanical engineer at the Massachusetts Institute of Technology. "This is a rather everyday system, but it has almost any of the complexities that one could care for in any physical system."

Dr. Kramer said he no longer remembered exactly why he and Dr. Lobkovsky began recording crinkling plastic. But he said they became intrigued by a fact that most people take for granted: Smooth, unwrinkled plastic has only one "stable" configuration, since no matter how it is bent (as long as it is not creased), the plastic will fall flat again. But crushed and wrinkled plastic has innumerable stable states: Deform it, and it holds the shape or something close to it.

So the researchers recorded the noises emitted during the slow stretching of a common plastic that they had crushed and recrushed many times. "As far as we know, no one had ever looked at this sound before," Dr. Kramer said. They were surprised to find an irregular series of extremely brief pops with unpredictable energies that varied by a factor of hundreds of thousands.

In creating a theory of wrappers, the researchers found a parallel in the rolling of a boulder over a hilly, irregular landscape. Different wrapper conformations correspond to the boulder's rolling into different depressions in the landscape. And like a theatergoer gradually opening a wrapper, someone might push the boulder very slowly up a hill. Nevertheless, at some point it will tumble quickly down the other side—and, in the case of the wrapper, emit that irritating snap.

As homely as that process is, it has much in common with a large drug molecule that behaves differently in the body, depending on what shape it takes, said Dr. James Kakalios, a physicist at the University of Minnesota. "If we understood that, it would have a revolutionary impact on everybody's life," since a new generation of therapeutic drugs could be engineered, he said.

But what does the research say about how to unwrap a piece of hard candy when, horribly, a tickle in the throat arrives during the final, fading bars of Tchaikovsky's "Pathetique"?

"Unwrap it as quickly as possible and get it over with," Dr. Kramer said.

—June 1, 2000

Scientists Bring Light to Full Stop, Hold It, Then Send It on Its Way

By JAMES GLANZ

Researchers say they have slowed light to a dead stop, stored it and then released it as if it were an ordinary material particle.

The achievement is a landmark feat that, by reining in nature's swiftest and most ethereal form of energy for the first time, could help realize what are now theoretical concepts for vastly increasing the speed of computers and the security of communications.

Two independent teams of physicists have achieved the result, one led by Dr. Lene Vestergaard Hau of Harvard University and the Rowland Institute for Science in Cambridge, Massachusetts, and the other by Dr. Ronald L. Walsworth and Dr. Mikhail D. Lukin of the Harvard-Smithsonian Center for Astrophysics, also in Cambridge.

Light normally moves through space at 186,300 miles a second. Ordinary transparent media like water, glass and crystal slow light slightly, an effect that causes the bending of light rays that allows lenses to focus images and prisms to produce spectra.

Using a distantly related but much more powerful effect, the Walsworth-Lukin team first slowed and then stopped the light in a medium that consisted of specially prepared containers of gas. In this medium, the light became fainter and fainter as it slowed and then stopped. By flashing a second light through the gas, the team could essentially revive the original beam.

The beam then left the chamber carrying nearly the same shape, intensity and other properties it had when it entered. The experiments led by Dr. Hau achieved similar results with closely related techniques.

"Essentially, the light becomes stuck in the medium, and it can't get out until the experimenters say so," said Dr. Seth Lloyd, an associate professor of mechanical engineering at the Massachusetts Institute of Technology who is familiar with the work.

Dr. Lloyd added, "Who ever thought that you could make light stand still?"

He said the work's biggest impact could come in futuristic technologies called quantum computing and quantum communication. Both concepts rely heavily on the ability of light to carry so-called quantum information, involving particles that can exist in many places or states at once.

Quantum computers could crank through certain operations vastly faster than existing machines; quantum communications could never be eavesdropped upon. For both these systems, light is needed to form large networks of computers. But those connections are difficult without temporary storage of light, a problem that the new work could help solve.

A paper by Dr. Walsworth, Dr. Lukin and three collaborators—Dr. David Phillips, Annet Fleischhauer and Dr. Alois Mair, all at Harvard-Smithsonian—is scheduled to appear in the Jan. 29 issue of *Physical Review Letters.*

Citing restrictions imposed by the journal *Nature,* where her report is to appear, Dr. Hau refused to discuss her work in detail.

Two years ago, however, *Nature* published Dr. Hau's description of work in which she slowed light to about 38 miles an hour in a system involving beams of light shone through a chilled sodium gas.

Dr. Walsworth and Dr. Lukin mentioned Dr. Hau's new work in their paper, saying she achieved her latest results using a similarly chilled gas. Dr. Lukin cited her earlier work, which Dr. Hau produced in collaboration with Dr. Stephen Harris of Stanford University, as the inspiration for the new experiments.

Those experiments take the next step, stopping the light's propagation completely.

"We've been able to hold it there and just let it go, and what comes out is the same as what we sent in," Dr. Walsworth said. "So it's like a freeze-frame."

Dr. Walsworth, Dr. Lukin and their team slowed light in a gas form of rubidium, an alkaline metal element.

The deceleration of the light in the rubidium differed in several ways from how light slows through an ordinary lens. For one thing, the light dimmed as it slowed through the rubidium.

Another change involved the behavior of atoms in the gas, which developed a sort of impression of the slowing wave.

This impression, actually consisting of patterns in a property of the atoms called their spin, was a kind of record of the light's passing and was enough to allow the experimenters to revive or reconstitute the original beam.

Both Dr. Hau's original experiments on slowing light, and the new ones on stopping it, rely on a complex phenomenon in certain gases called electromagnetically induced transparency, or EIT.

This property allows certain gases, like rubidium, that are normally opaque to become transparent when specially treated.

For example, rubidium would normally absorb the dark red laser light used by Dr. Walsworth and his colleagues, because rubidium atoms are easily excited by the frequency of that light.

But by shining a second laser, with a slightly different frequency, through the gas, the researchers rendered it transparent.

The reason is that the two lasers create the sort of "beat frequency" that occurs when two tuning forks simultaneously sound slightly different notes.

The gas does not easily absorb that frequency, so it allows the light to pass through it; that is, the gas becomes transparent.

But another property of the atoms, called their spin, is still sensitive to the new frequency. Atoms do not actually spin but the property is a quantum-mechanical effect analogous to a tiny bar magnet that can be twisted by the light.

As the light passes through, it alters those spins, in effect flipping them. Though the gas remains transparent, the interaction serves as a friction or weight on the light, slowing it.

Using that technique, Dr. Hau and Dr. Harris in the earlier experiment slowed light to a crawl. But they could not stop it, because the transparent "window" in the gas became increasingly narrower, and more difficult to pass through, as the light moved more and more slowly.

In a recent theoretical advance, Dr. Lukin, with Dr. Susanne Yelin of Harvard-Smithsonian and Dr. Michael Fleischhauer of the University of Kaiserslautern in Germany, discovered a way around this constraint.

They suggested waiting for the beam to enter the gas container, then smoothly reducing the intensity of the second beam.

The three physicists calculated that this procedure would narrow the window, slowing the first beam, but also "tune" the system so that the beam always passes through.

The first beam, they theorized, should slow to an infinitesimally slow speed, finally present only as an imprint on the spins, with no visible light remaining. Turning the second beam back on, they speculated, should reconstitute the first beam.

The new experiments bore those ideas out.

"The light is actually brought to a stop and stored completely in the atoms," Dr. Harris said. "There's no other way to do that. It's been done—done very convincingly, and beautifully."

—January 18, 2001

With Little Evidence,
String Theory Gains Influence

By JAMES GLANZ

Scientists have come up with theories in the shower, on barren mountains, while driving to work and even in their sleep. But what all theories have in common is that their predictions are eventually tested in experiments, where nature determines which inspirations are right and which are wrong.

Then there is string theory, the ambitious, profoundly mathematical attempt to knit together all of physics—from gravity to quantum mechanics to subatomic forces—into a single sublime formalism. Though string theorists first suspected they might be onto a "theory of everything" in the mid-1980s, and the field is the hottest area in theoretical physics, string theorists have yet to devise a make-or-break laboratory test for their ideas.

In part, that is because the theorized strings cannot be observed directly; they are thought to be vibrating entities smaller than a trillionth of a trillionth the size of an atom. Different string vibrations somehow correspond to different particles in nature, but scientists have yet to develop more than fragments of what they presume will ultimately be a complete theory.

Nevertheless, string theorists are already collecting the spoils that ordinarily go to the experimental victors, including federal grants, prestigious awards and tenured faculty positions. Less than a decade ago, there were hardly any jobs for string theorists, said Dr. David Gross, director of the Kavli Institute for Theoretical Physics at the University of California at Santa Barbara.

"Nowadays," Dr. Gross said, "if you're a hotshot young string theorist, you've got it made."

Dr. Gross has no problems with that success; he was one of string theory's early developers. But some physicists are dismayed by the dominance of a theory that has yet to prove itself experimentally.

"I think the whole theory is a long shot," said Sir Roger Penrose, a physicist at Oxford University. He said he had nothing against an interesting long shot but that string theory had "taken over at the expense of all other areas."

Dr. John Baez, a scientist in the mathematics department at the University of California at Riverside, who studies a different approach to unification based more directly on relativity theory, said, "String theorists keep saying that they're succeeding."

"The rest of us can wonder whether they are walking along the road to triumph," Dr. Baez said, "or whether in 20 years they'll realize that they were walking into this enormous, beautiful, mathematically elegant cul-de-sac."

A number of physicists discussed the question at a conference in Santa Barbara this month to honor Dr. Gross on his 60th birthday.

Their thoughts revealed how, in lean experimental times, physicists rely on their aesthetic senses, purely mathematical clues, suggestive connections with established theories, and a Houdini-like taste for escaping roadblocks. While physics has always called these tricks into play, researchers are relying on them as never before in the case of string theory, simply because it attempts to reach so far into the unknown.

Still, said Dr. Jeffrey Harvey of the University of Chicago, scientists in the field are confident their ideas are based in reality; they have, as he put it, "the feeling that string theory is something we discover rather than invent."

String theory has constantly changed since it first emerged several decades ago, and even its ardent adherents concede that they still do not understand more than what Dr. Gross called "the tail of the tiger," or a few suggestive parts of what is believed to be a complete theory. Until recently the physical crux of the theory was thought to be vibrating, 10-dimensional loops of string, roughly a billion trillion times smaller than a proton. Different modes of vibration of the strings (made of what, no one is sure) represented different particles in nature.

Now physicists believe the ultimate objects are 11-dimensional membranes. Either way, the extra dimensions beyond the usual four would be curled up so as to be nearly imperceptible. And because the vibrations would include the graviton, the particle thought to transmit gravity, as well as particles involved in the strong and weak nuclear forces and electromagnetism, string theory offered the prospect of unifying physics.

But with that aesthetic attraction came deep problems. First, in the 1980s, it seemed that the strings had a basic inability to cope with known differences between particles and their mirror images, and other such broad facts of nature. But closer study showed that, contrary to all expectation, various terms in the theory canceled each other, fixing the problem.

A decade later, the field again narrowly escaped when what seemed to be several different string theories all turned out to be different facets of the same underlying theory —a fortunate development, because physicists would not know what to do with more than one "ultimate" theory of the universe. Those episodes tell physicists they are on to something important, said Dr. Edward Witten, a physicist at the Institute for Advanced Study in Princeton, New Jersey.

"If it would turn out that string theory, which has led to so many miraculous-looking discoveries over so many decades, has nothing to do with nature, to me this would be a remarkable cosmic conspiracy," Dr. Witten said.

At the same time, string theory's strong influence on several branches of pure mathematics have provided a good test of the "intellectual horsepower" of the idea, said Dr. Stephen Shenker, a Stanford physicist. He said the wider impact suggested both that physicists had not missed some crucial mathematical error and that the theory was rich enough to be an encompassing theory of nature.

All those developments, Dr. Gross said, "do convince people that there's something here."

But despite these checks, said Dr. Stephen W. Hawking, the University of Cambridge physicist, who also attended the conference, physicists should also keep in mind that they have been wrong before and that string theory still contains many mysteries.

"For most of the last hundred years, we have thought that the theory of everything was just 'round the corner," Dr. Hawking said. "We keep making new discoveries, but I don't think we can yet say the end is in sight."

—March 13, 2001

Quantum Stew:
How Physicists Are Redefining Reality's Rules

By GEORGE JOHNSON

Struggling to understand the strange implications of modern physics, readers in the 1930s and '40s turned to a popular children's book for adults called *Mr. Tompkins in Wonderland* by the physicist George Gamow. In a series of dreams, Mr. Tompkins finds himself in surreal surroundings where the constants of nature have been changed so that matter behaves in ways that defy common sense.

In one dream, a number known as Planck's constant, which governs the intensity of quantum theory's perplexing effects, is cranked up so high that ordinary objects behave like elementary particles, which have the curious ability to act like both hard little kernels and ethereal waves. Things as large as billiard balls suddenly behave like electrons, spreading out all over the table, following many different paths at once. A visit to a quantum pool hall leaves poor Mr. Tompkins feeling drunk. The reason "quantum elephantism" doesn't really happen, the book explains, is that Planck's constant is extremely small, affecting only the tiniest objects—electrons and photons but not billiard balls.

But nothing involving quantum theory is ever so clear-cut. Recent experiments are demonstrating that quantum weirdness is not limited to the atomic realm. In late September 2001, a team of Danish physicists reported that a phenomenon called "quantum entanglement"—the "spooky action at a distance" that troubled Einstein—can affect not just individual particles but clusters of trillions of atoms. And last week, the Nobel Prize in physics was awarded for experiments showing how quantum mechanics can be exploited to make a couple of thousand atoms crowd together into a single superatom—what the scientists called "a kind of smeared-out, overlapping stew."

Experiment by experiment, the abstractions of quantum theory are taking on substance, impinging on phenomena closer to home. Physicists are developing a new finesse—getting a feel for quantum mechanics by playing with atoms the way their predecessors mastered Newtonian physics by fooling around with swinging pendulums or marbles rolling down inclined planes.

The practice is paying off with a deeper understanding of reality's rules. In Mr. Tompkins's time, the difference between the mysterious quantum realm and the hard-edged world of everyday life was assumed to be simply a matter of size.

Much beyond the magnitude of an atom, as quantum effects faded, objects took on definite positions in space and time. In recent years the situation has revealed itself as somewhat more subtle. Whether an object is dominated by quantum fuzziness has less to do with how big it is than with how well it can be shielded from outside disturbances—tiny vibrations, bombarding air molecules or even particles of light.

Larger things are indeed harder to isolate from the roiling environment—hence the predictable behavior of billiard balls. But with their delicate touch, physicists are steadily bringing the quantum ambiguities further into the macroscopic domain.

Consider the case of quantum entanglement. A subatomic particle can spin clockwise or counterclockwise like a top—but with a quantum twist. As long as it remains isolated from its environment, it lingers in a state of limbo, rotating both clockwise and counterclockwise at the same time. Only when it is measured or otherwise disturbed does it randomly snap into focus, assuming one state or the other. *And* becomes *either/or.*

Stranger still, two subatomic particles can be linked so that they must rotate in opposite directions. Force one to spin clockwise and the other instantly begins spinning counterclockwise, no matter how far they are separated in space.

In the past, experimenters had entangled two photons this way, and last year, in a major leap, they quantum-mechanically tethered four atoms together. The recent excitement came when physicists at the University of Aarhus in Denmark reported in the Sept. 27, 2001 issue of *Nature* that they had briefly entangled two clouds consisting of trillions of cesium atoms. In one cloud, most of the atoms were spinning one way; in the other cloud most were spinning, mirrorlike, in the opposite direction.

Correlating groups of atoms this way may find a use in quantum computers, devices where calculations are performed using single atoms or particles as counters. (Think of them as quantum abacus beads.) Theoreticians have proved that a quantum computer, if one can be built, could solve problems now considered impossible.

The experiments that won this year's Nobel in physics involved synchronizing atoms in a different but equally counterintuitive way.

Because of their quantum nature, atoms (like the particles they are made of) act like waves. The slower they move, the more stretched-out they become, dropping in pitch like a musical note sliding down the scale. Take a rarefied gas—atoms darting around in a container—and cool it so that the motion becomes slower and slower. Each atom's wavelength will widen until finally, as the temperature nears absolute zero, they all overlap, forming an exotic substance called a Bose-Einstein condensate. Imagine 2,000 billiard balls merging into one.

It is impossible for us denizens of the macro world to really picture such a state. We would have to have grown up in a universe with different constants, like the ones in Mr. Tompkins's dreams. As Gamow put it in his preface, "Even a primitive savage in such a world would be acquainted with the principles of relativity and quantum theory, and would use them for his hunting purposes and everyday needs."

As for developing quantum instincts, physicists are working their way up to the level of savages, striking sparks, building their first fires.

—October 16, 2001

String Theory, at 20, Explains It All (or Not)

By DENNIS OVERBYE

They all laughed 20 years ago.

It was then that a physicist named John Schwarz jumped up on the stage during a cabaret at the Physics Center in Aspen, Colorado, and began babbling about having discovered a theory that could explain everything. By prearrangement men in white suits swooped in and carried away Dr. Schwarz, then a little-known researcher at the California Institute of Technology.

Only a few of the laughing audience members knew that Dr. Schwarz was not entirely joking. He and his collaborator, Dr. Michael Green, now at the University of Cambridge, had just finished a calculation that would change the way physics was done. They had shown that it was possible for the first time to write down a single equation that could explain all the laws of physics, all the forces of nature— the proverbial "theory of everything" that could be written on a T-shirt.

And so emerged into the limelight a strange new concept of nature, called string theory, so named because it depicts the basic constituents of the universe as tiny wriggling strings, not point particles.

"That was our first public announcement," Dr. Schwarz said recently.

By uniting all the forces, string theory had the potential of achieving the goal that Einstein sought without success for half his life and that has embodied the dreams of every physicist since then. If true, it could be used like a searchlight to illuminate some of the deepest mysteries physicists can imagine, like the origin of space and time in the big bang and the putative death of space and time at the infinitely dense centers of black holes.

In the last 20 years, string theory has become a major branch of physics. Physicists and mathematicians conversant in strings are courted and recruited like star quarterbacks by universities eager to establish their research credentials. String theory has been celebrated and explained in best-selling books like *The Elegant Universe,* by Dr. Brian Greene, a physicist at Columbia University, and even on popular television shows.

Last summer in Aspen, Dr. Schwarz and Dr. Green (of Cambridge) cut a cake decorated with "20th Anniversary of the First Revolution Started in Aspen," as they and other theorists celebrated the anniversary of their big breakthrough. But even as they ate cake and drank wine, the string theorists admitted that after

20 years, they still did not know how to test string theory, or even what it meant.

As a result, the goal of explaining all the features of the modern world is as far away as ever, they say. And some physicists outside the string theory camp are growing restive. At another meeting, at the Aspen Institute for Humanities, only a few days before the string commemoration, Dr. Lawrence Krauss, a cosmologist at Case Western Reserve University in Cleveland, called string theory "a colossal failure."

String theorists agree that it has been a long, strange trip, but they still have faith that they will complete the journey.

"Twenty years ago no one would have correctly predicted how string theory has since developed," said Dr. Andrew Strominger of Harvard. "There is disappointment that despite all our efforts, experimental verification or disproof still seems far away. On the other hand, the depth and beauty of the subject, and the way it has reached out, influenced and connected other areas of physics and mathematics, is beyond the wildest imaginations of 20 years ago."

In a way, the story of string theory and of the physicists who have followed its siren song for two decades is like a novel that begins with the classic "What if?"

What if the basic constituents of nature and matter were not little points, as had been presumed since the time of the Greeks? What if the seeds of reality were rather teeny tiny wiggly little bits of string? And what appear to be different particles like electrons and quarks merely correspond to different ways for the strings to vibrate, different notes on God's guitar?

It sounds simple, but that small change led physicists into a mathematical labyrinth, in which they describe themselves as wandering, "exploring almost like experimentalists," in the words of Dr. David Gross of the Kavli Institute for Theoretical Physics in Santa Barbara, California.

String theory, the Italian physicist Dr. Daniele Amati once said, was a piece of 21st-century physics that had fallen by accident into the 20th century.

And, so the joke went, would require 22nd-century mathematics to solve.

Dr. Edward Witten of the Institute for Advanced Study in Princeton, New Jersey, described it this way: "String theory is not like anything else ever discovered. It is an incredible panoply of ideas about math and physics, so vast, so rich you could say almost anything about it."

The string revolution had its roots in a quixotic effort in the 1970s to understand the so-called "strong" force that binds quarks into particles like protons and neutrons. Why were individual quarks never seen in nature? Perhaps because they were on the ends of strings, said physicists, following up on work by Dr. Gabriele Veneziano of CERN, the European research consortium.

That would explain why you cannot have a single quark—you cannot have a string with only one end. Strings seduced many physicists with their mathematical elegance, but they had some problems, like requiring 26 dimensions and a plethora of mysterious particles that did not seem to have anything to do with quarks or the strong force.

When accelerator experiments supported an alternative theory of quark behavior, known as quantum chromodynamics, most physicists consigned strings to the dustbin of history.

But some theorists thought the mathematics of strings was too beautiful to die.

In 1974 Dr. Schwarz and Dr. Jöel Scherk from the École Normale Supérieure in France noticed that one of the mysterious particles predicted by string theory had the properties predicted for the graviton, the particle that would be responsible for transmitting gravity in a quantum theory of gravity, if such a theory existed.

Without even trying, they realized, string theory had crossed the biggest gulf in physics. Physicists had been stuck for decades trying to reconcile the quirky rules known as quantum mechanics, which govern atomic behavior, with Einstein's general theory of relativity, which describes how gravity shapes the cosmos.

That meant that if string theory was right, it was not just a theory of the strong force; it was a theory of all forces.

"I was immediately convinced this was worth devoting my life to," Dr. Schwarz recalled. "It's been my life work ever since."

It was another 10 years before Dr. Schwarz and Dr. Green (Dr. Scherk died in 1980) finally hit pay dirt. They showed that it was possible to write down a string theory of everything that was not only mathematically consistent but also free of certain absurdities, like the violation of cause and effect that had plagued earlier quantum gravity calculations.

In the summer and fall of 1984, as word of the achievement spread, physicists around the world left what they were doing and stormed their blackboards, visions of the Einsteinian grail of a unified theory dancing in their heads.

"Although much work remains to be done, there seem to be no insuperable obstacles to deriving all of known physics," one set of physicists, known as the Princeton string quartet, wrote about a particularly promising model known as heterotic strings. (The quartet consisted of Dr. Gross; Dr. Jeffrey Harvey and Dr. Emil Martinec, both at the University of Chicago; and Dr. Ryan M. Rohm, now at the University of North Carolina.)

The Music of Strings

String theory is certainly one of the most musical explanations ever offered for nature, but it is not for the untrained ear. For one thing, the modern version of the theory decreed that there are 10 dimensions of space and time.

To explain to ordinary mortals why the world appears to have only four dimensions—one of time and three of space—string theorists adopted a notion first bruited about by the German mathematicians Theodor Kaluza and Oskar Klein in 1926. The extra six dimensions, they said, go around in sub-submicroscopic loops, so tiny that people cannot see them or store old *National Geographics* in them.

A simple example, the story goes, is a garden hose. Seen from afar, it is a simple line across the grass, but up close it has a circular cross section. An ant on the hose can go around it as well as travel along its length. To envision the world as seen by string theory, one only has to imagine a tiny, tiny six-dimensional ball at every point in space-time.

But that was only the beginning. In 1995, Dr. Witten showed that what had been five different versions of string theory seemed to be related. He argued that they were all different manifestations of a shadowy, as-yet-undefined entity he called "M theory," with *M* standing for "mother," "matrix," "magic," "mystery," "membrane" or even "murky."

In M theory, the universe has 11 dimensions—10 of space and one of time— and it consists not just of strings but also of more extended membranes of various dimension, known generically as "branes."

This new theory has liberated the imaginations of cosmologists. Our own universe, some theorists suggest, may be a four-dimensional brane floating in some higher-dimensional space, like a bubble in a fish tank, perhaps with other branes—parallel universes—nearby. Collisions or other interactions between the branes might have touched off the big bang that started our own cosmic clock ticking or could produce the dark energy that now seems to be accelerating the expansion of the universe, they say.

Toting Up the Scorecard

One of string theory's biggest triumphs has come in the study of black holes. In Einstein's general relativity, these objects are bottomless pits in space-time, voraciously swallowing everything, even light, that gets too close, but in string theory they are a dense tangle of strings and membranes.

In a prodigious calculation in 1995, Dr. Strominger and Dr. Cumrun Vafa, both of Harvard, were able to calculate the information content of a black hole, matching a famous result obtained by Dr. Stephen Hawking of University of Cambridge using more indirect means in 1973. Their calculation is viewed by many people as the most important result yet in string theory, Dr. Greene said.

Another success, Dr. Greene and others said, was the discovery that the shape, or topology, of space, is not fixed but can change, according to string theory. Space can even rip and tear.

But the scorecard is mixed when it comes to other areas of physics. So far, for example, string theory has had little to say about what might have happened at the instant of the big bang.

Moreover, the theory seems to have too many solutions. One of the biggest dreams that physicists had for the so-called theory of everything was that it would specify a unique prescription of nature, one in which God had no choice, as Einstein once put it, about details like the number of dimensions or the relative masses of elementary particles.

But recently theorists have estimated that there could be at least 10^{100} different solutions to the string equations, corresponding to different ways of folding up the extra dimensions and filling them with fields—gazillions of different possible universes.

Some theorists, including Dr. Witten, hold fast to the Einsteinian dream, hoping that a unique answer to the string equations will emerge when they finally figure out what all this 21st-century physics is trying to tell them about the world.

But that day is still far away.

"We don't know what the deep principle in string theory is," Dr. Witten said.

For most of the 20th century, progress in particle physics was driven by the search for symmetries—patterns or relationships that remain the same when we swap left for right, travel across the galaxy or imagine running time in reverse.

For years physicists have looked for the origins of string theory in some sort of deep and esoteric symmetry, but string theory has turned out to be weirder than that.

Recently it has painted a picture of nature as a kind of hologram. In the holographic images often seen on bank cards, the illusion of three dimensions is created on a two-dimensional surface. Likewise, string theory suggests that in nature all the information about what is happening inside some volume of space is somehow encoded on its outer boundary, according to work by several theorists, including Dr. Juan Maldacena of the Institute for Advanced Study and Dr. Raphael Bousso of the University of California at Berkeley.

Just how and why a three-dimensional reality can spring from just two dimensions, or four dimensions can unfold from three, is as baffling to people like Dr. Witten as it probably is to someone reading about it in a newspaper.

In effect, as Dr. Witten put it, an extra dimension of space can mysteriously appear out of "nothing."

The lesson, he said, may be that time and space are only illusions or approximations, emerging somehow from something more primitive and fundamental about nature, the way protons and neutrons are built of quarks.

The real secret of string theory, he said, will probably not be new symmetries, but rather a novel prescription for constructing space-time.

"It's a new aspect of the theory," Dr. Witten said. "Whether we are getting closer to the deep principle, I don't know."

As he put it in a talk in October 2004, "It's plausible that we will someday understand string theory."

Tangled in Strings

Critics of string theory, meanwhile, have been keeping their own scorecard. The most glaring omission is the lack of any experimental evidence for strings or even a single experimental prediction that could prove string theory wrong—the acid test of the scientific process.

Strings are generally presumed to be so small that "stringy" effects should show up only when particles are smashed together at prohibitive energies, roughly 10^{19} billion electron volts. That is orders of magnitude beyond the capability of any particle accelerator that will ever be built on Earth. Dr. Harvey of Chicago said he sometimes woke up thinking, What am I doing spending my whole career on something that can't be tested experimentally?

This disparity between theoretical speculation and testable reality has led some critics to suggest that string theory is as much philosophy as science, and that it has diverted the attention and energy of a generation of physicists from other perhaps more worthy pursuits. Others say the theory itself is still too vague and that some promising ideas have not been proved rigorously enough yet.

Dr. Krauss said, "We bemoan the fact that Einstein spent the last 30 years of his life on a fruitless quest, but we think it's fine if a thousand theorists spend 30 years of their prime on the same quest."

The Other Quantum Gravity

String theory's biggest triumph is still its first one, unifying Einstein's lordly gravity that curves the cosmos and the quantum pinball game of chance that lives inside it.

"Whatever else it is or is not," Dr. Harvey said in Aspen, "string theory is a theory of quantum gravity that gives sensible answers."

That is no small success, but it may not be unique.

String theory has a host of lesser-known rivals for the mantle of quantum gravity, in particular a concept called loop quantum gravity, which arose from work by Dr. Abhay Ashtekar of Penn State and has been carried forward by Dr. Carlo Rovelli of the University of Marseille and Dr. Lee Smolin of the Perimeter Institute for Theoretical Physics in Waterloo, Ontario, among others.

Unlike string theory, loop gravity makes no pretensions toward being a theory of everything. It is only a theory of gravity, space and time, arising from the applications of quantum principles to the equations of Einstein's general relativity. The adherents of string theory and of loop gravity have a kind of Microsoft-Apple kind of rivalry, with the former garnering a vast majority of university jobs and publicity.

Dr. Witten said that string theory had a tendency to absorb the ideas of its critics and rivals. This could happen with loop gravity. Dr. Vafa; his Harvard colleagues, Dr. Sergei Gukov and Dr. Andrew Neitzke; and Dr. Robbert Dijkgraaf of the University of Amsterdam report in a recent paper that they have found a connection between simplified versions of string and loop gravity.

"If it exists," Dr. Vafa said of loop gravity, "it should be part of string theory."

Looking for a Cosmic Connection

Some theorists have bent their energies recently toward investigating models in which strings could make an observable mark on the sky or in experiments in particle accelerators.

"They all require us to be lucky," said Dr. Joesph Polchinski of the Kavli Institute.

For example, the thrashing about of strings in the early moments of time could leave fine lumps in a haze of radio waves filling the sky and thought to be the remains of the big bang. These might be detectable by the Planck satellite being built by the European Space Agency for a 2007 launching date, said Dr. Greene.

According to some models, Dr. Polchinski has suggested, some strings could be stretched from their normal submicroscopic lengths to become as big as galaxies or more during a brief cosmic spurt known as inflation, thought to have happened a fraction of a second after the universe was born.

If everything works out, he said, there will be loops of string in the sky as big as galaxies. Other strings could stretch all the way across the observable universe. The strings, under enormous tension and moving near the speed of light, would wiggle and snap, rippling space-time like a tablecloth with gravitational waves.

"It would be like a whip hundreds of light-years long," Dr. Polchinski said.

The signal from these snapping strings, if they exist, should be detectable by the Laser Interferometer Gravitational-Wave Observatory, which began science observations two years ago, operated by a multinational collaboration led by Caltech and the Massachusetts Institute of Technology.

Another chance for a clue will come in 2007 when the Large Hadron Collider is turned on at CERN in Geneva and starts colliding protons with seven trillion volts of energy apiece. In one version of the theory—admittedly a long shot—such collisions could create black holes or particles disappearing into the hidden dimensions.

Everybody's favorite candidate for what the collider will find is a phenomenon called supersymmetry, which is crucial to string theory. It posits the existence of a whole set of ghostlike elementary particles yet to be discovered. Theorists say they have reason to believe that the lightest of these particles, which have fanciful names like photinos, squarks and selectrons, should have a mass-energy within the range of the collider.

String theory naturally incorporates supersymmetry, but so do many other theories. Its discovery would not clinch the case for strings, but even Dr. Krauss of Case Western admits that the existence of supersymmetry would be a boon for string theory.

And what if supersymmetric particles are not discovered at the new collider? Their absence would strain the faith, a bit, but few theorists say they would give up.

"It would certainly be a big blow to our chances of understanding string theory in the near future," Dr. Witten said.

Beginnings and Endings

At the end of the Aspen celebration talk turned to the prospect of verification of string theory. Summing up the long march toward acceptance of the theory, Dr. Stephen Shenker, a pioneer string theorist at Stanford, quoted Winston Churchill: "This is not the end, not even the beginning of the end, but perhaps it is the end of the beginning."

Dr. Shenker said it would be great to find out that string theory was right.

From the audience Dr. Greene piped up, "Wouldn't it be great either way?"

"Are you kidding me, Brian?" Dr. Shenker responded. "How many years have you sweated on this?"

But if string theory is wrong, Dr. Greene argued, wouldn't it be good to know, so physics could move on? "Don't you want to know?" he asked.

Dr. Shenker amended his remarks. "It would be great to have an answer," he said, adding, "It would be even better if it's the right one."

—December 7, 2004

A Giant Takes on Physics's Biggest Questions

By DENNIS OVERBYE

The first thing that gets you is the noise.

Physics, after all, is supposed to be a cerebral pursuit. But this cavern—almost measureless to the eye, stuffed as it is with an Eiffel Tower's worth of metal, eight-story wheels of gold fan-shaped boxes, thousands of miles of wire and fat ductlike coils—echoes with the shriek of power tools, the whine of pumps and cranes, beeps and clanks from wrenches, hammers, screwdrivers and the occasional falling bolt. It seems like no place for the studious.

The physicists, wearing hard hats, kneepads and safety harnesses, are scrambling like Spider-Man over this assembly, appropriately named Atlas, ducking under waterfalls of cables and tubes and crawling into hidden room-size cavities stuffed with electronics.

They are getting ready to see the universe born again.

Again and again and again—30 million times a second, in fact.

Starting sometime next summer if all goes according to plan, subatomic particles will begin shooting around a 17-mile underground ring stretching from the European Center for Nuclear Research, or CERN, near Geneva, into France and back again—luckily without having to submit to customs inspections.

Crashing together in the bowels of Atlas and similar contraptions spaced around the ring, the particles will produce tiny fireballs of primordial energy, re-creating conditions that last prevailed when the universe was less than a trillionth of a second old.

Whatever forms of matter and whatever laws and forces held sway Back Then—relics not seen in this part of space since the universe cooled 14 billion years ago—will spring fleetingly to life, over and over again in all their possible variations, as if the universe were enacting its own version of the *Groundhog Day* movie. If all goes well, they will leave their footprints in mountains of hardware and computer memory.

"We are now on the endgame," said Lyn Evans, of CERN, who has been in charge of the Large Hadron Collider, as it is called, since its inception. Call it the Hubble Telescope of Inner Space. Everything about the collider sounds, well, large—from the 14 trillion electron volts of energy with which it will smash together protons, its cast of thousands and the $8 billion it cost to build, to the 128 tons of liquid helium needed to cool the superconducting magnets that keep the particles whizzing around their track and the three million DVDs worth of data it will spew forth every year.

The day it turns on will be a moment of truth for CERN, which has spent 13 years building the collider, and for the world's physicists, who have staked their credibility and their careers, not to mention all those billions of dollars, on the conviction that they are within touching distance of fundamental discoveries about the universe. If they fail to see something new, experts agree, it could be a long time, if ever, before giant particle accelerators are built on Earth again, ringing down the curtain on at least one aspect of the age-old quest to understand what the world is made of and how it works.

"If you see nothing," said a CERN physicist, John Ellis, "in some sense then, we theorists have been talking rubbish for the last 35 years."

Fabiola Gianotti, a CERN physicist and the deputy spokeswoman for the team that built the Atlas, said, "Something must happen."

The accelerator, Dr. Gianotti explained, would take physics into a realm of energy and time where the current reigning theories simply do not apply, corresponding to an era when cosmologists think that the universe was still differentiating itself, evolving from a primordial blandness and endless potential into the forces and particles that constitute modern reality.

She listed possible discoveries, like a mysterious particle called the Higgs, which is thought to endow other particles with mass, new forms of matter that explain the mysterious dark matter waddling through the cosmos and even new dimensions of space-time.

"For me," Dr. Gianotti said, "it would be a dream if, finally, in a couple of years in a laboratory we are going to produce the particle responsible for 25 percent of the universe."

Halfway around the ring stood her rival of sorts, Jim Virdee from Imperial College London, wearing a hard hat at the bottom of another huge cavern. Dr. Virdee is the spokesman, which is physics-speak for *leader,* of another team, some 2,500 strong, with another giant detector, the poetically named

Compact Muon Solenoid detector, which was looming over his shoulder like a giant cannon.

The prospect of discovery, Dr. Virdee said, is what sustained him and his colleagues over the 16 years it took to develop their machine. Without such detectors, he said, "this field which began with Newton just stops."

"When we started, we did not know how to do this experiment and did not know if it would work," he said. "Twenty-five hundred scientists can work together. Our judge is not God or governments, but nature. If we make a mistake, nature will not hesitate to punish us."

Game of Cosmic Leapfrog

The advent of the CERN collider also cements a shift in the balance of physics power away from American dominance that began in 1993, when Congress canceled the Superconducting Supercollider, a monster machine under construction in Waxahachie, Texas. The supercollider, the most powerful ever envisioned, would have sped protons around a 54-mile racetrack before slamming them together with 40 trillion electron volts.

For decades before that, physicists in the United States and Europe had leapfrogged one another with bigger, more expensive and, inevitably, fewer of these machines, which get their magic from Einstein's equation of mass and energy. The more energy that these machines can pack into their little fireballs, the farther back in time they can go, closer and closer to the big bang, the smaller and smaller things they can see. Recalling those times, Dr. Evans said: "There was a nice equilibrium across the Atlantic. People used to come and go."

Now, Dr. Evans said, "The center of gravity has moved to CERN."

The most powerful accelerator now operating is the trillion-electron-volt Tevatron, colliding protons and their antimatter opposites, antiprotons, at the Fermi National Accelerator Laboratory in Batavia, Ill. But it is scheduled to shut down by 2010.

CERN was born amid vineyards and farmland in the countryside outside Geneva in 1954 out of the rubble of postwar Europe. It had a twofold mission that included rebuilding European science and having European countries work together.

Today, it has 20 countries as members. Yearly contributions are determined according to members' domestic economies, and a result is a stable annual budget of about a billion Swiss francs. The vineyards and cows are still there, but so are strip malls and shopping centers.

It was here that the World Wide Web was born in the early 1990s, but the director-general of CERN, Robert Aymar, joked that the lab's greatest fame was as a locus of conspiracy in the novel *Angels and Demons,* by the author of *The DaVinci Code,* Dan Brown. The lab came into its own scientifically in the early '80s, when Carlo Rubbia and Simon van der Meer won the Nobel Prize by colliding protons and antiprotons there to produce the particles known as the W and Z bosons, which are responsible for the so-called weak nuclear force that causes some radioactive decays.

Bosons are bits of energy, or quanta, that, according to the weird house rules of the subatomic world, transmit forces as they are tossed back and forth in a sort of game of catch between matter particles. The Ws and Zs are closely related to photons, which transmit electromagnetic forces, or light.

The lab followed up that triumph by building a 17-mile-long ring, the Large Electron-Positron collider, or LEP, to manufacture W and Z particles for further study. Meanwhile, the United States abandoned plans for an accelerator named Isabelle to leapfrog to the giant supercollider in Texas.

Even before that supercollider was canceled, in 1993, however, CERN physicists had been mulling building their own giant proton collider in the LEP tunnel.

In 1994, after the supercollider collapse gave its own collider a clear field, the CERN governing council gave its approval. The United States eventually agreed to chip in $531 million for the project. CERN also arranged to borrow about $400 million from the European Investment Bank. Even so, there was a crisis in 2001 when the project was found to be 18 percent over budget, necessitating cutting other programs at the lab. The collider's name comes from the word *hadron,* which denotes subatomic particles like protons and neutrons that feel the "strong" nuclear force that binds atomic nuclei.

Whether the Europeans would have gone ahead if the United States had still been in the game depends on whom you ask. Dr. Aymar, who was not there in the '90s, said there was no guarantee then that the United States would succeed, even if it did proceed.

"Certainly in Europe the situation of CERN is such that we appreciate competition," he said. "But we assume that we are the leader and we have every intention to remain the leader. And we'll do everything which is needed to remain the leader."

To match the American machine, however, the Europeans, with a much smaller tunnel—17 miles instead of 54—had to adopt a riskier design, in particular by doubling the strength of their magnets.

"In this business, society is prepared to support particle physics at a certain level," Dr. Evans said. "If you want society to accept this work, which is not cheap, you have to be really innovative."

Cocktail Party Physics

The payoff for this investment, physicists say, could be a new understanding of one of the most fundamental of aspects of reality, namely the nature of mass.

This is where the shadowy particle known as the Higgs boson, aka the God particle, comes in.

In the Standard Model, a suite of equations describing all the forces but gravity, which has held sway as the law of the cosmos for the last 35 years, elementary particles are born in the big bang without mass, akin to Adam and Eve being born without sin.

Some of them (the particles, that is) acquire their heft, so the story goes, by wading through a sort of molasses that pervades all of space. The Higgs process, named after Peter Higgs, a Scottish physicist who first showed how this could work in 1964, has been compared to a cocktail party where particles gather their masses by interaction. The more they interact, the more mass they gain.

The Higgs idea is crucial to a theory that electromagnetism and the weak force are separate manifestations of a single so-called electroweak force. It shows how the massless bits of light called photons could be long-lost brothers to the heavy W and Z bosons, which would gain large masses from such cocktail party interactions as the universe cooled.

The confirmation of the theory by the Nobel-winning work at CERN 20 years ago ignited hopes among physicists that they could eventually unite the rest of the forces of nature.

Moreover, Higgs-like fields have been proposed as the source of an enormous burst of expansion, known as inflation, early in the universe, and, possibly, as the secret of the dark energy that now seems to be speeding up the expansion of the universe. So it is important to know whether the theory works and, if not, to find out what does endow the universe with mass.

But nobody has ever seen a Higgs boson, the particle that personifies this molasses. It should be producible in particle accelerators, but nature has given confusing clues about where to look for it. Measurements of other exotic particles suggest that the Higgs's mass should be around 90 billion electron volts, the unit of choice in particle physics. But other results, from the LEP collider at CERN before it shut down in 2000, indicate that the Higgs must weigh more than 114 billion electron volts. By comparison, an electron is half a million electron volts, and a proton is about 2,000 times heavier.

"We've nearly ruled out the Standard Model, if you want to say it that way," said John Conway, a Fermilab physicist. The new collider was specifically designed to hunt for the Higgs particle, which is key both to the Standard Model and to any greater theory that would supersede it.

Theorists say the Higgs or something like it has to show up simply because the Standard Model breaks down and goes kerflooey at energies exceeding one trillion electron volts. If you try to predict what happens when two particles collide, you end up with nonsense, explained Dr. Ellis of CERN, a senior theorist with the long white hair and a bushy beard to prove it.

"There is either a violation of probability or some new physics," Dr. Ellis said.

Nima Arkani-Hamed of Harvard said he would bet a year's salary on the Higgs.

"If the Higgs or something like it doesn't exist," Dr. Arkani-Hamed said, "then some very basic things like quantum mechanics are wrong."

A result, Dr. Gianotti said, is "Either we find the Higgs boson, or some stranger phenomenon must happen."

Nightmares

If the CERN experimenters find the Higgs, Nobel Prizes will flow like water. But just finding the elusive particle will not be enough to satisfy the theorists, who profess to be haunted by a much deeper problem; namely, why the putative particle is not millions of times heavier than it appears to be.

When they try to calculate the mass of the Higgs particle using the Standard Model and quantum mechanics, they get what Dr. Ellis called "a very infinite answer."

Rather than a trillion electron volts or so, quantum effects push the mass all the way up to *10 quadrillion* trillion electron volts, known as the Planck energy, where gravity and the other particle forces are equal.

The culprit is quantum weirdness, one principle of which is that anything that is not forbidden will happen. That means the Higgs calculation must include the effects of its interactions with all other known particles, including so-called virtual particles that can wink in and out of existence, which shift its mass off the scale.

As a result, if the Standard Model is valid for all energies, said Joesph Lykken, a Fermilab theorist, "then you are in deep doodoo trying to explain why the Higgs mass isn't a quadrillion times bigger than it needs to be."

Another way to put it is to ask why gravity is so much weaker than the other forces—the theory wants them all to be equal.

Theorists can rig their calculations to have the numbers come out right, but it feels like cheating. "What we have to do to equations is crazy," Dr. Arkani-Hamed said.

One solution that has been proposed is a new principle of nature called supersymmetry that, if true, would be a bonanza for the CERN collider.

It posits a relation between the particles of matter like electrons and quarks and particles that transmit forces like photons and the W boson. For each particle in one category, there is an as-yet-undiscovered superpartner in the other category.

"Supersymmetry doubles the world," Dr. Arkani-Hamed said.

These superpartners cancel out all the quantum effects that make the Higgs mass skyrocket. "Supersymmetry is the only known way to manage this," Dr. Lykken said.

Because Higgs bosons are expected to be produced very rarely, it could take at least a year or more for physicists to confirm their discovery at the collider. But some supersymmetric particles, if they exist, should be produced abundantly and could thus pop out of the data much sooner. "Suppose a gluino exists at 300 billion electron volts," Dr. Arkani-Hamed said, referring to a putative superpartner. "We could know the first day if they exist."

For several years, supersymmetry has been a sort of best bet to be the next step beyond the Standard Model, which is undefeated in experiments but has enormous gaps. The Standard Model does not include gravity or explain why, for example, the universe is matter instead of antimatter or even why particles have the masses they do.

In the end, Michelangelo Mangano, a theorist at CERN, said, "The standard model prediction can't be the end of the story."

Supersymmetry also fixes a glitch in the age-old dream of explaining all the forces of nature as manifestations of one primordial force. It predicts that at a high enough energy, all the forces—electromagnetic, strong and weak—have identical strengths.

"If supersymmetry is right, unification works," Dr. Ellis said.

But there is no direct evidence for any of the thousands of versions of super-symmetry that have been proposed. Indeed, many theorists are troubled that its effects have not already shown up in precision measurements at accelerators.

"It doesn't smell good," Dr. Arkani-Hamed said. Physicists say the best indirect evidence for supersymmetry comes from the skies, where the galaxies have been found to be swaddled by clouds of invisible dark matter, presumably unknown particles left over from the big bang. "Dark matter is a very physical argument." Dr. Ellis said. "If you take astrophysics seriously, there has to be some unseen stuff out there."

On the menu of discoveries, there is always None of the Above. As Dr. Gianotti put it: "Nature has chosen another solution. This will be great."

There are indeed other potential solutions that go by the name of Technicolor or the Little Higgs. But what if the collider sees nothing?

That, Dr. Ellis said, would be interesting for the theorists, who would have to retool and try to think even deeper thoughts about quantum mechanics and relativity, but bad for the experimentalists. Without any results, they would be unlikely to obtain financing for the next big machine planned, the $7 billion International Linear Collider.

A worse nightmare, several theorists said, would be seeing just the Higgs, but nothing else. That would leave them where they are now, stuck in the Standard Model, with no answer to their embarrassing fine-tuning problem, no dark matter and no clue to a better theory.

To add to the confusion, according to the Standard Model, the Higgs can have only a limited range of masses without severe damage to the universe. If it is too light, the universe will decay. If it is too heavy, the universe would have blown up already. According to Dr. Ellis, there is a magic value between 160 billion and 180 billion electron volts that would ensure a stable universe and require no new physics at all.

But that would leave theorists with nothing more to do and a world in which basic questions would remain forever unanswered.

Dr. Ellis said, "I can't believe God would push the button on a theory like that."

But, he conceded, "For the ILC, a boring Higgs is better than nothing."

Sunken Cathedrals

There was more than birds singing and trees blooming outside the main CERN cafeteria in March to suggest that springtime for physics was approaching.

Some 300 feet beneath the warming grass, the magnets that are the guts of the collider, thick as tree trunks, long as boxcars, weighing in at 35 tons apiece, were strung together like an endless train stretching away into the dim lamplight and around a gentle curve.

A technician on his way to a far sector of the collider ring bicycled past.

"When you fold in the technology combined with the scale," said Peter Limon, a Fermilab physicist on duty here, "I don't think anything on Earth or in space that we know about beats it."

Running through the core of this train, surrounded by magnets and cold, were two vacuum pipes, one for protons going clockwise, the other counterclockwise. Traveling in tight bunches along the twin beams, the protons will cross each other at four points around the ring, 30 million times a second. During each of these violent crossings, physicists expect that about 20 protons, or the parts thereof—quarks or gluons—will actually collide and spit fire. It is in vast caverns at those intersection points that the knee-padded and hard-hatted physicists are assembling their detector, or "sunken cathedrals" in the words of a CERN theorist, Alvaro de Rujula, to capture the holy fire. Two of the detectors are specialized. One, called Alice and led by Jürgen Schukraft of CERN, is designed to study a sort of primordial fluid, called a quark-gluon plasma, that is created when the collider smashes together lead nuclei.

The other, LHCb, is led by Tatsuya Nakada of CERN and the Swiss Federal Institute of Technology in Lausanne. It is designed to hunt for subtle differences in matter and antimatter that could help explain how the universe, which was presumably born with equal amounts of both, came to be dominated by matter.

The other two, the aforementioned Atlas and Compact Muon Solenoid, or CMS for short, are the designated rival workhorses of the collider, designed expressly to capture and measure every last spray of particle and spark of energy from the proton collisions.

The rivals represent complementary strategies for hunting the Higgs particle, which is expected to disintegrate into a spray of lesser particles. Exactly which particles depends on how massive the Higgs really is.

One telltale signature of the Higgs and other subatomic cataclysms is a negatively charged particle known as a muon, a sort of heavy electron that comes flying out at nearly the speed of light. Physicists measure muon momentum by seeing how much their paths bend in a magnetic field.

It is the need to have magnets strong enough and large enough to produce measurable bending, physicists say, that determines the gigantic size of the detectors.

The Compact Muon Solenoid, built by Dr. Virdee's group, weighs 12,000 tons, the heaviest instrument ever made. It takes its name from a massive super-conducting electromagnet that produces a powerful field running along the path of the protons.

Conversely, the magnetic field on Atlas wraps like tape around the proton beam. The Atlas collaboration has been led from its start by Peter Jenni of CERN. At 150 feet long and 80 feet high, Atlas is bigger than its rival, but it is much lighter, about 7,000 tons, about as much as the Eiffel Tower. The physicists like to joke that if you threw it in the ocean in a plastic bag it would float.

The two detectors have much in common, including "onion layers" of instruments to measure different particles and the ability to cope with harsh radiation and vast amounts of data. Dr. Virdee compared the central CMS detector, made of strips of silicon that record the passage of charged particles, to a 60-megapixel digital camera taking 40 million pictures a second. "We have to time everything to the nanosecond," he said.

To manage this onslaught the teams' computers have to perform triage, and winnow those events to a couple hundred per second. That is dangerous, Dr. Gianotti said, "because we are looking for something rare." The Higgs occurs once in every trillion events, she said.

Contending Armies

The competition between Atlas and the CMS is in keeping with a long tradition of having rival teams and rival detectors at big experiments to keep each other honest and to cover all the bets. As Dr. Mangano put it, "If you screw it up, others are here to crucify you."

At the Fermilab Tevatron, the teams, several hundred strong, are called CDF and D0. In the glory years 20 years ago at CERN, they were called UA1 and UA2. Over the years, as the machines have grown, so have the groups that built them, from teams to armies, 1,800 people from 34 countries for Atlas and 2,520 from 37 countries for the CMS. The other two experiments—Alice with 1,000 scientists, and LHCb with 663—are only slightly smaller.

Robert Cousins of UCLA and CMS joked that he was old enough so that after 25 years in the business "half my friends are on Atlas, the others on CMS." Dr. Jenni said all 1,800 Atlas scientists would have their names on the first papers out of the collider, adding: "The people who work in the pit make as important a physics contribution as those who end up in front of the computers. This is a big step in energy. It's new territory, and that's in the end why everyone is excited."

At the end of the day, Dr. Mangano said, unless there is a major problem both machines will perform. "It will come down to sociology," he said. "How quickly can they analyze the data? How do you manipulate and analyze the data? The process of understanding is long."

There could be new phenomena, he added, new particles that theorists have not thought of.

Dr. Mangano pointed out that it had been a long time since high-energy physicists had made a fundamental discovery. And back then, when Dr. Rubbia was doing his Nobel work, there were well-defined theories of what would be found. Now, everything will be new.

"There are many students who have never seen data," Dr. Mangano said. "I don't know how much longer we can keep going like that."

What comes out of the Large Hadron Collider, he said, "will determine the future of the field."

Dr. Arkani-Hamed said the tension was keeping him awake at night. "Nobody knows how this is going to go," he said. "That's what makes it so cool. The experiment itself is so spectacular."

Sipping an espresso in his office, Dr. Mangano refused to consider the possibility of failure. "It's like saying, 'Suppose you drive into a tree on the way home,'" he said. "Let's hope we get home safely and we see something."

—May 15, 2007

Physicists Find Elusive Particle
Seen as Key to Universe

By DENNIS OVERBYE

Signaling a likely end to one of the longest, most expensive searches in the history of science, physicists said that they had discovered a new subatomic particle that looks for all the world like the Higgs boson, a key to understanding why there is diversity and life in the universe.

Like Omar Sharif materializing out of the shimmering desert as a man on a camel in *Lawrence of Arabia,* the elusive boson has been coming slowly into view since last winter, as the first signals of its existence grew until they practically jumped off the chart.

"I think we have it," said Rolf-Dieter Heuer, the director-general of CERN, the multinational research center headquartered in Geneva. The agency is home to the Large Hadron Collider, the immense particle accelerator that produced the new data by colliding protons. The findings were announced in Aspen, Colorado by two separate teams. Dr. Heuer called the discovery "a historic milestone."

He and others said that it was too soon to know for sure, however, whether the new particle is the one predicted by the Standard Model, the theory that has ruled physics for the last half-century. The particle is predicted to imbue elementary particles with mass. It may be an impostor as yet unknown to physics, perhaps the first of many particles yet to be discovered.

That possibility is particularly exciting to physicists, as it could point the way to new, deeper ideas, beyond the Standard Model, about the nature of reality.

For now, some physicists are simply calling it a "Higgs-like" particle.

"It's something that may, in the end, be one of the biggest observations of any new phenomena in our field in the last 30 or 40 years," said Joe Incandela, a physicist of the University of California at Santa Barbara, and a spokesman for one of the two groups reporting new data on Wednesday.

At the Aspen Center for Physics, a retreat for scientists, bleary-eyed physicists drank champagne in the wee hours as word arrived via webcast from CERN. It was a scene duplicated in Melbourne, Australia, where physicists had gathered for a major conference, as well as in Los Angeles, Chicago, Princeton, New York, London and beyond—everywhere that members of a curious species have dedicated their lives and fortunes to the search for their origins in a dark universe.

In Geneva, 1,000 people stood in line all night to get into an auditorium at CERN, where some attendees noted a rock-concert ambience. Peter Higgs, the University of Edinburgh theorist for whom the boson is named, entered the meeting to a sustained ovation.

Confirmation of the Higgs boson or something very much like it would constitute a rendezvous with destiny for a generation of physicists who have believed in the boson for half a century without ever seeing it. The finding affirms a grand view of a universe described by simple and elegant and symmetrical laws—but one in which everything interesting, like ourselves, results from flaws or breaks in that symmetry.

According to the Standard Model, the Higgs boson is the only manifestation of an invisible force field, a cosmic molasses that permeates space and imbues elementary particles with mass. Particles wading through the field gain heft the way a bill going through Congress attracts riders and amendments, becoming ever more ponderous.

Without the Higgs field, as it is known, or something like it, all elementary forms of matter would zoom around at the speed of light, flowing through our hands like moonlight. There would be neither atoms nor life.

Physicists said that they would probably be studying the new particle for years. Any deviations from the simplest version predicted by current theory—and there are hints of some already—could begin to answer questions left hanging by the Standard Model. For example, what is the dark matter that provides the gravitational scaffolding of galaxies?

And why is the universe made of matter instead of antimatter?

"If the boson really is not acting standard, then that will imply that there is more to the story—more particles, maybe more forces around the corner," Neal Weiner, a theorist at New York University, wrote in an e-mail. "What that would be is anyone's guess at the moment."

The announcement was also an impressive opening act for the Large Hadron Collider, the world's biggest physics machine, which cost $10 billion to build and began operating only two years ago. It is still running at only half-power.

Physicists had been icing the champagne ever since last December. Two teams of about 3,000 physicists each—one named Atlas, led by Fabiola Gianotti, and the other CMS, led by Dr. Incandela—operate giant detectors in the collider, sorting the debris from the primordial fireballs left after proton collisions.

Last winter, they both reported hints of the same particle. They were not able, however, to rule out the possibility that it was a statistical fluke. Since then, the collider has more than doubled the number of collisions it has recorded.

The results announced in Aspen capped two weeks of feverish speculation and Internet buzz as the physicists, who had been sworn to secrecy, did a breakneck analysis of about 800 trillion proton-proton collisions over the last two years.

Up until last weekend, physicists at the agency were saying that they themselves did not know what the outcome would be. Expectations soared when it was learned that the five surviving originators of the Higgs boson theory had been invited to the CERN news conference.

The December signal was no fluke, the scientists said. The new particle has a mass of about 125.3 billion electron volts, as measured by the CMS group, and 126 billion, according to Atlas. Both groups said that the likelihood that their signal was a result of a chance fluctuation was less than one in 3.5 million, "five sigma," which is the gold standard in physics for a discovery.

On that basis, Dr. Heuer said that he had decided to call the Higgs result a "discovery." He said, "I know the science, and as director-general I can stick out my neck."

Dr. Incandela's and Dr. Gianotti's presentations were repeatedly interrupted by applause as they showed slide after slide of data presented in graphs with bumps rising like mountains from the sea.

Dr. Gianotti noted that the mass of the putative Higgs, apparently one of the heaviest subatomic particles, made it easy to study its many behaviors. "Thanks, nature," she said.

Gerald Guralnik, one of the founders of the Higgs theory, said he was glad to be at a physics meeting "where there is applause, like a football game."

Asked to comment after the announcements, Dr. Higgs seemed overwhelmed. "For me, it's really an incredible thing that's happened in my lifetime," he said.

Dr. Higgs was one of six physicists, working in three independent groups, who in 1964 invented what came to be known as the Higgs field. The others were Tom Kibble of Imperial College, London; Carl Hagen of the University of Rochester; Dr. Guralnik of Brown University; and François Englert and Robert Brout, both of Université Libre de Bruxelles.

One implication of their theory was that this cosmic molasses, normally invisible, would produce its own quantum particle if hit hard enough with the right amount of energy. The particle would be fragile and fall apart within a millionth of a second in a dozen possible ways, depending upon its own mass.

Unfortunately, the theory did not describe how much this particle should weigh, which is what made it so hard to find, eluding researchers at a succession of particle accelerators, including the Large Electron-Positron Collider at CERN, which closed down in 2000, and the Tevatron at the Fermi National Accelerator Laboratory, or Fermilab, in Batavia, Illinois, which shut down in 2011.

Along the way the Higgs boson achieved a notoriety rare in abstract physics. To the eternal dismay of his colleagues, Leon Lederman, the former director of Fermilab, called it the "God particle," in his book of the same name, written with Dick Teresi. (He later said that he had wanted to call it the "goddamn particle.")

Finding the missing boson was one of the main goals of the Large Hadron Collider. Both Dr. Heuer and Dr. Gianotti said they had not expected the search to succeed so quickly.

So far, the physicists admit, they know little about their new boson. The CERN results are mostly based on measurements of two or three of the dozen different ways, or "channels," by which a Higgs boson could be produced and then decay.

There are hints, but only hints so far, that some of the channels are overproducing the boson while others might be underproducing it, clues that maybe there is more at work here than the Standard Model would predict.

"This could be the first in a ring of discoveries," said Guido Tonelli of CERN.

In an e-mail, Maria Spiropulu, a professor at the California Institute of Technology who works with the CMS team of physicists, said: "I personally do not want it to be Standard Model anything—I don't want it to be simple or symmetric or as predicted. I want us all to have been dealt a complex hand that will send me (and all of us) in a (good) loop for a long time."

Nima Arkani-Hamed, a physicist at the Institute for Advanced Study in Princeton, said: "It's a triumphant day for fundamental physics. Now some fun begins."

—*July 4, 2012*

CHAPTER 2

The Practical Atom

Wireless Signals across the Ocean

Guglielmo Marconi announced on December 1901 in Newfoundland the most wonderful scientific development of recent times. He stated that he had received electric signals across the Atlantic Ocean from his station in Cornwall, England.

Marconi explains that before leaving England he made his plans for trying to accomplish this result, for, while his primary object was to communicate with Atlantic liners in midocean, he also hoped to receive wireless messages across the Atlantic.

The Marconi station in Cornwall is a most powerful one. An electric force a hundred times greater than at the ordinary stations is generated there. Before he left England, Marconi arranged that the electrician in charge of the station, which is located at Poldhu, should begin sending signals daily after a certain date, which Marconi was to cable to him upon perfecting the arrangements in Newfoundland. Marconi arrived in St. John's, Newfoundland eight days ago. He selected Signal Hill, at the entrance to the harbor, as an experimenting station, and moved his equipment there. Last Monday he cabled to the Poldhu station orders to begin sending signals at 3 p.m. daily and to continue them until 6 p.m., these hours being respectively 11:30 a.m. and 2:30 p.m., St. John's, Newfoundland time.

During these hours Marconi elevated a kite, with the wire by means of which signals are sent or received. He remained at the recorder attached to the receiving apparatus, and, to his profound satisfaction, signals were received by him at intervals, according to the program arranged previously with the operator at Poldhu. These signals consisted of the repetition at intervals of the letter *S*, which in Marconi's code is made by three dots, or quick strokes. This signal was repeated so frequently, and so perfectly in accord with the detailed plan arranged to provide safeguards against the possibility of a mistake that Marconi was satisfied that it was a genuine transmission from England.

Again the following day, during the same hours, the kite was elevated and the same signals were renewed.

This made the assurance so complete that Marconi cabled word of his success to his principals in England, and also made it known to the governor of Newfoundland, Sir Cavendish Boyle, who apprised the British Cabinet of the result of the experiments.

Marconi, though satisfied of the genuineness of the signals and that he has succeeded in his attempts to establish communication across the Atlantic without the use of wires, emphasizes the fact that the system is yet only in an embryonic stage. He says, however, that the possibility of its ultimate development is demonstrated by the success of the present experiments with incomplete and imperfect apparatus, as the signals can only be received by the most sensitively adjusted apparatus, and he is working under great difficulties owing to the conditions prevailing here. The Cornwall coast is 1,700 miles from St. John's.

In view of the success attending these trials, Marconi will for the present disregard the matter of communicating with transatlantic steamers. He will return to England next week, and will conduct the experiments from Poldhu. He explains that the greater electrical power there will enable him to send more effective signals. He will undertake this work himself, leaving assistants here to erect a mast and receive the signals as he forwards them. It is not possible to send return signals from here until a powerful electric battery is installed.

Premier John Bond of Newfoundland offers to Marconi every facility within the power of the colonial government for the carrying out of his plans.

Marconi intends to build a large, fully equipped, experimental station near St. John's, besides the Lloyd station at Cape Race. The former will have the same equipment as the Poldhu station, and will play the same part on this side of the Atlantic as Poldhu does on the other side. It is expected that the St. John's station will communicate with New York on one side, and Cornwall on the other, being midway between the two. This establishment will probably cost about $60,000, and is intended to perform the same work as a modern cable station.

Marconi announces that he will remain in England until after the coronation of King Edward next summer, and that he hopes to send the "news" of that event across the Atlantic by the wireless method, so as to prove the capability of the system for such purposes. He will probably in the meantime equip all vessels of the leading lines of steamers with his apparatus.

—*December 15, 1901*

Light-Energy Ideas Told by Millikan

"Of the many jolts which the physicist has received in the rapid discoveries which have been made the past quarter century, the worst has come in the field of photo-electricity," said Professor Robert A. Millikan in his final address in the Lee De Forest course at Yale.

Dr. Millikan, who is a Nobel Prize winner in physics, is director of the Norman Bridge Laboratory at Pasadena, where he has performed revolutionary experiments in spectroscopy. The feature of his experiments, dwelt upon in his closing address, was that they form a complete check on the Einstein corpuscular theory.

For the first time since the course was established three years ago, Dr. De Forest attended. Professor Hubert M. Turner of the Dunham Electrical Laboratory at Yale said in introducing Professor Millikan:

"When Dr. De Forest developed the three-element electron tube 20 years ago no one had any idea of the many uses to which it would be put. Now the photo-electric cell is very much in the same position as was the audion at that time.

"We find that the photo-electric cell is now being used by Dr. De Forest in his phono-film or 'talking movie'; by C. Francis Jenkins in his transmission of pictures by radio, and by General Electric, Westinghouse and other laboratories in their general research work. It appears likely that it will be in general use in the near future."

Professor Millikan said that in 1900 Leonard found that when light was allowed to fall on a metal, energy was emitted and was independent of the intensity of the light. He assumed that the light falling on the atoms of the metal simply released a spring inside it and let off the energy. Dr. Millikan called this the trigger theory.

In 1905 Einstein, with the disregard for the past which is so characteristic of him, assumed that the radiant energy was something bulletlike. The next 10 years were spent in experimental checks and Einstein received the Nobel Prize in 1920 because his work in this field had been verified.

Dr. Millikan's work at Chicago was the first quantitative proof of this theory, which is a corpuscular theory and cannot be reached on any other concept. Bohr threw away the corpuscular concept because no concept is necessary; the energy is equal to the Bohr constant times the incident frequency of the light, minus the ionizing potential of the metal. There was no work more important than the determination of the conditions under which this equation held.

In 1903 Dr. Millikan began his work on photo electricity and found that the effect was independent of temperature. The previous theory would require that the photo-electric effect be dependent upon temperature and it was a great surprise when he raised the temperature of a piece of aluminum from room temperature to 220 degrees and found no change in the number of electrons emitted. Since that time, these results have been questioned, but there is no doubt that under most conditions there is no thermal effect.

The bulletlike theory of Niels Bohr was very hard to reconcile with other phenomena, but experiments were given a complete check, and no experimental work of the last 10 years has been more conclusive than this, which has set up the validity of this concept for all frequencies higher than the critical frequency.

Dr. Millikan then discussed the much-debated question of the direction of emission of the electrons with respect to the incident beam of light. He told of some experiments of Lockbridge, which have not yet been published, in which the direction of the electrons can be measured closer than five degrees.

Recently, German scientists have said that five irregular tracks which are seen in apparatus such as Lockbridge's are optical illusions; but recent experiments show that these irregular paths are followed by the electrons.

Watson, using newly developed apparatus, allowing X-rays to fall on a fine tungsten wire, found that the electrons are emitted in all directions, but that the maximum number are emitted at an angle between 65 and 75 degrees with the incident beam. In order to get these results, it is necessary to hold the current in the magnetic coil constant for a week at a time. This means adjusting it every two hours.

The results completely overthrow the work of Frank Bubb, in that he predicted a band of 20 degrees width in a direction opposite to that of the incident beam in which there would be no electrons, and these experiments show that there is no difference in the amount of radiation either in the direction of the beam or opposite to it.

Leaving the field of photo electricity, Dr. Millikan showed that he had discovered that the same effect on temperature exists when electrons are drawn out of tungsten wire by means of an electric field. He attributed the success of his recent experiments to the high potential gradients which he was able to get by using very fine wires and fairly low voltage differences. The total number of electrons which are obtained in this way is greater than the number contained in the tungsten wires, thus showing that they must be conduction electrons and the same as those obtained in the photo-electric effect. These are independent of temperature, in the case of tungsten, until 1,100 degrees absolute is reached.

All these considerations indicate the corpuscle theory. The experiments in interference and radio indicate that the wave theory should be applied. The results seem to be that it will be necessary for the physicist to use both theories in attempting to find one picture that will cover all cases. Each can now be applied in its own separate field, and until someone comes forward with a theory that will apply equally well to both it will be necessary to use the two.

—April 11, 1926

Compton to Strive for Atomic Energy

An extensive study of the problem of releasing atomic energy, the solution of which would create a limitless reservoir of power and would bring undreamed-of changes to civilization, will begin at the University of Chicago shortly, Dr. Arthur H. Compton, Nobel Prize physicist of that institution, has announced. Dr. Compton will supervise the experiments, which are to continue for several years.

A primary goal of the study will be to produce an extremely high voltage in electrons in X-rays. If this voltage can be raised to between 10 million and 20 million volts pressure, it is likely that the experiment will be successful, and the door pointing the way to the release of atomic energy will have been thrown open, in the opinion of Dr. Compton.

The achievement of this tremendous voltage, which has never been attained in any laboratory on Earth, would help to approach the high temperatures existing normally on the sun, where atomic energy is continuously being released, Dr. Compton explained.

Temperatures inside the Sun

"On the sun, the electron and proton particles of the atom constantly coalesce to produce the photon, which is radiated away into space," Dr. Compton said. "The temperatures in the sun's interior are probably as high as 40 million degrees and it is quite impossible to achieve any such temperature here. But it is possible that the essential characteristics of the vital action of coalescence may be due to the high speed of electrons, and in that case it is not a vain hope to anticipate making electrons and protons coalesce on Earth to release the huge energies which they contain."

Because of the purely experimental stage of the work, Dr. Compton said he is unwilling to divulge publicly the exact nature of the experiments to be undertaken at the University of Chicago. He explained that it would be years before the undertaking could succeed, and that "tremendous" technical difficulties would have to be overcome. Among these technical difficulties, he said, were the production of a tube capable of withstanding the great pressure it is hoped to create, and the arrangement of new and intricate equipment of many kinds.

The combination of the electron and proton, the two invisible particles that pursue their orbits inside the tiny atom, is one of three possible ways known for the release of atomic energy. The experiments at the University of Chicago are to attack the coalescense problem, as it seems the most likely way to find a solution, Dr. Compton explained. The other two methods, he said, consist of a study of radioactivity and the cosmic rays.

"If the key to harnessing the power of atomic energy is ever discovered, our present civilization will undergo a very radical change," Dr. Compton said. "There is enough atomic energy in a teaspoonful of ordinary water to provide all the energy to run New York City, with all of its transit systems, factories and the life of the metropolis in general."

Dr. Compton revealed the forthcoming study of atomic energy in an address called "Do Things Have a Beginning and an End?" at the College of the City of New York, Twenty-third Street and Lexington Avenue. More than two thousand persons heard Dr. Compton's talk, the last of a series on the "Nature of Things."

Discusses Age of the Earth

Touching on the age of the Earth, Dr. Compton asserted that radioactive material gives an approximately accurate basis for concluding that our globe is about 2 billion years old. The oldest minerals found on the Earth are over 1.5 billion years old, he said. Life in some form began on this planet about 1 billion years ago, he added.

"The best modern evidence of science points to the belief that man is no more than 2 million years old," he said.

The sun probably attained the ripe age of some 5,000,000,000,000 years, Dr. Compton asserted. So vast, however, is the amount of the sun's energy that, although the Earth's parent body is radiating away 400,000 tons of matter a second, there has been no noticeable decrease in the solar energy.

"If the sun's energy were other than radioactive or atomic, our parent globe would be completely burned up in less than 10,000 years," he said. The noted physicist said the best evidence tends to show that the universe began about 1,000,000,000,000,000 years ago.

—March 28, 1931

Radar—I

By Hanson W. Baldwin

In the war in the air over Western Europe, radar—the "seeing eye" of radio—is playing a major role.

Radar (an abbreviation of "radio direction and ranging"), as it is called in the United States, or radio locator as it is known in Britain, is of high importance not only in air war but in every phase of modern combat.

The device was of major significance in the Battle of Britain. Today radio locator stations dot the coasts of nearly every combatant country, night fighters are equipped with radar to guide them to enemy bombers, and the equipment of a man-of-war is not complete without one or several radars.

Radar enabled one of our modern battleships in a night action in the Solomons to locate, fire at and straddle on the first salvo a Japanese battleship eight miles away.

The Japanese ship was never actually seen by our men until after she had been hit and was afire. Radar also played a role in various British naval victories in the Mediterranean.

Radar is also of major importance in controlling the fire of antiaircraft guns, and at night has replaced or supplemented searchlights.

Basic Development Prewar

Probably it is the most important single technical development of this war.

Its basic development, however, was largely prewar; in fact, it was discovered years ago and, contrary to popular impression, the Scotchman, Sir Robert Alexander Watson-Watt, though one of the first, was not the only inventor who grasped the principles of radar and helped to develop it into a practical operating instrument. The July 1943 issue of the magazine *Radio News* reports that "some of the basic fundamentals of radar were covered in patents of a now unknown Frenchman."

Actually, discovery and development of the device was apparently more or less simultaneous and parallel in three or four countries. It is likely that early British, American, German—and perhaps French and Japanese researches and developments proceeded independently each of the other. The Italians were slow to adopt radar, but now have it, probably the German version.

In this country the development began in 1922 when two scientists, Dr. A. Hoyt Taylor and Leo C. Young of the Naval Aircraft Radio Laboratory, discovered the basic principle of radar. In 1931 the Radio Division of the Bureau of Engineering ordered the investigation of the use of radio for the detection of aircraft and surface vessels.

Battleship Test in 1938

In 1932 the War Department was informed of the possibilities shown by these investigations. In 1935 $100,000 was appropriated by the House for research. In 1936 a practical demonstration of the device was given, and in 1938 probably the first ship to be equipped with radar, the old battleship *New York*, received an experimental installation. Vice Admiral Alfred W. Johnson said then that "The equipment is one of the most important radio developments since the advent of radio itself."

The British radio location system, largely developed under the inspiration and influence of Watson-Watt, was the first large-scale practical installation in the Allied countries, although Germany is believed to have had a number of radar stations in operation when the war started, or shortly thereafter.

Today, the Allied nations have pooled ideas, and development and research knowledge, with the result that United Nations's radars, manufactured in greatest volume in this country, are plainly better than those Japanese radars which we have captured, and are believed to be equal, or perhaps superior, to the German device.

—*July 12, 1943*

Radar—II

By HANSON W. BALDWIN

Radar, one of the most important technical developments of this war, is a radio means of locating, and measuring the distance to, a target—either air or surface.

Radar and radio locators transmit, with the use of electronic equipment, ultra-high-frequency radio impulses, which continue on a straight path until they strike an object, when they are reflected, or "bounced back," to the transmission point.

The time interval between transmission of the original impulse and reception of the "echo" can be measured and translated into the distance to the object. The bearing of the object can also be determined. Indeed, radio locators, with the aid of cathode ray tubes and translucent screens, can actually depict a representation of the target—whether plane or surface ship.

Thus the target does not have to be seen. Night, clouds, low visibility, fog make no difference to the "eye" of radar.

Radar is not foolproof, is not without error and has its limitations. The waves travel in a straight line and hence the range is roughly limited—dependent on the height of the transmitter above the sea—by the curvature of the Earth's surface. It is of little use in valleys or in areas hemmed in by high mountains. It is best mounted high on masts or on mountain peaks away from tall buildings.

Skill in Use Is Required

Radar requires much training and considerable skill to operate and a great deal more skill to maintain it and keep it calibrated. It can obviously give very little warning of planes that are "hedge-hopping," that is, keeping very close to the ground. This is the technique adopted by most of the recent German daylight raiders against Britain.

Moreover, radar is simply a means of locating an enemy, not of destroying that enemy. Today, because of radar, the means of locating the enemy are usually more accurate and certain than the means of destroying the enemy.

Even if radar accurately trains an antiaircraft gun toward an unseen enemy plane, it is one thing to know that the plane definitely is at such and such a point in the heavens and another to hit that point. A correction for "lead"—the distance the plane will travel between the time the gun is fired and the time the shell reaches the plane's altitude—must be estimated and ballistic corrections for temperature, wind, the number of times the gun has been fired and the like must be considered. Errors in these calculations—human errors, not radar errors—may be made, with the result that the shell will miss. This is true in any gunnery, for gunnery is not an exact science. Moreover, planes often fly above the effective range of the available antiaircraft guns.

The same general considerations are also true of the other uses of radar—on convoy escort vessels, for example, for use against submarines. The radio waves transmitted by radar do not penetrate water; hence radar is no good against submerged submarines. Used at night against submarines on the surface, it is very effective, but detection of the submarine does not mean that the submarine is destroyed. It may submerge to safety before the escort vessel can reach the spot indicated.

Radar in night fighters helps to guide the planes to enemy bombers, but again it does not solve the problem of destroying those bombers. Location is not destruction.

Signals Mark Friendly Warcraft

Radio locators on the ground, at sea and in the air, therefore, are a method of information of the enemy's plane or ship movements. Our own ship and plane movements will, of course, be picked up by radar, and to avoid firing at our own ships and planes various identification signals and systems have to be arranged. This can be done without too much difficulty.

Mass air raids make the use of radar, particularly when controlling antiaircraft guns, more difficult, for a great many planes "saturating" a small airspace might present an almost insuperable problem in exact fire control. When this saturation point is reached, however, antiaircraft guns can switch from individual fire against specific plane targets to barrage fire, using radar only for the approximate altitude and range.

For each trick of the offense, there is a new trick of the defense, and there is no doubt that radar and radio locators have tremendously helped the intelligence systems of the defense. But it is one thing to know what the enemy is doing, one thing to predict an air raid, and another thing to do anything about it, another thing to stop it.

The offense, in other words, particularly in the air, still has the edge on the defense and will probably continue to have it until the means of destroying the enemy—antiaircraft guns, night fighters, escort ships, etc.—become relatively certain and as exact as the means of locating the enemy—radar.

That has not yet occurred and may never occur. If it does, some other new device of the offense will be built to nullify this defensive advantage.

It is still true today, despite radar and despite antiaircraft guns and defense fighters, that in nearly any air raid some planes will get through. Even in our lop-sided air victory over Guadalcanal recently, when the Japanese lost 94 planes to our 6, some of the enemy planes bombed and damaged their objectives.

Radar cannot prevent this. What it has done is to supply us with sufficient warning to alert the means of destruction—the defending planes, antiaircraft guns and the like—thus enabling us, not to prevent raids, but to reduce damage and to exact some toll from the enemy.

—*July 13, 1943*

Dec. 2, 1942—The Birth of the Atomic Age

By WILLIAM L. LAURENCE

Mr. Laurence, science writer for The New York Times, *served as special consultant to the War Department to study the development of the atomic bomb and to explain this new force to the public. He was present when the first atomic bomb was exploded at Alamogordo, New Mexico, and saw from a plane the explosion of the bomb dropped on Nagasaki, which brought about Japan's surrender. He also witnessed the explosion at Bikini.*

Dec. 2, 1946 will be the first official anniversary of the atomic age, commemorating that fateful December day when man lighted the first atomic fire on this planet, the first fire that did not have its origin in the sun. For it was on that day—Dec. 2, 1942—at the gloomy squash court underneath the west stands of Stagg Field on the University of Chicago campus, that man succeeded at last in operating an atomic furnace, the energy of which came from the vast cosmic reservoir supplying the sun and the stars with their radiant heat and light—the nucleus of the atoms of which the material universe is constituted.

To understand what took place four years ago, it is necessary to review briefly the events that led up to it. It was known that to produce an atomic bomb it would be necessary to start a chain reaction in a mass of atoms. As one atom split, it would have to set off the trigger that would split a second atom.

Early in 1939 came the epoch-making discovery that when uranium atoms were split by bombarding them with neutrons—so called because they are electrically neutral parts of atoms—each atom gave off at least one more neutron. This neutron, it was reasoned, could in turn split another atom of uranium, and the process could go on indefinitely under the proper conditions. In theory, all of the essentials for a chain reaction were fulfilled. If such a reaction were started, calculations showed, it would liberate vast amounts of energy—3 million times more energy than that given out by equal masses of coal, 20 million times more explosive force than equal masses of TNT. In 1939 this meant the threat of the most destructive weapon the world had ever dreamed of.

Soon came another important discovery. It was established that the uranium that undergoes fission was not the ordinary abundant type of the element, of atomic weight 238, but the much rarer type, having the atomic weight 235. (In nature uranium always contains one part of U-235 to 140 parts of U-238.)

Why, it may be asked, didn't ordinary uranium explode?

First, because U-238 simply absorbs a good portion of the neutrons liberated in the process of splitting U-235.

Second, because a number of impurities present in a natural mixture of uranium also have a strong appetite for neutrons.

Third, because many of the neutrons escape from the surface of the mass of uranium like steam bubbling off hot water.

It was therefore recognized that before a chain reaction could be achieved, it would be necessary to do three things: First of all, the U-235 would have to be separated from the U-238; second, all the other impurities would have to be removed; third, it would be necessary to get a "critical mass" of the concentrated U-235, that is, a size large enough to retain most of the neutrons within the system. This critical mass, the minimum necessary to start an atomic explosion, was soon determined on theoretical grounds to be somewhere between 1 and 100 kilograms.

How could these three objectives be achieved? U-238 and U-235 could not be separated by chemical means because they are different forms (isotopes) of the same element, and the physical means then available were so slow that it would have taken 1,000 separation devices 1,000 years, to produce about 30 grams. Obviously, another approach had to be found—and quickly. The Germans were known to be at work on an atomic bomb and there was then good reason to fear that they would develop it first.

The solution suggested itself independently and almost simultaneously to Prof. Enrico Fermi and Dr. Leo Szilard, both of whom were then working at Columbia University. It had by that time been established that U-238 would absorb neutrons only if they were moving rather fast. The slow-speed, low-energy neutrons would not penetrate the nucleus. But it was exactly those slow neutrons that split U-235. What was needed was some method of slowing down neutrons.

Elements that slow down neutrons are known as "moderators." One of the most efficient of these is "heavy water," in which the hydrogen is double the weight of ordinary hydrogen. But the only plant in the world producing "heavy water" on a large scale was situated in Norway and was being used by the Nazis in their atomic bomb project. (This plant, by the way, was destroyed in 1943 by members of the Norwegian underground in one of the great epics of the war, as a result of which the Nazis were left far behind in the race for atomic weapons.)

Another "moderator" is graphite, the soft carbon used in pencils. Drs. Fermi and Szilard decided that graphite, because of its availability, would be the most suitable moderator for their purpose. It was their idea to build a huge spherical lattice of graphite bricks in which small lumps of the natural uranium mixture would be embedded at regular intervals. They named the structure a "pile," a name that has stuck to the gigantic descendants of the original, at Oak Ridge, Tennessee, and at Hanford, Washington.

If such a structure were large enough, they reasoned, most of the neutrons, born through the fission of U-235, would remain inside the "pile," to produce further fissions and thus maintain a chain reaction. The neutrons would move slowly because the graphite moderator would slow them down.

Two formidable obstacles stood in their way. To build such a pile it was first of all necessary to get large quantities of metallic uranium in a form purer than any uranium then in existence; otherwise, the impurities would absorb neutrons and stop the reaction. Only very small quantities of metallic uranium were being produced at that time, and these were of doubtful purity. Second, the graphite bricks had to be of a similar degree of purity, if they were to be free of neutron-absorbing substances. No graphite of such purity was then in existence.

Early in 1942, the researches on the uranium-graphite pile which had been carried out at Columbia University were shifted to the University of Chicago under the direction of Prof. Arthur H. Compton. Having learned that the Westinghouse Company had been producing small amounts of metallic uranium, Dr. Compton telephoned Dr. Harvey C. Rentschler, Westinghouse research director.

"How soon can Westinghouse supply three tons of metallic uranium?" Dr. Compton casually asked.

Dr. Rentschler was aghast. The total output of pure uranium metal up to that time had been a few grams. On being informed that uranium was necessary for a vital secret war project, he went to work. By November 1942, the three tons were delivered.

Meantime, other companies entered the picture, and new and simpler processes for purifying uranium ore and graphite were developed. By Nov. 7, 1942, a total of 12,400 pounds of pure uranium metal had been collected at the west stands of the Chicago squash court, and many more tons of highly purified uranium oxide, as well as tons of the purest graphite ever produced. The stage was set.

Actual work on the first self-sustaining chain-reaction pile began on Nov. 7, 1942. Many preliminary experiments had enabled the pile-builders to figure out the proper shape and dimensions that would yield the most efficient results. The structure was to be a sphere. The graphite was cut in square bricks and built up in layers. At the corners of the graphite bricks in each alternate layer were placed the uranium lumps, those of the pure metal being placed in the center of the pile. A timber framework resting on the squash court floor supported the structure, which was planned to consist of sixteen layers.

Success or failure hinged on the "multiplication factor" of the neutrons. If, for example, each neutron that split an atom of U-235 liberated another neutron which in turn would split another atom, the fission process would maintain itself. In that case the multiplication factor, designated by the scientists as K, would be equal to one. In other words, the effective birth rate would be equal to the death rate and the chain-reaction would go on.

If, on the other hand, only 99 fission-producing neutrons were born for every 100 that caused fission, then the multiplication factor would be less than one and the reaction would die out. If the multiplication factor was slightly greater than one, so much the better, for this would give an added safety factor. The scientists referred to such a multiplication factor as the "Great God K."

As the pile grew, precautions had to be taken to prevent a runaway chain reaction, which would have caused untold catastrophe. In any mass of uranium, fission occurs spontaneously as stray neutrons smash into the parts of U-235. Therefore, an elaborate system of controls was devised. These consisted of a series of boron steel rods and strips of cadmium inserted through slots in the pile. Boron and cadmium have an enormous capacity for devouring neutrons, so that the number of neutrons in the pile could be controlled by either pushing or pulling the boron rods and cadmium strips in and out of the pile. From the very beginning the cadmium strips and boron rods were placed in "retard" position to make sure that "Great God K" did not make a surprise appearance.

By the night of Dec. 1, 1942, eleven layers of graphite-uranium bricks had been piled up. Late that evening there were signs that the goal was near. But Dr.

Fermi, with true scientific imperturbability, decided to call it a day. Early the next morning the atomic "bricklayers" were back on the job. It was one of the coldest days of the winter. The squash court was badly heated. But the bricklayers worked on, oblivious of the gloom and cold.

Dr. Walter H. Zinn, then on leave from the College of the City of New York, was master of ceremonies on that Dec. 2. Present were Drs. Fermi, Szilard and Compton; Drs. Samuel K. Allison, Herbert L. Anderson, George Weil, Eugene Wigner, Norman Hilberry, Volney C. Wilson and John Marshall. There was one young woman in the group, Leona Woods, who later became Mrs. John Marshall.

Present also was Dr. Crawford H. Greenewalt, a member of the board of directors of the du Pont Company, which later built and operated the giant piles at Oak Ridge, Tennessee, and at Hanford, Washington. It was largely on the basis of what Dr. Greenewalt saw later that afternoon that the du Pont Company agreed to undertake to build these plants.

As the 12th layer was completed, everyone present became aware that one of the great moments in history was near. As the cadmium strips were cautiously pulled out, the instruments registering the multiplication rate of the neutrons began clicking louder and louder. These clicks were the heralds of the atomic age.

As the clicks grew more frequent, extra precautions were decided on. Two of the young physicists in the group, Dr. Alvin C. Graves of the University of Texas and Harold V. Lichtenberger of Millikan College, Decatur, Illinois, were selected to serve in what their colleagues called the "suicide brigade." They stood silently on a high platform overlooking the pile, each holding a bucket filled with a solution of cadmium, ready to pour it should the "Great God K" show any sign of becoming rambunctious. For two hours they stood thus tensely waiting for a signal that never came, hoping all the while that human nerves and muscles would be equal to the task.

Slowly all the cadmium strips but one were pulled out. Then, as the "suicide brigade" stood on the alert, the last one was pulled out to the proper calculated distance.

The scientists had previously figured out that if the number of neutrons per second reached a count of more than 1,600 it would mean a multiplication factor greater than one. Tensely and silently they stood around the neutron counters. Click, click, click. Twelve hundred, 1,400, 1,600. Sixteen hundred and one. The atomic "baby" had emitted its first lusty cry.

Dr. Greenewalt rushed back to a conference room in Eckhert Hall, where his colleagues had been debating since morning as to whether the du Pont Company should go into the building and operation of the fantastic plants suggested by the Manhattan District.

"Gentlemen," Dr. Greenewalt said, his eyes popping, "there is no need for further discussion."

At the same time, Dr. Compton held a short long-distance telephone conversation with President James Bryant Conant of Harvard.

"The Italian navigator has arrived in the New World and found the continent much smaller than he thought it was," said Dr. Compton.

"I hope the natives received him kindly," replied Dr. Conant.

In its final appearance that day the "pile," actually a manmade model of a living star, was a three-quarter-complete sphere, flat at the top, a shape geometers know as an oblate spheroid, resembling a giant doorknob. A 13th layer was added for luck, and the scientists called it a day.

The atomic power output that day was at the rate of only one-half watt, which corresponds to the splitting of one-half of a millionth of a gram of U-235 per hour. The force of fission had been so well controlled that there was no danger at all of fire or explosion. Later the rate was increased to 200 watts. There was still a long way to go to the giant plants with an output of millions of kilowatts, at the same time transmuting the useless U-238 into huge quantities of fissionable plutonium, the manmade element that destroyed Nagasaki. Furthermore, what had been demonstrated was a controlled chain reaction with slow neutrons, whereas for use as an explosive it is necessary to produce an uncontrolled chain reaction with fast neutrons.

But while there was still much work ahead, the road was clear. It was a straight line from then on to New Mexico, Hiroshima, Nagasaki and victory. It marked the end of an era and the beginning of a new one, with incalculable potentialities for good and for evil.

—*December 1, 1946*

Drama of the Atomic Bomb
Found Climax in July 16 Test

By WILLIAM L. LAURENCE

The Atomic Age began at exactly 5:30 Mountain War Time on the morning of July 16, 1945, on a stretch of semi-desert land about 50 airline miles from Alamagordo, New Mexico, just a few minutes before the dawn of a new day on this Earth.

At that great moment in history, ranking with the moment in the long ago when man first put fire to work for him and started on his march to civilization, the vast energy locked within the hearts of the atoms of matter was released for the first time in a burst of flame such as had never before been seen on this planet, illuminating Earth and sky for a brief span that seemed eternal with the light of many super-suns.

The elemental flame, first fire ever made on Earth that did not have its origin in the sun, came from the explosion of the first atomic bomb. It was a full-dress rehearsal preparatory to use of the bomb over Hiroshima and Nagasaki—and other Japanese military targets had Japan refused to accept the Potsdam Declaration for her surrender.

The rehearsal marked the climax in the penultimate act of one of the greatest dramas in our history and the history of civilized man—a drama in which our scientists, with the Army Corps of Engineers as director, were working against time to create an atomic bomb ahead of our German enemy.

The collapse of Germany marked the end of the first act of this drama. The successful completion of our task, in the greatest challenge by man against nature so far, brought down the curtain on the second act.

The grand finale came three weeks afterward over the skies of Japan with a swift descent of the curtain on the greatest war in history.

The atomic flash in New Mexico came as a great affirmation of the prodigious labors of our scientists during the past four years, in which they managed to "know the unknowable and unscrew the inscrutable."

It came as the affirmative answer to the until-then-unanswered question: "Will it work?"

With the flash came a delayed roll of mighty thunder, heard, just as the flash was seen, for hundreds of miles. The roar echoed and reverberated from the distant hills and the Sierra Oscuro Range nearby, sounding as though it came from some supramundane source as well as from the bowels of the Earth.

The hills said "yes" and the mountains chimed in "yes." It was as if the Earth had spoken and the suddenly iridescent clouds and sky had joined in one mighty affirmative answer. Atomic energy—yes.

It was like the grand finale of a mighty symphony of the elements, fascinating and terrifying, uplifting and crushing, ominous, devastating, full of great promise and great forebodings.

I watched the birth of the Era of Atomic Power from the slope of a hill in the desert land of New Mexico, on the northwestern corner of the Alamogordo Air Base, about 125 miles southwest of Albuquerque. The hill, named Compania Hill for the occasion, was 20 miles to the northwest of Zero, the code name given to the spot chosen for lighting the first atomic fire on this planet. The area embracing Zero and Compania Hill, 24 miles long and 18 miles wide, had the code name Trinity.

Caravan of Scientists by Night

I joined a caravan of three buses, three automobiles and a truck carrying radio equipment at 11 p.m. Sunday, July 15, at Albuquerque. There were about 90 of us in that strange caravan, traveling silently and in utmost secrecy through the night on probably as unusual an adventure as any in our day.

With the exception of your correspondent, the caravan consisted of scientists from the highly secret atomic bomb research and development center in the mesas and canyons of New Mexico, 25 miles northwest of Santa Fe, where we solved the secret of translating the fabulous energy of the atom into the mightiest weapon ever made by man. It was from there that the caravan set out at 5:30 that Sunday afternoon for its destination, 212 miles to the south.

These were the "mesa-men" on the march, dwellers in the "caves" in the interior of atoms, pioneer explorers of vast new continents in hitherto forbidden realms of the cosmos, builders of the civilization of tomorrow.

Here on trails hallowed by pioneers of other days, who opened new frontiers and did not rest until they conquered a continent, "covered wagons" were rolling again through the night on their way to open still newer frontiers of a continent that has no limits in space.

The caravan wound its way slowly over the tortuous roads overlooking the precipitous canyons of northern New Mexico, passing through Espanola, Santa

Fe and Bernalillo, arriving at Albuquerque at about 10 p.m. Here it was joined by Sir James Chadwick, who won the Nobel Prize and knighthood for his discovery of the neutron, the key that unlocks the atom; Professor Ernest O. Lawrence of the University of California, master atom-smasher, who won the Nobel Prize for his discovery of the cyclotron; Professor Edwin M. McMillan, also of the University of California, one of the discoverers of plutonium, the new atomic energy element, and several others from the atomic bomb center, who, with your correspondent, had arrived during the afternoon.

The night was dark with black clouds and not a star could be seen. Occasionally, a bolt of lightning would rend the sky and reveal for an instant the flat semi-desert landscape, rich with historic lore of past adventure. We, too, were headed for adventure, Argonauts on the way to a Golden Fleece richer by far than Jason ever found. We were on the road to the baffled golden Seven Cities of Cibola, sought in vain by Francisco Coronado on the trails not too far away from the area we were traversing.

We rolled along on U.S. Highway 85, running between Albuquerque and El Paso, through sleeping ancient Spanish-American towns, their windows dark, their streets deserted—towns with music in their names, Las Lunas, Belen, Bernardo, Alamillo, Socorro, San Antonio.

At San Antonio we turned east and crossed "the bridge on the Rio Grande with the detour in the middle of it." We traveled $10\frac{1}{2}$ miles eastward on U.S. Highway 380, where we turned south on a specially built dirt road, running for 25 miles to the Base Camp at Trinity.

The end of our trail was reached after we had covered about $5\frac{1}{5}$ miles on the dirt road. Here we saw the first signs of life since we had left Albuquerque about three hours earlier, a line of silent men dressed in helmets. A little farther ahead a detachment of military police examined our special credentials.

We descended and looked around us. The night was still pitch-black save for an occasional flash of lightning in the eastern sky, outlining for a brief instant the range of Sierra Oscuro directly ahead of us. We were in the middle of the New Mexico desert, miles away from nowhere, not a sign of life, not even a blinking light on the distant horizon. This was to be our caravansary until the zero hour.

From a distance to the southeast the beam of a searchlight probed the clouds. This gave us our first sense of orientation. The bombing test site, Zero, was a little to the left of the searchlight beam, 20 miles away. With the darkness and the waiting in the chill of the desert, the tension became almost unbearable.

Directions for Observers' Safety

We gathered around in a circle to listen to directions on what we were to do at the time of the "shot," directions read aloud by the light of a flashlight:

At a short signal of the siren at minus five minutes to Zero, "all personnel whose duties did not specifically require otherwise" were to prepare "a suitable place to lie down on."

At a long signal of the siren at minus two minutes to Zero "all personnel whose duties did not specifically require otherwise" were to "lie prone on the ground immediately, the face and eyes directed toward the ground and with the head away from Zero."

"Do not watch for the flash directly," the directions read, "but turn over after it has occurred and watch the cloud. Stay on the ground until the blast wave has passed [two minutes].

"At two short blasts of the siren, indicating the passing of all hazard from light and blast, all personnel will prepare to leave as soon as possible.

"The hazard from blast is reduced by lying down on the ground in such a manner that flying rocks, glass and other objects do not intervene between the source of blast and the individual. Open all car windows.

"The hazard from light injury to eyes is reduced by shielding the closed eyes with bended arms and lying facedown on the ground. If the first flash is viewed, a 'blind spot' may prevent you from seeing the rest of the show.

"The hazard from ultraviolet light injuries to the skin is best overcome by wearing long trousers and shirts with long sleeves."

David Dow, assistant to the scientific director of the Atomic Bomb Development Center, handed each of us a flat piece of colored glass used by arc welders to shield their eyes. Dr. Edward Teller of George Washington University cautioned us against sunburn.

Someone produced sunburn lotion and passed it around.

It looked eerie seeing a number of our highest-ranking scientists seriously rubbing sunburn lotion on their faces and hands in the pitch blackness of the night, 20 miles away from the expected flash. These were the men who, more than anybody, knew the potentialities of atomic energy on the loose. It gave one an inkling of their confidence in their handiwork.

The bomb was set on a structural steel tower 100 feet high. Nine miles away to the southwest was the Base Camp. This was G.H.Z. for the scientific high command, of which Professor Kenneth T. Bainbridge of Harvard University was field commander.

Here were erected barracks to serve as living quarters for the scientists, a mess hall, a commissary, a Post Exchange and other buildings. Here the vanguard of the atomists, headed by Prof. J. Robert Oppenheimer of the University of California, scientific director of the atomic bomb project, lived like soldiers at the front, supervising the enormously complicated details involved in the epoch-making tests.

Here early that Sunday afternoon gathered Maj. Gen. Leslie R. Groves, commander-in-chief of the atomic bomb project; Brig. Gen. T. F. Farrell, hero of World War I, General Groves's deputy; Prof. Enrico Fermi, Nobel Prize winner and one of the leaders in the project; President James Bryant Conant of Harvard; Dr. Vannevar Bush, director of the Office of Scientific Research and Development; Dean Richard C. Tolman of the California Institute of Technology; Prof. R. F. Bacher of Cornell; Col. Stafford L. Warren, University of Rochester (New York) radiologist; and a host of other leaders in the atomic bomb program.

At the Base Camp was a dry, abandoned reservoir, about 500 feet square, surrounded by a mound of earth about eight feet high. Within this mound bulldozers dug a series of slit trenches, each about three feet deep, seven feet wide and about 25 feet long.

At a command over the radio at zero minus one minute, all observers at Base Camp, about 150 of the "Who's Who" in science and the armed forces, lay down "prone on the ground" in their preassigned trenches, "face and eyes directed toward the ground and with the head away from Zero."

Three other posts had been established, south, north and west of Zero, each at a distance of 10,000 yards (5.7 miles). These were known respectively, as South-10,000, North-10,000 and West-10,000, or S-10, N-10 and W-10.

Here the shelters were much more elaborate, wooden structures, their walls reinforced by cement, buried under a massive layer of earth.

S-10 was the control center. Here Professor Oppenheimer, as scientific commander-in-chief, and his field commander, Professor Bainbridge, issued orders and synchronized the activities of the other sites.

Here the signal was given and a complex of mechanisms was set in motion that resulted in the greatest burst of energy ever released by man on Earth up until that time.

No switch was pulled, no button pressed, to light this first cosmic fire on this planet.

At 45 seconds to zero, set for 5:30, young Dr. Joseph L. McKibben of the University of California, at a signal from Professor Bainbridge, activated a master robot that set off a series of other robots. Moving "electronic fingers" writ and moved on, until at last strategically spaced electrons moved to the proper place at the proper split second.

The 45 seconds passed and the moment was zero.

At our observation post on Compania Hill the atmosphere had grown tenser as the zero hour approached. We had spent the first part of our stay partaking of an early-morning picnic breakfast that we had taken along with us. It had grown cold in the desert and many of us, lightly clad, shivered. Occasionally, a drizzle came down, and the intermittent flashes of lightning made us turn apprehensive glances toward Zero.

We had had some disturbing reports that the test might be called off because of the weather. The radio we had brought along for communication with Base Camp kept going out of order, and when we had finally repaired it, some loud band would drown out the news we wanted to hear.

We knew there were two specially equipped B-29 Superfortresses high overhead to make observations and recordings in the upper atmosphere, but we could neither see nor hear them. We kept gazing through the blackness.

Suddenly, at 5:29:50, as we stood huddled around our radio, we heard a voice ringing through the darkness, sounding as though it had come from above the clouds:

"Zero minus 10 seconds!"

A green flare flashed out through the clouds, descended slowly, opened, grew dim and vanished into the darkness.

The voice from the clouds boomed out again:

"Zero minus three seconds!"

Another green flare came down. Silence reigned over the desert. We kept moving in small groups in the direction of Zero. From the east came the first faint signs of dawn.

And just at that instant there rose from the bowels of the Earth a light not of this world, the light of many suns in one.

It was sunrise such as the world had never seen, a great green supersun climbing in a fraction of a second to a height of more than 8,000 feet, rising ever higher until it touched the clouds, lighting up Earth and sky all around with a dazzling luminosity.

Up it went, a great ball of fire about a mile in diameter, changing colors as it kept shooting upward, from deep purple to orange, expanding, growing bigger, rising as it was expanding, an elemental force freed from its bonds after being chained for billions of years.

For a fleeting instant the color was unearthly green, such as one sees only in the corona of the sun during a total eclipse.

It was as though the Earth had opened and the skies had split. One felt as though he had been privileged to witness the Birth of the World—to be present at the moment of Creation when the Lord said: Let there be light.

On that moment hung eternity. Time stood still. Space contracted into a pinpoint.

To another observer, Prof. George B. Kistiakowsky of Harvard, the spectacle was "the nearest thing to Doomsday that one could possibly imagine."

"I am sure," he said, "that at the end of the world—in the last millisecond of the Earth's existence—the last man will see what we saw!"

A great cloud rose from the ground and followed the trail of the Great Sun.

At first it was a giant column that soon took the shape of a supramundane mushroom. For a fleeting instant it took the form of the Statue of Liberty magnified many times.

Up it went, higher, higher, a giant mountain born in a few seconds instead of millions of years, quivering convulsively.

It touched the multicolored clouds, pushed its summit through them, kept rising until it reached a height of 41,000 feet—12,000 feet higher than the Earth's highest mountain.

All through this very short but extremely long time interval not a sound was heard. I could see the silhouettes of human forms motionless in little groups, like desert plants in the dark.

The newborn mountain in the distance, a giant among pigmies against the background of the Sierra Oscuro range, stood leaning at an angle against the clouds, a vibrant volcano spouting fire to the sky.

Roar Reverberations over Desert

Then out of the great silence came a mighty thunder. For a brief interval the phenomena we had seen as light repeated themselves in terms of sound.

It was the blast from thousands of blockbusters going off simultaneously at one spot.

The thunder reverberated all through the desert, bounced back and forth from the Sierra Oscuros, echo upon echo. The ground trembled under our feet as in an earthquake.

A wave of hot wind was felt by many of us just before the blast and warned us of its coming.

The Big Boom came about 100 seconds after the Great Flash—the first cry of a newborn world. It brought the silent, motionless silhouettes to life, gave them a voice.

A loud cry filled the air. The little groups that hitherto had stood rooted to the Earth like desert plants broke into a dance, the rhythm of primitive man dancing at one of his fire festivals at the coming of spring.

They clapped their hands as they leaped from the ground—earthbound man symbolizing a new birth in freedom—the birth of a new force that for the first time gives man means to free himself from the gravitational pull of the Earth that holds him down.

The dance of the primitive man lasted but a few seconds, during which an evolutionary period of about 10,000 years had been telescoped. Primitive man was metamorphosed into modern man—shaking hands, slapping each other on the back, laughing like happy children.

The sun was just rising above the horizon as our caravan started on its way back to Albuquerque and Los Alamos. It rose to see a new thing under the sun, a new era in the life of man.

We looked at it through our dark lenses to compare it with what we had seen.

"The sun can't hold a candle to it!" one of us remarked.

—September 26, 1945

Atomic Bombing of Nagasaki
Told by Flight Member

By WILLIAM L. LAURENCE

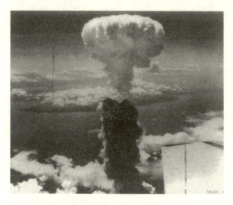

We are on our way to bomb the mainland of Japan. Our flying contingent consists of three specially designed B-29 "Superfortresses," and two of these carry no bombs. But our lead plane is on its way with another atomic bomb, the second in three days, concentrating in its active substance an explosive energy equivalent to 20,000 and, under favorable conditions, 40,000 tons of TNT.

We have several chosen targets. One of these is the great industrial and shipping center of Nagasaki, on the western shore of Kyushu, one of the main islands of the Japanese homeland.

I watched the assembly of this manmade meteor during the past two days, and was among the small group of scientists and Army and Navy representatives privileged to be present at the ritual of its loading in the "Superfortress" last night, against a background of threatening black skies torn open at intervals by great lightning flashes.

It is a thing of beauty to behold, this "gadget." Into its design went millions of man-hours of what is without doubt the most concentrated intellectual effort in history. Never before had so much brainpower been focused on a single problem.

This atomic bomb is different from the bomb used three days ago with such devastating results on Hiroshima.

I saw the atomic substance before it was placed inside the bomb. By itself it is not at all dangerous to handle. It is only under certain conditions, produced in the bomb assembly, that it can be made to yield up its energy, and even then it gives only a small fraction of its total contents—a fraction, however, large enough to produce the greatest explosion on Earth.

The briefing at midnight revealed the extreme care and the tremendous amount of preparation that had been made to take care of every detail of the mission, to make certain that the atomic bomb fully served the purpose for which it was intended. Each target in turn was shown in detailed map and in aerial photographs. Every detail of the course was rehearsed—navigation, altitude, weather, where to land in emergencies. It came out that the Navy had submarines and rescue craft, known as Dumbos and Superdumbos, stationed at various strategic points in the vicinity of the targets, ready to rescue the fliers in case they were forced to bail out.

The briefing period ended with a moving prayer by the chaplain. We then proceeded to the mess hall for the traditional early morning breakfast before departure on a bombing mission.

A convoy of trucks took us to the supply building for the special equipment carried on combat missions. This included the "Mae West," a parachute, a lifeboat, an oxygen mask, a flak suit and a survival vest. We still had a few hours before takeoff time, but we all went to the flying field and stood around in little groups or sat in jeeps talking rather casually about our mission to the Empire, as the Japanese home islands are known hereabouts.

In command of our mission is Maj. Charles W. Sweeney, 25, of 124 Hamilton Avenue, North Quincy, Massachusetts. His flagship, carrying the atomic bomb, is named *The Great Artiste*, but the name does not appear on the body of the great silver ship, with its unusually long, four-bladed, orange-tipped propellers. Instead it carried the number 77, and someone remarks that it was "Red" Grange's winning number on the gridiron.

Bombardier an 8th A. F. Veteran

Major Sweeney's copilot is First Lieut. Charles D. Albury, 24, of 252 Northwest Fourth Street, Miami, Florida. The bombardier, upon whose shoulders rests the responsibility of depositing the atomic bomb square on its target, is Capt. Kermit K. Beahan of 1004 Telephone Road, Houston, Texas, who is celebrating his 27th birthday on Aug. 9, 1945, the day of the bombing.

Captain Beahan has the awards of the Distinguished Flying Cross, the Air Medal and one Silver Oak Leaf Cluster, the Purple Heart, the Western Hemisphere Ribbon, the European Theatre Ribbon and two battle stars. He participated in the first Eighth Air Force heavy bombardment mission against the Germans from England on Aug. 17, 1942, and was on the plane that transported Gen. Dwight D. Eisenhower from Gibraltar to Oran at the beginning of the North African invasion. He has had a number of hair-raising escapes in combat.

The navigator on *The Great Artiste* is Capt. James F. Van Pelt Jr., 27, of Oak Hill, W. Va. The flight engineer is M/Sgt. John D. Kuharek, 32, of 1054 Twenty-second Avenue, Columbus, Nebraska; S/Sgt. Albert T. De Hart of Plainview, Texas, who celebrated his 30th birthday Aug. 8, 1945, is the tail gunner; the radar operator is S/Sgt. Edward K. Buckley, 32, of 529 East Washington Street, Lisbon, Ohio. The radio operator is Sgt. Abe M. Spitzer, 33, of 6551 Pelham Parkway, North Bronx, New York; Sgt. Raymond Gallagher, 23, of 572 South Mozart Street, Chicago, is assistant flight engineer.

The lead ship is also carrying a group of scientific personnel, headed by Comdr. Frederick L. Ashworth, USN, one of the leaders in the development of the bomb. The group includes Lieut. Jacob Beser, 24, of Baltimore, Maryland, an expert on airborne radar.

The other two Superfortresses in our formation are instrument planes, carrying special apparatus to measure the power of the bomb at the time of explosion, high-speed cameras and other photographic equipment.

Our "Superfortress" is the second in line. Its commander is Capt. Frederick C. Bock, 27, of 300 West Washington Street, Greenville, Michigan. Its other officers are Second Lieut. Hugh C. Ferguson, 21, of 247 Windermere Avenue, Highland Park, Michigan, pilot; Second Lieut. Leonard A. Godfrey, 24, of 72 Lincoln Street, Greenfield, Massachusetts, navigator; and First Lieut. Charles Levy, 26, of 1954 Spencer Street, Philadelphia, bombardier.

The enlisted personnel of this "Superfort" are: T/Sgt. Roderick F. Arnold, 28, of 130 South Street, Rochester, Michigan, flight engineer; Sgt. Ralph D. Curry, 20, of 1101 South Second Avenue, Hoopeston, Illinois, radio operator; Sgt. William C. Barney, 22, of Columbia City, Indiana, radar operator; Corp. Robert J. Stock, 21, of 415 Downing Street, Fort Wayne, Indiana, assistant flight engineer; and Corp. Ralph D. Belanger, 19, of Thendara, New York, tail gunner.

The scientific personnel of our "Superfortress" includes S/Sgt. Walter Goodman, 22, of 1956 Seventy-fourth Street, Brooklyn, New York, and Lawrence Johnson, graduate student at the University of California, whose home is at Hollywood, California.

The third "Superfortress" is commanded by Maj. James Hopkins, 1311 North Queen Street, Palestine, Texas. His officers are Second Lieut. John E. Cantlon, 516 North Takima Street, Tacoma, Washington, pilot; Second Lieut. Stanley C.

Steinke, 604 West Chestnut Street, West Chester, Pennsylvania., navigator; and Second Lieut. Myron Faryna, 16 Elgin Street, Rochester, New York, bombardier.

The crew are Tech. Sgt. George L. Brabenec, 9717 South Lawndale Avenue, Evergreen, Illinois; Sgt. Francis X. Dolan, 30-60 Warren Street, Elmhurst, Queens, New York; Corp. Richard F. Cannon, 160 Carmel Road, Buffalo, New York; Corp. Martin G. Murray, 7356 Dexter Street, Detroit, Michigan, and Corp. Sidney J. Bellamy, 529 Johnston Avenue, Trenton, New Jersey.

On this "Superfortress" are also two distinguished observers from Britain, whose scientists played an important role in the development of the atomic bomb. One of these is Group Capt. G. Leonard Cheshire, famous Royal Air Force pilot, who is now a member of the British military mission to the United States. The other is Dr. William G. Denny, professor of applied mathematics, London University, one of the group of eminent British scientists that has been working at the "Y-Site" near Santa Fe, New Mexico, on the enormous problems involved in taming the atom.

Group Captain Cheshire, whose rank is the equivalent to that of colonel in the United States Army Air Forces, was designated as an observer of the atomic bomb in action by Winston Churchill when he was still prime minister. He is now the official representative of Prime Minister Clement R. Attlee.

In Storm Soon After Takeoff

We took off at 3:50 this morning and headed northwest on a straight line for the Empire. The night was cloudy and threatening, with only a few stars here and there breaking through the overcast. The weather report had predicted storms ahead part of the way but clear sailing for the final and climactic stages of our odyssey.

We were about an hour away from our base when the storm broke. Our great ship took some heavy dips through the abysmal darkness around us, but it took these dips much more gracefully than a large commercial airliner, producing a sensation more in the nature of a glide than a "bump," like a great ocean liner riding the waves, except that in this case the airwaves were much higher and the rhythmic tempo of the glide much faster.

I noticed a strange eerie light coming through the window high above the navigator's cabin and as I peered through the dark all around us I saw a startling phenomenon. The whirling giant propellers had somehow became great luminous disks of blue flame. The same luminous blue flame appeared on the plexiglass windows in the nose of the ship, and on the tips of the giant wings it looked as though we were riding the whirlwind through space on a chariot of blue fire.

It was, I surmised, a surcharge of static electricity that had accumulated on the tips of the propellers and on the dielectric material in the plastic windows. One's thoughts dwelt anxiously on the precious cargo in the invisible ship ahead of us. Was there any likelihood of danger that this heavy electric tension in the atmosphere all about us might set it off?

I expressed my fears to Captain Bock, who seems nonchalant and imperturbed at the controls. He quickly reassures me:

"It is a familiar phenomenon seen often on ships. I have seen it many times on bombing missions. It is known as St. Elmo's Fire."

On we went through the night. We soon rode out the storm and our ship was once again sailing on a smooth course straight ahead, on a direct line to the Empire.

Our altimeter showed that we were traveling through space at a height of 17,000 feet. The thermometer registered an outside temperature of 33 degrees below zero Centigrade, about 30 below Fahrenheit. Inside our pressurized cabin, the temperature was that of a comfortable air-conditioned room, and a pressure corresponding to an altitude of 8,000 feet. Captain Bock cautioned me, however, to keep my oxygen mask handy in case of emergency. This, he explained, might mean either something going wrong with the pressure equipment inside the ship or a hole through the cabin by flak.

The first signs of dawn came shortly after 5 o'clock. Sergeant Curry, who had been listening steadily on his earphones for radio reports, while maintaining a strict radio silence himself, greeted it by rising to his feet and gazing out the window.

"It's good to see the day," he told me. "I get a feeling of claustrophobia hemmed in this cabin at night."

He is a typical American youth, looking even younger, than his 20 years. It takes no mind reader to read his thoughts.

"It's a long way from Hoopeston, Illinois," I find myself remarking.

"Yep," he replies, as he busies himself decoding a message from outer space.

"Think this atomic bomb will end the war?" he asks hopefully.

"There is a very good chance that this one may do the trick," I assure him, "but if not, then the next one or two surely will. Its power is such that no nation can stand up against it very long."

This was not my own view. I had heard it expressed all around a few hours earlier, before we took off. To anyone who had seen this manmade fireball in action, as I had less than a month ago in the desert of New Mexico, this view did not sound overoptimistic.

By 5:50 it was real light outside. We had lost our lead ship, but Lieutenant Godfrey, our navigator, informs me that we had arranged for that contingency. We have an assembly point in the sky above the little island of Yakushima, southeast of Kyushu, at 9:10. We are to circle there and wait for the rest of our formation.

Our genial bombardier, Lieutenant Levy, comes over to invite me to take his front-row seat in the transparent nose of the ship and I accept eagerly. From that vantage point in space, 17,000 feet above the Pacific, one gets a view of hundreds of miles on all sides, horizontally and vertically. At that height the vast ocean below and the sky above seem to merge into one great sphere.

I was on the inside of that firmament, riding above the giant mountains of white cumulous clouds, letting myself be suspended in infinite space. One hears the whirl of the motors behind one, but it soon becomes insignificant against the immensity all around and is before long swallowed by it. There comes a point where space also swallows time and one lives through eternal moments filled with an oppressive loneliness, as though all life had suddenly vanished from the Earth and you are the only one left, a lone survivor traveling endlessly through interplanetary space.

My mind soon returns to the mission I am on. Somewhere beyond these vast mountains of white clouds ahead of me lies Japan, the land of our enemy. In about four hours from now, one of its cities, making weapons of war for use against us, will be wiped off the map by the greatest weapon ever made by man. In one-tenth of a millionth of a second, a fraction of time immeasurable by any clock, a whirlwind from the skies will pulverize thousands of its buildings and tens of thousands of its inhabitants.

Our weather planes ahead of us are on their way to find out where the wind blows. Half an hour before target time we will know what the winds have decided.

Does one feel any pity or compassion for the poor devils about to die? Not when one thinks of Pearl Harbor and of the Death March on Bataan.

Captain Bock informs me that we are about to start our climb to bombing altitude. He manipulates a few knobs on his control panel to the right of him and I alternately watch the white clouds and ocean below me and the altimeter on the bombardier's panel. We reached our altitude at 9 o'clock. We were then over

Japanese waters, close to their mainland. Lieutenant Godfrey motioned to me to look through his radar scope. Before me was the outline of our meeting point. We shall soon meet our lead ship and proceed to the final stage of our journey.

We reached Yakushima at 9:12 and there, about 4,000 feet ahead of us, was *The Great Artiste* with its precious load. I saw Lieutenant Godfrey and Sergeant Curry strap on their parachutes and I decided to do likewise.

We started circling. We saw little towns on the coastline, heedless of our presence. We kept on circling, waiting for the third ship in our formation.

It was 9:56 when we began heading for the coastline. Our weather scouts had sent us coded messages, deciphered by Sergeant Curry, informing us that both the primary target as well as the secondary were clearly visible.

The winds of destiny seemed to favor certain Japanese cities that must remain nameless. We circled around them again and again and found no opening in the thick umbrella of clouds that covered them. Destiny chose Nagasaki as the ultimate target.

We had been circling for some time when we noticed black puffs of smoke coming through the white clouds directly at us. There were fifteen bursts of flak in rapid succession, all too low. Captain Bock changed his course. There soon followed eight more bursts of flak, right up to our altitude, but by this time were too far to the left.

We flew southward down the channel and at 11:33 crossed the coastline and headed straight for Nagasaki about 100 miles to the west. Here again we circled until we found an opening in the clouds. It was 12:01 and the goal of our mission had arrived.

We heard the prearranged signal on our radio, put on our arc-welder's glasses and watched tensely the maneuverings of the strike ship about half a mile in front of us.

"There she goes!" someone said. Out of the belly of *The Great Artiste* what looked like a black object went downward.

Captain Bock swung around to get out of range; but even though we were turning away in the opposite direction, and despite the fact that it was broad daylight in our cabin, all of us became aware of a giant flash that broke through the dark barrier of our arc-welder's lenses and flooded our cabin with intense light.

We removed our glasses after the first flash, but the light still lingered on, a bluish-green light that illuminated the entire sky all around. A tremendous blast wave struck our ship and made it tremble from nose to tail. This was followed by four more blasts in rapid succession, each resounding like the boom of cannon fire hitting our plane from all directions.

Observers in the tail of our ship saw a giant ball of fire rise as though from the bowels of the Earth, belching forth enormous white smoke rings. Next they saw a giant pillar of purple fire, 10,000 feet high, shooting skyward with enormous speed.

By the time our ship had made another turn in the direction of the atomic explosion the pillar of purple fire had reached the level of our altitude. Only about 45 seconds had passed. Awestruck, we watched it shoot upward like a meteor coming from the Earth instead of from outer space, becoming ever more alive as it climbed skyward through the white clouds. It was no longer smoke, or dust, or even a cloud of fire. It was a living thing, a new species of being, born right before our incredulous eyes.

At one stage of its evolution, covering millions of years in terms of seconds, the entity assumed the form of a giant square totem pole, with its base about three miles long, tapering off to about a mile at the top. Its bottom was brown, its center was amber, its top white. But it was a living totem pole, carved with many grotesque masks grimacing at the Earth.

Then, just when it appeared as though the thing has settled down into a state of permanence, there came shooting out of the top a giant mushroom that increased the height of the pillar to a total of 45,000 feet. The mushroom top was even more alive than the pillar, seething and boiling in a white fury of creamy foam, sizzling upward and then descending earthward, a thousand Old Faithful geysers rolled into one.

It kept struggling in an elemental fury, like a creature in the act of breaking the bonds that held it down. In a few seconds it had freed itself from its gigantic stem and floated upward with tremendous speed, its momentum carrying into the stratosphere to a height of about 60,000 feet.

But no sooner did this happen when another mushroom, smaller in size than the first one, began emerging out of the pillar. It was as though the decapitated monster was growing a new head.

As the first mushroom floated off into the blue, it changed its shape into a flowerlike form, its giant petal curving downward, creamy white outside, rose-colored inside. It still retained that shape when we last gazed at it from a distance of about 200 miles.

—September 9, 1945

Visit to Hiroshima Proves
Its World's Most-Damaged City

By WILLIAM H. LAWRENCE

The atomic bomb still is killing Japanese at a rate of 100 daily in flattened, rubble-strewn Hiroshima, where the secret weapon harnessing the power of the universe itself as a destructive agent was used for the first time on Aug. 6, 1945.

I was among the first few foreigners to reach the site of this historic bombing and walked for nearly two hours today through streets where the stench of death still pervades and survivors or relatives of the dead, wearing gauze patches over their mouths, still probe among the ruins for bodies or possessions.

This is the world's most damaged city, worse than Warsaw or Stalingrad, which held the record in Europe. Fully four square miles, constituting 60 percent of the city, are absolutely leveled and the houses and buildings in the rest of the city are irreparably damaged.

Japanese announced that the death toll had passed 53,000, an increase of 20,000 in the figure reported Aug. 20, and it was predicted the final count would exceed 80,000 dead.

On Aug. 20, the latest date for which Japanese official detailed statistics are available, the casualties were 33,000 dead, 30,000 missing, 13,960 seriously wounded and 43,500 listed as wounded "not so seriously."

This accounted for approximately one-third of Hiroshima's prewar population of 343,000, but in addition it was stated that most of the other persons in the city suffered minor wounds that were not considered serious enough for medical treatment in view of the great shortage of doctors to deal with this disaster.

Japanese doctors told us they were helpless to deal with burns caused by the bomb's great flash or with the other physical ailments caused by the bomb. Some said they thought that all who had been in Hiroshima that day would die as a result of the bomb's lingering effects.

They told us that persons who had been only slightly injured on the day of the blast lost 86 percent of their white blood corpuscles, developed temperatures of 104 degrees Fahrenheit, their hair began to drop out, they lost their appetites, vomited blood and finally died.

The bomb fell about 8:15 a.m. on a clear day, just after the "all clear" signal ending an air raid alert had been sounded, and many of Hiroshima's residents were in the streets when the sky above them was lighted by a brilliant flash that seared everything below it.

Most of the deaths and destruction occurred in a fraction of a second, although fires smoldered for more than a day in the ruins of wooden and stone houses crumpled inward. Bodies of men, women and children were thrown about the streets and the cries of the terrified wounded filled the air.

A witness of the bombing said, "Everything had been scorched to the ground, everything that still lived was waiting to die."

So terrible was the blast that every wounded person thought he had been hit by an individual bomb, and it was not until hours later that it was recognized that a new weapon of undreamed-of power had been utilized against them.

We were told that the bomb descended by parachute from three Superfortresses and had exploded about 150 feet above the ground.

While the Japanese still are staggered by the initial impact of this weapon and its lingering effects, their scientists still have not figured out whether the blast will have continuing harmful effects upon all who live in Hiroshima.

They already have banned the drinking of city water because of chemical changes caused by the radioactive rays of the bomb, and they asked us whether we thought residing in the bombed area would be harmful.

U.S. Scientists Plan Study

We told them that we did not know, but that a party of American scientists who had worked on development of the bomb soon would come to Hiroshima to study its effects and to make recommendations.

[American atomic bomb experts have arrived in Japan to check the effects of the bombs on Hiroshima and Nagasaki, an American Broadcasting Company correspondent reported Tuesday from Japan. He quoted one of the experts as saying that "effects of the bomb could well disintegrate the white corpuscles in the human bloodstream," but that Japanese reports that emanations from the bombed area damaged the health of visitors three weeks later might be exaggerated.]

Of Hiroshima's prewar population of 343,000, we were told that about 120,000 lived in the rubble of the city and its suburbs. Most of these have suffered injuries and it has been noted that those who spent more than a few hours in the bombed area daily suffer from severe headaches and general physical disorders. It is not believed that a stay of a few hours will affect a person.

As a war correspondent in Europe and the Pacific, I have never looked upon such scenes of death and destruction. It was enough to take your breath away when standing in the center of the area where the bomb fell. You could see nothing but rubble and the seared walls of a few earthquake-proof buildings that remained upright.

Steel was twisted and tile was burned into dust. The wood was charred and torn into small fragments. Air-raid shelters were crushed in.

The damage in Hiroshima is greater than that in Nagasaki, which I saw from a low level in an airplane, but there were indications that the Nagasaki bomb was in some ways more powerful. It appeared that most of the buildings of Nagasaki disintegrated, leaving no rubble to mark the damaged area. Only in a few places were piles of stone, steel, galvanized iron and wood, typical of bomb damage in any city.

Although our Japanese naval guides from Kure, who showed us through the city, told us without emotion that the residents of Hiroshima hated Americans, whom they regarded as the cruelest people in the world, this party of war correspondents and photographers attached to the United States Strategic Air Forces—the outfit that dropped the bomb—was able to walk at will through the streets without being molested.

A 22-year-old American-born Japanese naval lieutenant interpreter, with whom I walked, stopped persons on the streets for us to ask them "whether they had been in the city the day the bomb fell." A deaf old man, recognizing us as Americans, came up and shook hands "with each of us and made the sign of the Cross to tell us that he was a Christian." Through the interpreter he told us his family had perished in the raid.

Our interpreter, who asked us not to insist that he give us his name, said that he was born in Sacramento, California, and that his father still lived in the United States. He came to Japan with his mother about 10 years ago and lives in Kure, the battered Japanese naval base.

We walked past large granite buildings from which stone fragments still were dropping and peered inside a roofless stone structure that serves as the emergency headquarters for three banks, to which an imperial messenger had just delivered relief funds, for which residents were standing in line.

Surprisingly, the streetcars which were not burned out, still operate and Japanese riding on them looked out with more curiosity than hostility at the tall white men in Army uniforms, studying the devastation their country had caused.

Some Bodies Still in Ruins

It was a chilly, drizzly day, but hundreds were moving amid the rubble, from which most of the bodies had been removed and cremated, but a few still remain, giving off the awful, sickening odor of death.

Even the trees were killed by the bomb. Birds that looked like buzzards perched on the torn, leafless limbs.

Nobody was smiling. The patient, long-suffering Japanese, who believed they were winning the war up to the very day the emperor announced he had surrendered, moved slowly and quietly through the streets to carry out their personal business. There is no work for them here, except in cleaning up, and, as in all Japanese cities, there is little to eat.

In the rubble of destroyed stone and wooden houses we saw, occasionally, an unbroken bottle of sake that, somehow, had survived the blast, but nobody seemed to be drinking.

A visit to Hiroshima is an experience to leave one shaken by the terrible, incredible sights. Here is the final proof of what the mechanical and scientific genius of America has been able to accomplish in war through the invention of the airplane, especially the B-29 Superfortress and the atomic bomb. It should be the last evidence needed to convince any doubter of the need to retain and perfect our air offense and defense lest the fate of Hiroshima be repeated in Indianapolis or Washington or Detroit or New York.

Three Japanese newspapermen who interviewed us wanted to know the role of the atomic bomb in future warfare. We told them it was our purpose as one of the United Nations to make certain that peace is maintained throughout the world.

This has been the most unusual press trip this correspondent has ever participated in. We took off about 8 o'clock this morning from the Atsugi airfield near Tokyo in *The Headliner,* a Flying Fortress, piloted by Capt. Mark Magnan of Milwaukee, and flew through foul weather and dangerous mountain ranges toward Kure. Just over Kure there was an opening in the clouds and we were able to get down low enough to spot a short, 2,600-foot, naval fighter plane strip, into which Captain Magnan made an overwater approach, slamming on the brakes almost as soon as the wheels of the big bomber touched the ground. We stopped barely 50 feet short of stone obstacles that would have wrecked our plane at the edge of the field.

We were the first Americans in this battered Japanese naval base and the surprised airfield staff gaped at our big plane as it rolled to a stop. We explained our mission to them and they found two battered Ford automobiles that carried the reporters to the bombed-out granite headquarters of the naval base some miles away.

Japanese officers served tea to the correspondents, who sat beneath the covered driveway of the building entrance while Col. Chester W. Coltharp of the Far East Air Force and Lieut. Col. John R. McCrary of the Strategic Air Force went inside to explain our desires to the Japanese naval commander.

They were received by Vice Admiral Masao Kanazawa, five-feet-four inches tall, English-speaking former spokesman of the Japanese Naval Ministry, whose first words were: "It is all finished; it is good!"

Smiling and joking, the admiral spoke without apparent bitterness of the sad fate of the Japanese Navy, pointing outside his window to the anchorages where twisted, wrecked battleships, aircraft carriers and cruisers lay in ruins from the attacks of Navy and Army fliers. He said no big combat ship remained operational in Kure and that only about 10 destroyers were fit for sea duty.

Felt for 12 Miles Away

He said that although Kure is 12 miles from Hiroshima, he had felt the terrible explosion.

"It felt like a great wind, which made the trees of Kure sway back and forth," he said.

The admiral provided two Buicks and a Ford and assigned English-speaking officers, the interpreter from Sacramento and a Japanese naval surgeon, Dr. Taira, to accompany us into Hiroshima. We drove through ruins of Kure's urban district, 46 percent destroyed by a Superfortress incendiary raid on July 2, and moved swiftly along a paved highway, the sides of which were lined with boxes of ammunition, shells, crated airplane motors and other war supplies that had been moved out of the city for safety.

On a slippery bit of road, our car skidded and crashed into a crated motor and bounced off it into another, but none of the correspondents was injured.

On arriving in Hiroshima, the party of Americans split up to walk through the piles of debris and to talk with the city's residents. As we walked along, I chatted with the lieutenant from Sacramento and Dr. Taira. Neither was in

Hiroshima at the time the bomb exploded, but came to the city to aid the relief work a few hours after the blast.

Dr. Taira said searing burns were the principal cause of death and injury. He thought most of the people died within a fraction of a second. He compared the burns with those caused by the overuse of X-rays. He said he believed the report that the radioactivity created by the bomb would make sterile those who were not killed.

Discussing the destruction of white blood corpuscles in the human body, the doctor said he did not think that a stay of a few hours in the blasted city would affect newcomers.

After walking through the city for about two hours, we were taken to a modern undamaged building on the outskirts, where in the paneled former boardroom of the Eastern Oriental Manufacturing Company, motorcycle manufacturers, we were received by Hirokuni Dazai, who controlled the "Thought Police" in the Hiroshima prefecture. The "Thought Police" are similar to the Gestapo in Nazi Germany or the NKVD in the Soviet Union.

Dazai, who had returned to Hiroshima from Tokyo 40 minutes before the bomb exploded, provided the official casualty statistics and gave us an account of what happened to him.

He was wearing a white gauze bandage around his head. He said he had suffered a slight wound when his house collapsed upon him and his family. His wife was knocked unconscious, but his two children were only scratched.

He said he had not noticed the airplanes overhead and that the great flash that arced through the sky was his first knowledge that a bomb had fallen. He said he had believed hundreds of bombs had fallen when he felt the blast.

It was Dazai who sent the first report of the new bomb to Tokyo. His report undoubtedly played a major role in the emperor's decision four days later to advise the United Nations that Japan was willing to accept the Potsdam Declaration if he could keep his job.

Dazai said great fires kept relief parties out of the central part of the city for hours and interfered with land transport, including railways, so that it was almost impossible to move doctors in or take patients out.

We asked him his opinion of the use of this type of bomb. He replied that he believed we had in our possession the ability to destroy every living thing of the civilization established by the gods.

—September 5, 1945

Five Atomic Piles in Operation Here

The United States now has five atomic engines, or controlled nuclear chain-reaction piles in operation, some for research and some for production, Dr. Enrico Fermi, one of the scientists who harnessed the atom, has said.

Dr. Fermi, winner of the Nobel Prize and currently on the staff of the University of Chicago, spoke at a conference in the Museum of Science and Industry marking the fourth anniversary of man's first successful effort to control atomic energy. The historic event took place on a squash court under the stands at Stagg Field of the university, but it was not announced to the world until after the first atomic bomb was dropped on Hiroshima Aug. 6, 1945.

Two of the atomic piles are known to be at the Argonne National Laboratory in the Palos Hills forest preserve in Chicago. Professor Fermi did not disclose the location of the other three.

Another speaker at the conference was Maj. Gen. Leslie P. Groves, commanding general of the Manhattan Project, which directed development of the atomic bomb. He made a plea for adoption of the Baruch plan for international control of atomic energy, admitting that there were risks inherent in the plan, but declaring that the risks of not proceeding with it would be greater.

Hope Called Not Enough

"We all hope fervently that there will never be another war, but we cannot in these times of dark and brooding uncertainty base our future solely on a hope," General Groves said. "As we today commemorate the fourth anniversary of the most significant scientific event in our times, we face a future which may at times be obscured by the clouds of unbridled ambition, on the part of a few and of intolerance, distrust and ignorance. It is our duty to see that they are destroyed."

General Groves said peace could be attained only by international agreements fortified by penalties for violations. We must embrace international cooperation or international disintegration, he asserted.

Dr. John T. Tate, chairman of the council of representatives of 22 Midwestern universities which participate in the Argonne Laboratory, made a plea for more emphasis on research in fundamentals. He pointed out that while the controlled atomic chain reaction was first accomplished in the United States, the key scientific discoveries that made it possible were made in Europe.

Dr. Walter H. Zinn, director of the Argonne Laboratory, said no scientist wanted to have government control of science, but, nevertheless, the tremendous cost of atomic research made government support necessary. Materials for the first pile at the University of Chicago cost more than $1 million, he said.

Study of Peace Uses Urged

Intensive and advanced study of peacetime applications of atomic power, directed at ultimately freeing mankind from dependence on such diminishing resources as coal and oil, was called for today by Dean F. Ellis Johnson, engineer and educator, in a ceremony at the Hanford Engineer Works in Richland, Washington.

In dedicating a building for the training of nuclear scientists and engineers, Dean Johnson, former head of the School of Engineering at the University of Wisconsin, outlined a program for such work.

"It is of first importance to stimulate the best efforts of the younger scientists and technologists, and to attract ambitious and capable men into the field of nuclear engineering," Dean Johnson said.

The observance was part of a nationwide celebration of the fourth anniversary of the first large-scale release of atomic energy.

Threat to Coal Forecast

Dr. Arthur H. Compton, noted atomic scientist, declared on Dec. 2, 1945 that atomic power might become a serious competitor to the coal industry in the United States by about 1955, provided "political difficulties do not interfere."

His prediction was contained in a speech on the fourth anniversary of the loosing of nuclear fission in Chicago, made to a gathering of top-ranking world scientists at the Sorbonne in Paris.

Dr. Compton, chancellor of Washington University in St. Louis, also upheld the use of the atom bomb against Japan.

"Not to have used this newfound power to stop short the most disastrous war of history would have been unpardonable," he declared.

—December 3, 1946

Tiny Radios Made by Armed Services

By T. R. KENNEDY JR.

New radio-electronic equipment, in many cases saving as much as 75 percent in the size and weight of present-day Army and Navy apparatus, ranging from communications receivers to radars and servicing kits, was explained for the first time at the Navy exhibit at the convention of the Institute of Radio Engineers. The convention was held at the Commodore Hotel and Grand Central Palace in New York City.

The new "miniaturization" program is a joint project of the Army, Navy and civilian fields in electronics in which all radio devices are getting progressively smaller and smaller, it was explained. For instance, a complete Navy radio receiver formerly requiring 12,160 cubic inches of space has been compressed into a volume of 3,150 cubic inches without destroying its efficiency. Efficiencies in some cases have vastly improved, it was said.

This is being accomplished by new materials, new parts, vacuum tubes of the dwarf variety, and learning how to put them together more compactly.

A novel ten-tube, one-pound radio receiver, complete except for external battery supply, was shown by E. J. Nucci, electronics specialist of the Navy Bureau of Ships, Washington. It was sealed in a plastic case three inches by six inches by one inch thick, and was almost completely transparent. It was built to be discarded when trouble happens inside, he said.

Mr. Nucci demonstrated several instruments resembling oversized fountain pens, each one a compact and exact electrical meter designed to measure some function of an operating receiver. Each one, he said, was one of about 40 separate parts in a new technical radio-radar-electronics servicing kit now beginning to be carried on maintenance jobs by Army and Navy men, who call them their "radio lunch boxes." They weigh 25 pounds each. Eventually, Mr. Nucci said, the new kits would include a complete "oscilloscope," a device that aids radio men in aligning television receivers. The "scope," he went on, "may add two or three pounds."

Old Apparatus Is Heavy

Servicing apparatus now carried about by civilian radio technicians on home televising fixing jobs often weigh several hundred pounds and cost, if complete, about $500.

The whole new kit, with "scope, might eventually be less costly than the expensive and complicated, apparatus used today," Mr. Nucci predicted, and still be less in volume than a cubic foot.

A new scientific development, known as a semiconductor, which Bell Laboratories experts and others have developed to the point where they "look like a possible successor for many of the common uses of the ordinary vacuum tube," was described at one of the radio engineering technical sessions. The speakers were Dr. Walter H. Brattain and Dr. John Bardeen of the Laboratories, who said that recent tests have revealed the semiconductor useful as amplifiers over a range of frequencies up to "10 million cycles."

As made by the Laboratories, semiconductors are known as "transistors," and are about as large as the metal tips of shoelaces. In the transistor the small number of current-carrying electrons present can be controlled or varied 1,000-fold or more by changing the electronic structure of the material, Dr. Brattain said.

It is this or a similar property which makes a device amplify, he said.

—March 11, 1949

Ending of All Life by
Hydrogen Bomb Held a Possibility

By WILLIAM L. LAURENCE

The hydrogen bomb, if developed, could be rigged in such a way as to exterminate the entire world's population or most of it, four leading atomic scientists have warned.

This could be done simply by incorporating common substances in the hydrogen bomb, they declared. When detonated, the explosion would release tremendous quantities of neutrons, the most penetrating particles in nature. These, in turn, would enter into the nuclei of the incorporated element and make them intensely radioactive.

If an element such as cobalt were chosen, they pointed out, it would be transmuted into a radioactive element about 320 times as powerful in its radioactivity as radium. This deadly radioactive cobalt would be scattered into the atmosphere and carried by the westerly winds all over the surface of the Earth. Any living thing inhaling it, or even touched by it, would be doomed to certain death.

Other New Facts Brought Out

This hitherto unknown information about the potential horrors in store for humanity in the event of the development and use in warfare of the hydrogen bomb, as well as several other facts so far not fully realized by the public, were brought out at the University of Chicago Round Table Conference broadcast over the National Broadcasting Company's network. New York and vicinity will not hear this program until next week.

The participants in the discussion were Professors Hans A. Bethe of Cornell University, Frederick Seitz of the University of Illinois, and Leo Szilard and Dr. Harrison Brown of the University of Chicago. All four played major roles in the development of the wartime atomic bomb. Professor Szilard was a key figure in starting the government project.

The element best suited for the hydrogen bomb, they brought out, is deuterium, the double-weight isotope (twin) of simple hydrogen, which constitutes one part in 5,000 of all waters on Earth. The nuclei of deuterium, known as deuterons, are composed of one proton (positively charged elementary particle) and one neutron (neutral particle of about the same mass as the proton).

In a hydrogen bomb, two deuterons would be made to fuse by the intense temperatures developed by the detonation of an orthodox uranium-plutonium fission bomb, which would be used as the trigger for setting off the hydrogen bomb. In the fusing process, one out of every four deuterons would set free its neutron. This means that one-eighth of the total mass of heavy hydrogen exploded would be released as neutrons.

Neutrons Enter Air's Nitrogen

If allowed to go free into the air, these neutrons would enter the nitrogen in the atmosphere and transmute it into a radioactive form of carbon of atomic mass 14. The radioactive carbon, it was pointed out, loses half its radioactivity in 5,100 years. This is a rather weak type of radioactivity and would not be fatal unless a great many hydrogen bombs were exploded.

On the other hand, Professor Szilard and his colleagues pointed out, it would be possible to add to the contents of the bomb a substance that, when made radioactive by neutrons, has a very short lifetime, and is thus much more powerfully radioactive. A substance with a half-life of only five years, for example, such as cobalt, would be 1,000 times more deadly in its radioactivity than carbon 14.

"Assuming that we have a radioactive element that will last for five years," Professor Szilard said, "we just let it go into the air. During the following years it will gradually settle out and cover the whole Earth with dust. I have asked myself: 'How many neutrons, or how much heavy hydrogen, do we have to detonate to kill everybody on Earth by this particular method?' Well, I come out with about 50 tons of neutrons being needed to kill everybody, which means about 500 tons of heavy hydrogen.

"Who would want to kill everybody on Earth? Suppose we have a war, and suppose that we are at the point of winning the war against Russia, maybe after a struggle which lasts 10 years. Russia's rulers then can say: 'You come no further, you don't invade Europe, and you don't drop ordinary bombs outside or else we detonate our H-bombs and kill everybody.' Facing such a threat, I don't think we can go forward."

"Do you think," Dr. Brown asked, "that any nation would really be willing to kill all the people on Earth rather than suffer defeat themselves? Would we be willing to do it, for example?"

"I do not know whether we would be willing to do it," Professor Szilard replied, "and I don't know if Russia would be willing to do it, but I think that we might threaten to do it, and who will take the risk not to take that threat seriously?"

The discussion brought out that at our present state of knowledge it is not likely that a substance of short-lived radioactivity could be developed so that it would be possible for one nation to exterminate another and spare itself.

"We are faced with the possible ironical conclusion," Dr. Brown said in summing up this view, "that in this respect it becomes easier to kill all the people in the world than just a part of them." This, Professor Szilard assented, "is definitely so."

If we should build the hydrogen bomb, the scientists agreed, it would not entail just the cost of the bomb itself, but the expense of the defense measures we would be forced to take. These measures, the scientists said, would include the dispersal of the population of all our coastal cities, involving the transplantation of 30 to 60 million persons.

Tremendous Costs Seen

Our general defense measures, Dr. Szilard estimated, would reach $25 billion a year, while the dispersal proper would cost about $15 billion a year. The dispersal of the coastal cities would take about 10 years, Dr. Szilard said.

Dr. Bethe thought it would also be necessary to disperse our inland cities, since they, too, might be attacked by planes or "maybe by guided missiles."

To disperse all of our cities, Professor Szilard said, "we would probably have to spend something like $25 billion a year for 10 years."

"It certainly would mean a planned movement," he added. "It would mean controls much stricter than we ever had during the war time. It would be not a New Deal, but a Super-Super New Deal."

The dispersal, however, the four scientists agreed, would be of use only against the blast and flash burn effects of the hydrogen bomb. "The dispersal would be of no help at all against the effects of the radioactivity," they stressed, since that would go everywhere.

The size of the bomb, Professor Bethe said, would depend only on how much heavy hydrogen could be carried in a plane or in any other device that might be used to deliver the bomb. A bomb 1,000 times the power of the present atomic bomb would destroy an area 100 times greater, he said, adding: "If a bomb were exploded somewhere, then 10 miles away from it there would be almost complete destruction, and that would mean that a city as big as New York, that is, the biggest cities on Earth, would be destroyed by one single bomb."

Professor Seitz added that whereas the flash burn effect of the Hiroshima bomb extended only half a mile, the flash effect of the hydrogen bomb would be at least 30 times greater. This meant, he said, that people would suffer flash burns at least 20 miles away from the center of the detonation.

Professor Bethe said that by talking about the hydrogen bomb, "We unnecessarily gave the Russians the information that we consider it feasible, and the information that we are making it, which more or less forces them to do the same."

"We cannot predict whether the bomb can be made or not," Professor Bethe added. "On the other hand, from the decision which has been made, we must conclude that our experts believe that it is probable that we can make it. Even so, I think we must be prepared that it will take several years before the bomb has been completed."

He explained later that his "best guess" was that a minimum of three years would be required to complete the first model of the hydrogen bomb.

—February 27, 1950

Major Gains Seen in New Amplifiers

By WILLIAM L. LAURENCE

Important advances in the development of the transistor, a tiny amplifying device that can perform most of the functions of vacuum tubes and is expected to lead to a revolution in electronics, were announced in Philadelphia at the close of the annual meeting of the American Association for the Advancement of Science.

The report was presented by Jack A. Morton of the Bell Telephone Laboratories, where the transistor was invented three years ago.

Dr. William Shockley, also of the Bell laboratories, who initiated and directed the research leading to the invention of the original transistor, announced the invention of a second and radically new type of the instrument, which, he said, possessed "astonishing properties never before achieved in an amplifying device."

Meantime, Mr. Morton announced, the original type has been so developed that "it can be made as uniform in performance as vacuum tubes and can perform a wide variety of applications which now require commercial tubes of the vacuum type."

Progress in the development of the original type of transistor has been so successful, Mr. Morton reported, that it is expected to be put into trial use in the Bell System in about a year. The first application will be in the Bell System's nationwide toll dialing program, where transistors will be an integral part of equipment for automatically determining direct and alternate routes across the nation's telephone networks.

While transistors can do many of the things that vacuum tubes now do, they are expected to find their greatest use in applications where the use of vacuum tubes is now impractical; for example, in the complex switching mechanisms that are the basis of the dial telephone system.

They are also expected to have an important effect on the entire field of electronics, including applications in the electronic aspects of modern warfare.

The original transistor, known as a "point-contact" device, consists essentially of two hair-thin wires resting on a tiny speck of germanium, a semiconducting metallic element. These point contacts correspond to the terminals of a vacuum tube, but there is no glass envelope, no vacuum and no heating element to cause warm-up delay. The entire apparatus is housed in a metal cylinder about the size of a .22-caliber rifle shell, although it may be housed in a much smaller space for certain applications.

A "Nearly Ideal Amplifier"

The new transistor, meanwhile, is known as the "junction transistor." Dr. Shockley described it as "extremely efficient" and as "a nearly ideal amplifier for very low power applications."

The junction transistor is in the form of a small rectangular block, roughly the size of a kernel of corn. It has no point contacts but, instead, consists of a tiny rod-shaped piece of germanium, so treated that it embodies a thin electrically positive layer sandwiched between the two electrically negative ends.

The junction transistor, which derives its name from the two positive-negative junctions, differs markedly from the point-contact type, in which the contacts of the points play an essential role. This new form of transistor, Dr. Shockley said, is capable of amplifying 100,000 times. It occupies about $1/400$ of a cubic inch, in comparison with a typical subminiature vacuum tube, which occupies about $1/3$ of a cubic inch.

Surpass the "Older" Type

"Transistors of this type," Dr. Shockley said, "are much more efficient than the older type and consume far less power." The original type operates on less power than an ordinary flashlight bulb.

At the time of their invention in 1948, transistors were highly variable in their characteristics and of uncertain reliability, but these problems are now understood, Mr. Morton said, and it is expected that regular production can be started in the near future by the Western Electric Company, the manufacturing and supply unit of the Bell System.

"Transistors have been produced," he added, "which can withstand shock and vibration better than any known vacuum tube, and they are expected to have a service life considerably longer than that of commercial vacuum tubes in current use."

He also said that "transistors can now be designed for a great many specific functions."

Among those associated in the transistor work at Bell have been Dr. John Bardeen and Dr. Walter Brattain, the inventors of the point-contact transistor; Morgan Sparks and G. K. Teal, who built the first of the new type transistors, and R. L. Wallace Jr. and W. J. Pietenpol, who have been working on their development.

—*January 1, 1952*

Hydrogen Is Fused for Peace or War

By WILLIAM L. LAURENCE

The most spectacular event in atomic energy since the explosion of the first atomic bomb in the desert of New Mexico on the morning of July 16, 1945, took place at Eniwetok in the Marshall Islands, in the Central Pacific, at 7:15 a.m. Eniwetok time on Nov. 1, 1952. It was at that moment, which will surely go down as one of the great moments in man's history, that the hydrogen bomb made its first appearance on the world's stage.

It is virtually certain that the hydrogen bomb tested on that historic morning was a rather small-scale laboratory model, a so-called "test-tube" type.

However, this "baby" hydrogen bomb was all that was necessary to prove that super-weapons could be made of any desired power—weapons of an explosive force equal to that of 1,000 atomic bombs of the type that destroyed Hiroshima and Nagasaki.

Viewed as War-Deterrent

Such a weapon, exploding with a force equal to that of 20 million tons of TNT, could devastate an area of more than 300 square miles by blast and 1,200 square miles by fire. If encased in a shell of cobalt, it could produce a radioactive cloud equal to 5 million pounds of radium, spreading death and devastation over thousands of miles.

Because of its potential for destruction, such a weapon must be regarded as the greatest deterrent against aggressive war ever devised, and hence as a weapon to protect the free world against any totalitarian attempt to enslave it. What took place at Eniwetok may thus be looked upon in the future as a decisive turning point in history.

To produce large-size hydrogen bombs in quantity, three basic ingredients are necessary—vastly improved atomic fission bombs, five to 10 times the power of the Japanese models, to serve as triggers to set of the hydrogen fusion reaction; large quantities of the double-weight hydrogen, named deuterium; and relatively modest quantities of the triple-weight hydrogen, named tritium.

It would take only one ton of deuterium to produce an explosion equal to 20 million tons of TNT. This variant of hydrogen, which will constitute the principal explosive of the hydrogen bomb, can be obtained relatively cheaply in unlimited quantities from water. Tritium, however, no longer exists in nature, and hence will have to be re-created by modern alchemy. A plant for that purpose is now in an advanced stage of construction on the Savannah River near Aiken, South Carolina, and may begin operations sometime this year.

The United States already possesses the super-fission bomb trigger to generate the proper temperature, more than 100 million degrees Centigrade, necessary to fuse the nuclei of the deuterium atoms (deuterons), as well as vast stocks of deuterium. What is still lacked in quantity is tritium. As soon as this missing link becomes available, sometime in 1953, the stockpiling of large-scale hydrogen bombs can begin.

Only relatively small quantities of tritium will be required, as this element will serve only as a booster to speed up the fusion (explosion) of the deutrons.

Keel of Nautilus Laid

A second dramatic development in the field of atomic energy in 1952 was the laying by President Harry S. Truman at Groton, Connecticut, on June 14, of the keel of the *Nautilus*, the first atomic submarine. The vessel will be afloat in 1954.

It was officially announced at the same time that the first atomic-powered aircraft carrier may be expected "in about five years." A major effort is also being made to build an atomic-powered military airplane. It was also announced recently by the secretary of the Army that electricity from atomic energy (possibly for remote bases, port facilities and catastrophe areas) is being studied.

All authorities are now agreed that large-scale generation of electricity from atomic energy is possible, if cost is to be disregarded, but that it cannot yet compete, if power only is considered, with energy from conventional fuels.

On the other hand, nuclear reactors, in addition to producing vast quantities of heat that can be used to produce vast quantities of electricity, also can be designed to produce plutonium, one of the basic ingredients of the atomic fission bomb, which can also be used as the trigger for the hydrogen bomb. This tremendously important by-product thus offers the possibility of constructing dual-purpose atomic plants that would produce electricity for use by industry and plutonium for the national defense.

The Atomic Energy Commission made known in May 1951, that it would accept four proposals for "special studies of the practicability of business and industry building and operating reactors for the production of fissionable materials (plutonium) and power." These proposals were submitted by eight corporations, working in pairs.

These pairs were the following: The Monsanto Chemical Company and the Union Electric Company of Missouri; the Detroit Edison Company and the Dow Chemical Company; the Commonwealth Edison Company and the Public Service Company of Northern Illinois; the Pacific Gas and Electric Company and the Bechtel Corporation.

New Office Set Up

In May 1952, the AEC established a new office, under Dr. William Lee Davidson, to aid in the administration of the industrial participation program—involving studies of power generation—and to expand the areas in which all types of industry may find an interest in the national atomic "energy program."

By mid-1952 each of the four pairs of companies engaged in the industrial studies submitted to the commission some form of report. In April 1953 the Dow–Detroit Edison companies proposed that they undertake a further study to be jointly financed, partly by private and partly by public funds.

On Oct. 18, 1952, the Commission announced that 11 more companies would work with Dow–Detroit Edison in the same study program. Meanwhile, two additional companies—Pioneer Service and Foster Wheeler—were starting a separate power survey and four others have indicated interest.

A number of companies now have contracts with the Commission to perform work in the reactor development field. Some of these companies possess extensive knowledge concerning possibilities of atomic power for industrial purposes.

—*January 5, 1953*

Silicon "Battery" Represents a New Approach in Long Efforts to Harness Sun's Power

By WALDEMAR KAEMPFFERT

The scientists who attended the recent meeting of the National Academy of Sciences in Washington, D.C., saw a "solar battery" at work. With nothing but light as the source of power, a toy Ferris wheel was made to turn and enough power was supplied for a telephone conversation.

All this is of importance because on a single day the land areas of the temperate and tropical zones are flooded with more energy than the human race has utilized in the form of fuel, falling water and muscle since it came out of the trees a million years ago.

Scores of attempts have been made to utilize this energy. As far back as 1818 Augustin Mouchot's engine produced about one horsepower from a 20-square-yard reflector at the Paris exhibition. Since then similar engines ranging up to 50 horsepower have pumped water for irrigation in the United States, Egypt, Australia and Mexico.

John Ericsson, famous as the designer of the *Monitor,* the Civil War ironclad ship, spent $100,000 on solar engines, which derived their power from huge mirrors that focused the sun's rays on a boiler.

If these attempts at putting the sun to work have come to little, it is because solar radiation is spread over large areas. The capital investment in land and machinery is so large that solar energy has never paid off.

Attack by Transistor

This old problem has been attacked in a new way by the Bell Telephone Laboratories with the aid of the transistor, a device that was invented about six years ago by John Bardeen and Walter H. Brattain during the course of research directed by William Shockley. A transistor is a semiconductor. It may be of germanium or silicon. Silicon is better because it can stand more heat. Chemically pure germanium or silicon conducts virtually nothing. Actually they are good insulators. But introduce small quantities of impurities and they become semiconductors.

In the case of silicon, which is used by the Bell Telephone Laboratories, the impurities may be boron and phosphorus. The impurities form what is known as a "p-n junction," the *p* standing for "positive," the *n* for "negative." When light strikes the junction, a current flows, which means that the electrons in the crystal of silicon move about. A current is always a flow of electrons. As the electrons in the crystal move, blank spaces are left, technically called "holes." There is a perpetual movement of electrons and holes, the electrons seeking to fill holes and the holes appearing as electrons move out of them.

The silicon in the transistor is made in the form of wafers to obtain as large a surface as possible. Wafer is connected with wafer. A square foot of silicon wafer will generate about five watts on a clear day. A storage battery can be charged by day, and from this current can be obtained at night.

This "solar battery" has its possibilities in telephoning and as a power supply for low-power mobile equipment or as a sun-powered battery charger for amplifier stations along a rural telephone system. But there is no immediate prospect of drawing on the sun to supply a house with all the power that it needs.

Limited Use

This accomplishment of the Bell Laboratories is the outgrowth of research that has been carried on for years, chiefly for the purpose of clarifying the mysteries of conductors and semiconductors. Intensive research in microwave radio during the war led to marked improvement in rectifiers—devices that change alternating into direct current. Pure silicon proved to be of importance for the rectification of microwaves in radar. A method was developed for producing silicon in pure form. With this at its disposal, the Bell Laboratories could introduce impurities at will in the right amount and in the right way so as to obtain wafers that would be uniform in performance.

The "solar battery" has an efficiency of 6 percent—about half that of the average factory steam engine.

But what if there should be no sun? And what happens at night? A storage battery is charged by sunlight to supply power when needed.

—May 2, 1954

New "All-Transistor" Calculator
May Surpass Electronic Models

An experimental "all-transistor" calculator, which may surpass conventional electronic "robot-brained" machines, was exhibited in Poughkeepsie, New York, at a press preview of the new International Business Machines research laboratory.

The new model, with a computing unit about half the size of a comparable vacuum-tube unit and requiring only 5 percent as much power, employs 2,200 transistors. It ranks in capacity with IBM's type-604 electronic calculator, which uses 1,250 vacuum tubes.

The transistor, a semiconductor, can control the flow of electrons just as a vacuum tube does. Unlike a vacuum tube, it never burns out. And tiny crystals of the grayish-white metallic element germanium, sprouting hairlike wires, can perform many of the functions of much larger and more complicated electronic tubes.

W. W. McDowell, IBM vice president in charge of research and engineering, described the machine as representing another step toward the computers and data-processing machines of the future. He predicted it would soon be commercially acceptable.

The transistor model, believed to be the first fully operative computer of its kind, needs no bulky power supply or forced air cooling of components. Printed wiring has been incorporated into its design. The model contains 595 printed wiring panels, on which the transistors are mounted.

Crystals "Store" Information

In another phase of the company's research program, matrix crystals developed from barium titanate were shown. These pea-sized crystals can "store" 256 bits of information each and their employment in a calculator will permit greater capacity at a high speed and perhaps at a lower cost, the company said.

A major factor in a calculating machine, Mr. McDowell pointed out, is its "logic"—that is, its ability to add, subtract, multiply and divide. Recent developments, he noted, promise to increase the amount of "logic" tremendously and to increase also the amount of storage.

Two other phases of the research program deal with magnetic-core storage devices and a gas-tube counter. The cores, major "memory" components in a data-processing machine, are tiny, doughnut-shaped objects that can "remember" information indefinitely. Their use marks a transition from electronic tubes to semiconductors and solid-state material, the company said.

Counts to 2,000 a Second

The gas-tube counter, replacing the electro-mechanical counter in medium-speed machines, can count up to 2,000 a second. It operates on the principle of allowing the gas argon to glow within separate sections of the tube to designate digits.

Dr. James Rhyne Killian Jr., president of the Massachusetts Institute of Technology, will speak at the formal dedication of the new laboratory. Thomas J. Watson, chairman of the IBM board, will place the cornerstone.

The U-shaped, four-story structure, situated on a former twenty-seven-acre estate two miles east of IBM's manufacturing plant, adds 179,000 square feet to existing laboratory floor space. Its cost has been estimated at between $6 million and $7 million. The two wings contain 155 laboratories and offices for 600 engineers.

The interior has movable steel wall panels. The connecting wing houses an auditorium, reference library, lounge, shop, cafeteria, kitchen and receiving departments. The buildings are of steel, reinforced concrete and brick construction. Between them in an attractively landscaped area is a large pool.

—*October 8, 1954*

The Laser Lights Up the Future

By MAYA PINES

The hottest treasure hunt in the scientific world today involves a small device with the odd name of *laser* and such impressive potential that few companies in electronics, optics or space research dare to be left out of the race. The laser (an acronym for "light amplification by stimulated emission of radiation") stands for an entirely new conception of what light can do.

Its invention is comparable to the invention of the vacuum tube—with all the developments of radio, radar, TV and transistors yet to come.

With the laser's help, light stops being just something to see by. It becomes a powerful tool able to carry messages over gigantic distances, perform delicate surgery, make radar 10,000 times more precise, weld microscopic wires. When properly focused over a short distance, the narrow, intense beams of laser light have an even more startling property: They can vaporize any known material.

The first laser, a pencil-thin rod of synthetic ruby, went into operation in the summer of 1960. Since then more than 400 firms have rushed into the field and about 50 different types of lasers have been built in various parts of the country. An estimated $30 million was spent on laser research last year, much of it by the U.S. government.

Lasers make it possible to generate light in much the same fashion as radio, TV or radar waves, over which light has certain breathtaking advantages. Since laser light starts out in waves that are almost perfectly parallel, its rays never diverge seriously, regardless of distance. Furthermore, light waves are tens of thousands of times shorter than radio waves, which means that even a narrow band of visible light can hold trillions of cycles per second and thus transmit enormous amounts of information. It has been calculated that, under the right conditions, a single laser beam could carry as many messages—radio, telephone, teletypewriter and TV—as all communications channels in existence today.

Even now, radio and TV frequencies are crowded, but the present electronic traffic jam is nothing compared with that forecast for the near future. Communications experts believe that the message load will double within the next 10 years. This is why so much interest has centered on the laser's promise in the communications field, though its immediate uses there are still very few.

The major problem is that laser beams cannot penetrate clouds, snow or fog unless they use up a large part of their energy burning off whatever stands in the way. As long as lasers remain earthbound, therefore, drastic measures will have to be taken to bypass the weather. Right now, Bell Telephone Laboratories is studying the possibility of laying long underground vacuum pipes for this purpose, so that neither dust nor water vapor could interfere with the laser's beams. A cross-country network of such pipes, with relay stations every 30 miles or so to boost or bend the beams, may yet prove economically feasible.

Radar based on laser light is probably much closer to reality. One practical use will be in tracking Earth satellites. The time it takes laser beams to reach the satellite and come back, as well as the angles of the reflected light, will serve to calculate the satellite's position. Not only should this system prove vastly more accurate than current methods of tracking satellites by radio, but it may make it unnecessary for the satellites to broadcast radio signals of their own. Thus, even passive, or silent, satellites could be tracked without difficulty from a series of stations in generally cloudless areas, such as Arizona or New Mexico. Should clouds prevent the use of one such station, any other station within line of sight of the satellite could do the job. And because of the narrowness of the laser beam, radar of this type will be virtually jamproof.

Narrow as laser beams are to begin with, they can be focused with extraordinary precision by the use of simple lenses. For instance, a needle of red light shot up to the moon by MIT engineers last summer lit up a spot only two miles in diameter on the moon's surface; by contrast, the tightest microwave beam of the kind used for conventional radar would have spread over an area 500 miles wide.

Beside their narrowness, laser beams have another useful characteristic: the immense power and heat they can mobilize when focused on small areas at close range. A beam from a ruby laser can be millions of times hotter than the sun's surface. To demonstrate this power, researchers at General Electric have made laser light burn tiny holes in diamonds, the hardest substance known to man. The red flashes lasted only one two-thousandth of a second, but the target areas simply went up in puffs of blue smoke.

This searing heat can be put to work in ultradelicate surgery. Eye specialists at Columbia-Presbyterian Medical Center have used laser beams to destroy tiny tumors on a patient's retina instantly and painlessly. The light went through the eye's lens without harming it, then focused exactly where needed in the back of the eye and burned off the tumors. Although intense beams of ordinary light had been used for such operations before, the laser flash was 1,000 times briefer and delivered far less total heat, thus minimizing the danger involved. In experiments on rabbits, surgeons have shown that similar laser beams can weld a detached retina to the back of the eye. Eventually laser light may replace the finest scalpel for some kinds of surgery.

Microwelding and cutting with laser light have almost arrived in industry. With the demand for miniaturized equipment growing rapidly, the machine-tool industry is excited at the prospect of new means to cope with it. The first experimental lasers of this type have gone on sale. With their help, such jobs as cutting out almost invisible parts of tiny electrical circuits, or joining together infinitesimal wires, need no longer present formidable difficulties.

While many of the latest developments in laser techniques are secret, reports from scientific meetings indicate that the Russians have known the principles of lasers for several years. During an interview with an American industrialist last winter, Premier Khrushchev picked up a steel ruler from his desk, pointed to tiny holes in it which he said had been drilled by laser light, and boasted that Soviet scientists were well ahead of ours in this field.

Military men on both sides of the Iron Curtain undoubtedly are deeply interested in laser light's ability to vaporize the toughest materials. This science-fiction aspect of laser beams has led to much talk about their potential use as "death rays."

General Curtis E. LeMay, chief of staff of the Air Force, has referred to "beam-directed energy weapons." "Perhaps there will be weapons that strike with the speed of light," he speculated. "That kind of speed [186,300 miles per second] makes the 1,500-mile-an-hour I.C.B.M a relatively slow-moving target."

Fantastic as it seems, some researchers envision laser weapons orbiting the Earth on satellites. They would emit light beams so hot (over 1,000,000 degrees) that they could vaporize, or at least burn holes through, any intruding missiles. However, other scientists believe that such schemes are thoroughly impractical because of the enormous size and power of the device that would be required.

Conceivably, some kind of short-range ray gun might be developed for use in combat, just as in the comic strips. Rumors have it that the Government is secretly investigating such weapons. The fighting man of the future might then have to be equipped with a new version of medieval armor—a coat of mail made of some highly reflective material. And since the best mirrors available today can reflect as much as 98 percent of laser light, tomorrow's knights in shining armor might use their own shields as weapons to shoot back the rays and destroy their attackers.

Beyond the Earth's atmosphere, there is no limit to the effectiveness of laser light. In fact, lasers seem made to order for man's leap into space. Their beams can span great distances without interference and with a minimum amount of divergence. The most important use of the moon, or of a manmade space station, it has been suggested, will be as a communications center for laser-borne messages.

With an eye to space applications, the RCA Laboratories recently developed a sun-powered laser which, if placed on a satellite, could operate indefinitely.

The knotty problems of rendezvous in space may be spectacularly eased by a new kind of laser radar which measures speeds down to one 10,000th of an inch per second. According to the Sperry Gyroscope Company, which built a prototype of it, such radar would allow a huge spaceship to dock as lightly as a feather. This would avoid the danger of "a bump which could rip open the ship's skin" or a mere nudge which, in frictionless space, "could knock a space station so far out of orbit that a following vehicle seeking a landing site wouldn't be able to find it." Even the idea of transferring power over thousands of miles of space by means of laser light is becoming respectable nowadays. With good focusing, such power could be collected efficiently at the other end and converted to do useful work.

Going from outer space to underwater exploration, laser light in the blue-green range may be developed to probe the ocean floor for wrecks or submarines, reaching much farther than ordinary light.

Earthquakes and underground explosions may be detected more accurately when better seismographs, using lasers, are built.

Many other, more familiar devices are likely to be altered by the discovery of lasers. Automobile spark plugs that give off bursts of laser light are being talked about, as well as laser-light bulbs that never wear out. Better guidance instruments for ships and planes may result from a new, laser-based gyroscope. High-speed photography will reach new heights; even today, as a California professor has demonstrated, it is possible to take pictures at a rate of a half-million per second with the help of laser light.

Despite the extraordinary progress of lasers in less than three years, many scientists believe that their real impact will not be felt for another decade. By then it may be found that their greatest achievements lie in research on the frontiers of chemistry, biology or physics. For instance, laser light may produce hitherto unknown chemical reactions by heating only one ingredient in a chemical mixture. It may be used to spark small thermonuclear fires. It is already the most accurate measuring device known to man, and its future as a biological dissecting tool is full of promise.

As one of the laser's inventors, Dr. Arthur L. Schawlow, has remarked, "With the advent of the laser, man's control of light has reached an entirely new level. Indeed, one of the most exciting prospects for workers in the field is that this new order of control will open up uses for light that are as yet undreamed of."

—September 8, 1963

New Photo Technique Projects a World of Three-Dimensional Views

By WALTER SULLIVAN

Imagine a picture window set in what is actually the inside wall of a city apartment. In it one sees, in brilliant color and three dimensions, a garden abloom with spring flowers.

If one wishes to see beyond a statue in the foreground, one need only walk to one corner of the "window" and peer behind it, bringing into view more greensward and flowering shrubs. It is, in fact, a picture backed by solid masonry. Yet it is indistinguishable from the original view except that the bird on the branch is without song and the leaves do not move.

This is but one of the wonders that may come about—or have already done so—through a new technique known as holography. It has led to a race for the discovery of new applications reminiscent of what followed such electronic landmarks as the invention of the transistor and the laser.

The goals range from three-dimensional television and moving pictures, viewed without special glasses, to a variety of secret military applications. Soviet scientists have helped pioneer the field so it is likely that they, too, are highly active in this area.

In holography, a beam of coherent light—that is, an ordered succession of waves of one color or frequency—is split, for example, by a mirror. One part shines on the object to be photographed, the other is directed toward an emulsified plate.

Light waves reflected from the object meet the waves arriving from the light source. The interaction of the waves produces submicroscopic patterns of dark and light areas. These so-called interference patterns encode in the emulsion all the characteristics of the waves reflected from the object.

An interference pattern occurs when coherent waves of the same frequency meet one another. Where the wave crests from the two beams coincide, there are bright spots. Where the waves are out of phase with one another, there are varying shades of dark.

After the plate has been developed in the manner of ordinary photographic film and coherent light is applied, the waves of this light are altered by the interference pattern so that it reaches the eye with all the directional and brightness characteristics of the light reflected from the object.

Therefore, it is possible to "see" the object even though it is no longer present.

Among the applications of holography proposed in recent months or under development are the following:

- Radar systems enabling an airport traffic controller to look into a three-dimensional scope and watch all aircraft in the area, much as one watches fish in a glass-walled aquarium.
- Scanners that can map distant planets or enable engineers to see stress patterns in a whirring propeller.
- Computers that display their solution of engineering problems as three-dimensional images. The "object" that is displayed can be examined from one side or the other, as if it were really there, yet it exists only in the "imagination" of the computer.
- Photographs of fog or dust clouds that can be projected into space in three-dimensional form so that each particle can be examined, classified and counted by microscope. This makes possible for the first time the direct examination of such clouds.
- Machines that can process hundreds of thousands of pictures, picking out rapidly all patterns that conform to certain criteria. Such devices are already being used to find the signatures of oil-bearing geologic formations from explosion soundings of the Earth's interior. The same system can be used to screen electrocardiograms and fingerprint files or search aerial photographs for missile sites.
- Side-looking radars that enable aircraft flying offshore to map in detail cloud-covered installations along a coastline.
- Microscopes that can display directly the three-dimensional structure of proteins and other complex molecules formed from millions of atoms. Such a capability would revolutionize the development of new drugs.

Light Waves Captured

The principle of holography can be described in terms of a window. Holography, in a sense, captures and freezes all the light waves passing through the window. The waves are imprinted in the window glass (actually on a film or plate) in a manner that makes it possible to re-create them at will.

This process differs fundamentally from ordinary photography. In the latter process, a single two-dimensional image is thrown onto the film by the passing of light from the scene through a pinhole or lens system. The light that hits a given spot on the film originates in only one point of the scene. The brighter that point, the more its light exposes the film.

The result when the film is developed is a negative image that can readily be identified as a two-dimensional picture.

In holography every point on the film "sees" the whole scene, just as light from the entire landscape strikes every point on a windowpane. Ordinarily a film exposed in this manner, without benefit of lens or pinhole, shows no image at all. Every point on the film has been exposed to light from the entire scene and hence all points on the film are equally darkened.

What ordinary film does not record is the direction in which the light waves are moving. If there were some way to record this directional property, as well as color, there would be no need for lens and pinhole. The film could capture all the information passing through it.

This has become possible, in holography, because a way has been found to capture the wave motions of light. The light reflected from every point in a scene radiates from that point, just as waves radiate from the spot in a pond where a stone has fallen. These waves are moving at the speed of light, some 186,300 miles a second, and are unobservable in normal photography.

However, there is a way in which light waves can be "stopped." It was discovered in 1801 by Thomas Young in England. He found that if a pencil-like beam of light was split into two beams and if these beams, through mirrors, were brought back together on a screen, a pattern of dark and bright bands appeared.

The pattern is formed because the converging light waves in some places augment one another and in others they "fight" each other, reducing the brightness.

Young recognized that, to do this, the waves must be ordered, like those produced by a stone's plunge into water. In scientific terms they must be coherent. The waves produced in water by a wildly thrashing child are jumbled and chaotic. They are incoherent. Similarly, light waves from many sources—from many parts of a large lamp, for example—are incoherent. But from a point source they tend to be coherent. Hence they can be used to generate interference patterns.

The application of the interference principle is the essence of holography. If one man can be credited with its discovery, he is Dr. Dennis Gabor of the Imperial College of Science and Technology in London.

In a recent interview, he explained how the idea came to him in 1947 as he was waiting for a free tennis court at Rugby. During the 1920s and 1930s he worked in Berlin with the group that was pioneering the development of the electron microscope. In the postwar period, this device greatly enlarged man's knowledge of the very small, but the lens system and other components set limits on the degree of magnification.

Interference Patterns

Might there be some way to get around the lens problem? It occurred to Dr. Gabor that this might be possible through the capture, in a photograph, of the wave fronts of light impinging on that film. The key to his plan was the generation of interference patterns that would preserve on the photographic plate all the characteristics of light waves from the object under study.

As with Young's classic experiment, he allowed two sets of light waves to clash with each other. All the waves originated from a single point source—a mercury vapor lamp producing light essentially of one wavelength. Some of the waves passed through the target material (a microfilm inscribed with the names of three pioneers in optics—Huygens, Young and Fresnel). Other waves bypassed the target. The two sets of waves then encountered each other on the recording emulsion.

Whenever waves from any point in the target struck the emulsion; they met matching waves that had come directly from the lamp. Thus countless overlapping interference patterns were generated by waves arriving from every part of the target. The patterns were frozen into the photographic plate.

Once this plate had been developed it was possible to reconstruct the original image by directing a beam of coherent light through the plate.

Dr. Gabor called the resulting picture a hologram from the Greek word *holos*, meaning "whole," since it recorded the whole picture. His original goal was to achieve enormous magnification, rather than three-dimensional pictures.

In the electron microscope, the object under study is illuminated by a beam of electrons. Moving electrons have a wavelike motion whose wavelength is extremely short. Thus an electron microscope should show detail down to the scale of the larger atoms, but magnetic or electrostatic lenses in such instruments blur the image.

Dr. Gabor's holograms required no lens. If holograms made with electrons were illuminated with ordinary light, whose wavelengths are almost a million times greater, there should be a million-fold magnification.

For a variety of reasons, this original goal of holography has not yet come to fruition, and X-rays, rather than electrons, are being explored as a source of microscopic illumination. X-rays have wavelengths as short as those of electrons, but have not been used in microscopes because they do not respond to lenses. The problem is to generate a sufficiently intense and coherent beam of X-rays to generate holograms.

Dr. Gabor's work did not make much of a stir until 1962, when two scientists at the University of Michigan realized that the newly invented laser provided a source of light far better suited to holography than anything available to Dr. Gabor. The light from a laser is essentially of a single wavelength; it is highly coherent and extremely intense.

The two men in Michigan, Dr. Emmett Leith and Juris Upatnieks, saw that by employing a laser and a new geometry in their arrangement of light beams they could eliminate a major drawback in the Gabor-type hologram.

In projecting one of his holograms Dr. Gabor shone the illuminating beam directly into the eyes of the viewer. This not only made it hard to see the image but also would have presented a hazard if the intense beam of a laser had been used.

In the Leith-Upatnieks arrangement, the interference patterns were formed by splitting a laser beam.

One part of the beam illuminated the scene to be recorded. Waves reflected or scattered from the scene fell on the photographic plate. The other part of the beam, known as the reference beam, was diverted by mirrors so that it struck the plate at an angle. Interference between these two matching beams, one reflected from the scene and one unaltered by such reflection, generated the hologram on the surface of the emulsion.

Holograms produced by this method had almost unbelievable properties. When reilluminated by the laser, they produced an image that, to the viewer, seemed to hang in space. One could reach behind the film and thrust one's fingers into this three-dimensional picture.

Since the whole picture was encoded in each spot on the film, one could cover all but a small portion of the film and still obtain a picture, although much detail was lost, as well as the feeling of depth. Eyes operate like conventional cameras and sense depth only because they view the scene from two angles. To cover all but a small patch of the hologram was like covering one eye.

Soviet Proposal

Meanwhile, in 1962, a different form of holography was proposed by Dr. Y. N. Denisyuk of the Soviet Union. The same idea was suggested soon thereafter by Dr. P. J. van Heerden of the Polaroid Corporation and was first put into practice two years ago by Dr. George W. Stroke and A. E. Labeyrie at the University of Michigan and L. H. Lin and K. S. Pennington at Bell Telephone Laboratories.

The technique, which has helped make possible holograms visible under ordinary light as well as three-dimensional moving pictures, is based on the method first used for color photography. In 1891 Gabriel Lippmann in France made colored photos through an application of the interference principle.

He used a thick emulsion on his plates, which were backed with mercury, to form a reflecting surface. Light from the image passed through the emulsion and then was reflected back through it again. The reflected light waves interacted with incoming waves of their own wavelength, or color, to form layers of interference lines within the emulsion.

When ordinary white light fell on such plates after their development, the various wavelengths in the light responded to matching interference patterns in the emulsion, producing colors of the original shades. Although Lippman won a Nobel Prize for his achievement, the process was soon overshadowed by the far less cumbersome three-color method, which is the basis of modern color photography.

In the original application of this principle to holography, no reflecting surface was used. Instead the laser beam split, part of it shining through the back side of the photographic plate as the reference beam and part of it shining on the target in front of the plate. Light reflected back onto the plate from the target clashed with light from the reference beam, producing layered interference patterns inside the emulsion.

A remarkable property of a hologram made in this way is that, for viewing, it need not be illuminated by a laser beam. If ordinary white light, composed of all colors, shines onto the plate, only the wavelength of the laser that made the hologram will interact with the interference patterns. The rest of the light is lost. Hence, a bright light is needed to generate the image, which will be visible in a single color.

Holograms in full color have been made on thin emulsion film, using three lasers whose light represented three primary colors. Such holograms can be superimposed without getting in one another's way. The reason is that a hologram made with a reference beam striking at a certain angle can be displayed later only by an illuminating beam from that same angle. If beams from three lasers strike at widely separated angles, they will generate independent, but superimposed, holograms.

If after development the emulsion is illuminated by similar laser beams coming from the same directions, the three holograms in primary colors will appear and blend to form a full-color image in three dimensions.

A major advance has been the invention of color holograms that can be illuminated by ordinary light.

The thick emulsion process is used and the holograms are made with lasers emitting light in three primary colors. Three superimposed holograms are embedded in the emulsion, independent of one another because the interference pattern of each is sensitive only to the wavelength of light whereby it was formed.

When such an emulsion is illuminated by strong white (many-colored) light, only wavelengths in each of the primary colors interact with one of these patterns. Thus, holograms in the primary colors are generated and blend to form a picture.

The superimposition of holograms has opened the way for three-dimensional moving pictures.

Few Pictures Stored

In the emulsion holograms, only a few pictures can be stored. However, it's been calculated that, in an emulsion one centimeter thick, 10,000 holograms can be superimposed. If such a plate was rotated slowly in front of an illuminating beam, one after another of these pictures would spring into view.

Such three-dimensional movies of short duration have already been made, but the practical problems are formidable. One is finding a way to display holograms to a large audience. Dr. Gabor, in conjunction with CBS Laboratories, where he is a consultant, has taken out several patents on holographic movies.

To date scenes portrayed in holograms have largely been only a few feet in depth because laser beams do not retain their coherence—in other words they do not stay in step with themselves beyond that distance.

Thus, until recently the idea of a "picture window" showing a scene of great depth seemed beyond reach. Now two techniques that get around the depth limitation are under experimentation. Both offer the possibility of recording outdoor scenes holographically.

One method, developed by Dr. Robert V. Pole of International Business Machines, uses a "fly eye," each of whose hundreds of small lenses records on conventional film the scene as viewed from its location. This, in essence, is a way to freeze all the light coming through a window.

The result is a checkerboard of numerous, tiny photographs that make no sense to a human viewer because he does not have the multiple eyes of a fly. However, all these images can be thrown onto a film to form a hologram. While it displays the basic characteristics of other holograms, the images to date have been fuzzy, but Dr. Pole believes such drawbacks can largely be eliminated by better lens arrangements.

Another method produces holograms by passing light from the scene through a filter. This allows only one color, or wavelength, to enter the system. The image beam is then split to form two images of the scene. These are brought together in the emulsion in such a way that the light waves from each spot in the scene, being of the same wavelength, interfere with one another.

The result is a multitude of interference patterns, one for each part of the scene. Unfortunately, these patterns clash with one another in a manner that, to date, has limited the amount of detail that can be captured.

Probably the most startling hologram is the kind projected into space in front of the film. When a laser beam passes through a hologram, it generates two images. One is seen by looking through the hologram toward the light source, much as one looks through a window. Known as the virtual image, it resembles the original scene.

The second "real" image lies on the same side of the hologram as the observer. It is an inside-out image that can be thrown on a screen but is difficult to view with the eye, much as it is hard to see an image when looking into a slide projector.

Members of Dr. Leith's team at the University of Michigan have found a way to generate in front of the film a real image that appears normal. This is done by making a hologram from the real image of a previous hologram, somewhat as one might make a positive print from a negative one in conventional photography.

Nothing in the armatorium of the magician is more remarkable than the sight of a three-dimensional image hanging in space—in front of the film.

Among the highly varied applications of holography is what would appear a foolproof form of cryptography. If a plate of ground glass is placed between the target (for example, a top secret document) and the photographic plate, the light is scrambled in a manner that generates a seemingly meaningless hologram.

Study of Deformations

Only when the plate is illuminated by a beam passing through that same sheet of ground glass (or its twin), positioned precisely as before, are the light waves unscrambled and the image brought into view.

An application of holography to design problems involves the study of deformations in structures under stress. Before application of the stress, a hologram of the object is made. When developed and illuminated, this generates a three-dimensional image of the object as it was to begin with.

The object, left where it was relative to the hologram, is then squeezed. Wherever its shape is changed, even to a degree unobservable by other means, the light waves of its image clash with those in the hologram.

The battle between light waves from the two images, those of the unstressed device and those from its deformation, produces a series of interference lines visible to the unaided eye. These lines delineate areas of equal deformation (or stress) and can be photographed to guide engineers in their design work.

The presentation of holograms to the public seems to be in the offing. General Motors has on display at its Detroit headquarters a hologram showing a futuristic car. By lowering one's head, it is possible to look under the image of the car and see beyond.

Small holograms on film can be purchased for home display, but mass production methods are not yet satisfactory. The annual science book, *Science Year*, published by Field Enterprises, may include a hologram in its 1967 issue, and *Life* magazine has been exploring the possibility of publication. However, as an editor at *Life* put it the other day, the problem is "to find something that can be bound into a magazine and makes economic sense."

This remark is applicable to many current lines of holographic research. The prospects are exciting, but many applications are not yet economical.

—March 19, 1967

Nuclear Power Gain Reported, But Experts Express Doubts

By MALCOLM W. BROWNE

A professor at the University of Utah said that he and a British colleague had achieved nuclear fusion in a test cell simple enough to be built in a small chemistry laboratory.

The achievement, if verified, would represent a gigantic advance in the generation of nuclear energy. But experts elsewhere were skeptical.

Dr. B. Stanley Pons, chairman of the chemistry department at the university, said that he and Dr. Martin Fleischmann of the University of Southampton in England had conducted "a surprisingly simple experiment" in which the nuclei of heavy hydrogen atoms were forced to fuse at room temperature, producing energy. Most attempts in the United States, the Soviet Union and other countries to achieve controlled hydrogen fusion have assumed that hydrogen must be heated to a temperature near that of the sun before a useful fusion reaction can occur.

The scientists made the assertion in a statement issued at the University of Utah and at a news conference there.

Need for More Details

Spokesmen for the Lawrence Livermore National Laboratory in California, the Princeton Plasma Physics Laboratory in New Jersey and the University of Rochester Laboratory for Laser Energetics, where major fusion research projects have been conducted, declined to comment on Dr. Pons's assertion. These and other scientists said that the announcement issued by the University of Utah contained too few technical details to assess.

A spokesman for the university refused to describe the fusion experiment more fully, explaining that details were being withheld because of the need to resolve "patent problems."

The *Financial Times of London* yesterday published a diagram said to show the test cell used in the experiment. It showed two electrodes immersed in a container partly filled with heavy water (deuterium oxide). The negatively charged cathode, made of palladium metal, was identified as the site of the fusion reaction.

In his announcement, Dr. Pons said the experiment generated a great deal of heat as well as neutron radiation, and that the presumed fusion reaction produced more energy than it consumed. He and Dr. Fleischmann predicted that "the discovery will be relatively easy to make into a usable technology" but that "continued work is needed, first, to further understand the science, and secondly, to determine its value to energy economics."

While leading fusion experts elsewhere refrained from formal comment, several privately expressed doubts.

"The only cold process of which we are aware that could lead to hydrogen fusion," said one, "is called muon catalysis," in which short-lived particles called muons are used to neutralize the electric charges of neighboring hydrogen nuclei so that they can be fused. "Experiments at Los Alamos National Laboratory suggest that this technique could never be scaled up to generate power," the expert said.

Repeated efforts to reach Dr. Pons by telephone to respond to such doubts were unsuccessful. A university spokesman said Dr. Pons would be unavailable to respond to questions.

Conventional experiments aimed at developing hydrogen fusion focus on two techniques. In the first, electrically charged heavy hydrogen is compressed by powerful magnets and is thus heated to a high enough temperature to initiate fusion. In the other method, powerful laser beams are used to vaporize tiny glass capsules of heavy hydrogen. The implosion of the vaporized glass compresses and heats the heavy hydrogen to the fusion point.

—March 24, 1989

Claim of Achieving Fusion in Jar Gains Support in Two Experiments

By MALCOLM W. BROWNE

A recent claim by scientists in Utah that they had achieved nuclear fusion in a jar of water gained important support from two independent research groups that repeated part of the experiment.

Results reported by Texas A&M University and the Georgia Institute of Technology appear to have substantially strengthened the assertion by researchers at the University of Utah.

If the Utah findings are verified, they might have far-reaching importance as a new source of energy and would herald an entirely new field of scientific research. But even some of the scientists involved in the work cautioned that it was not yet clear whether their results were a laboratory curiosity or a break-through that would lead to a practical new power source.

Results of Two Experiments

In separate experiments that sought to confirm the Utah results, scientists at Texas A&M said that they had achieved a large energy increase but were not certain whether it was a result of nuclear fusion, and Georgia Tech researchers said that they had detected neutrons resulting from a fusion reaction but were not certain how much energy had been produced.

Unlike fission, in which atoms are split to produce energy, fusion involves the joining of the nuclei of hydrogen atoms to produce helium and large amounts of energy. Scientists have long sought to produce controlled fusion reactions that might provide a cheaper and somewhat safer source of energy than fission, which is used in nuclear power plants.

The Texas A&M and Georgia Tech findings, taken together, duplicated some of the most significant results first reported on March 23, 1989 by B. Stanley Pons of the University of Utah and Martin Fleischmann, a collaborator from the University of Southampton in Britain.

Technique Used in Utah

The two chemists said at a news conference in Utah that by passing an electrical current through a small container of heavy water (water in which hydrogen is replaced by its heavier isotope, deuterium), a fusion reaction was initiated in a palladium cathode.

They said the reaction produced substantially more heat energy than had been put into the cell, which suggested that a reaction of some kind was occurring and was yielding large gains in energy. They said tests at Salt Lake City also showed that the reaction was producing both neutrons and helium, neither of which would result from a simple chemical reaction. That would suggest that a nuclear fusion reaction took place, they said.

The Texas A&M and Georgia Tech announcements appeared to support that conclusion. .

At Texas A&M, Dr. Charles R. Martin, Dr. Kenneth N. Marsh and Dr. Bruce E. Gammon showed reporters an apparatus that they said had yielded from 20 percent to 80 percent more energy than had been put in.

The cell was enclosed in a water bath in which temperature was measured and maintained to within a small fraction of a degree. The water bath surrounded and collected heat produced by a cell similar to the one used at Utah.

"We have been unable so far to measure any other effect of the reaction," Dr. Martin said, "and we're certainly not confirming at this point that we're seeing nuclear fusion."

"We have not ruled out the possibility that this is simply a chemical reaction of some kind," Dr. Marsh said. "To do that will require a great deal more experimentation than we have yet been able to perform."

At the same time, the scientists said, it was very difficult to understand how a chemical reaction could have taken place between the materials in the cell that would have continuously produced the heat actually measured.

Reporters sought an assessment by the Texas A&M researchers as to how long it might take before cold fusion might become a significant source of power.

"Gigantic" Engineering Hurdles

"You haven't heard us say anything about fusion," Dr. Martin said. "A lot of alternative explanations have yet to be ruled out. Even if this should prove to be fusion, the engineering problems of harnessing fusion energy would be gigantic. It remains to be seen whether, in practice, fusion could create power more cheaply than other forms of fuel."

Other laboratories throughout the country have been trying to duplicate the Utah findings without success. Among those conducting the research are the Massachusetts Institute of Technology, Princeton University and the Lawrence Livermore National Laboratory in California.

The dream of cheap, plentiful fusion energy has prompted the federal government to pour billions of dollars into research over the years. Those research efforts have focused on two techniques: the use of powerful magnets to compress and heat electrically charged hydrogen, and the use of a powerful laser to implode a tiny capsule containing hydrogen.

While several laboratories believe they are on the verge of initiating fusion with the help of huge, costly machines, no existing fusion apparatus seems capable of producing more energy than it consumes.

If the Fleischmann-Pons apparatus proves to be as efficient as claimed, it would represent a major advance over previous attempts to achieve controlled fusion.

Dr. Martin said that research in cold fusion clearly warranted financing. Experiments that have been under way here for the past two weeks have been partly financed by the Office of Naval Research and the Electric Power Research Institute.

The experiment at Georgia Tech succeeded in measuring a flow of neutrons from a cell that was similar to the one used in Utah. In a fusion reaction, two deuterium (heavy hydrogen) atoms are joined together to create helium atoms, energy and neutron particles.

Dr. James Mahaffey, leader of the Georgia Tech group, commented in his announcement, "Our data convinced me that we are making neutrons in that vessel. There is no way to get neutrons unless something nuclear is going on."

He said the neutron counting device his group used measured an increase of 13-fold in the flow of neutrons emitted from the test cell when the current was switched on.

After the initial announcement at Utah, scientists at major laboratories conducting conventional fusion research expressed deep skepticism. It had been thought that the mutual repulsion of charged hydrogen atoms could be overcome only by duplicating temperatures and pressures common in the sun.

Many scientists remain skeptical, but a special discussion of the controversial assertions has been scheduled for a meeting in Dallas of the American Chemical Society. Vigorous debates are expected. Dr. Pons in Utah said he believed that fusion took place because atoms of deuterium were absorbed into the lattice structure of the palladium electrode when electrical current split heavy water into its constituents, deuterium and oxygen. The crystal lattice of the palladium metal holds the deuterium atoms so close to each other that many of them are able to join together, producing heat.

At the news conference at Georgia Tech, researchers expressed cautious optimism about the import of their replication work, which so far has cost $25,000 to perform.

"The Utah experiment was so fantastic and out of the mainstream of physics that simply reproducing it is significant," said Dr. Mahaffey, the leader of the five-member team.

Bill Livesay, a team member, added, "It remains to be seen whether this is a curiosity" limited to the laboratory, rather than paving the way for industrial-scale production of cheap energy.

The Georgia researchers said they took careful steps to be sure their measurements of neutrons, the "signature" of a fusion reaction, were from the experiment and not some outside source. The cathode of the electrolytic cell was shielded by graphite, tap water, paraffin, boron and lead, with the neutron counter being located in the tap water.

The researchers said they also did a control in which they took measurements of neutrons when the experiment was not running, so as to get a reading on "background" neutron counts.

During the experiment, they said, neutron counts averaged 600 counts per hour, substantially more that what might be expected if a nuclear reaction were not taking place.

In addition to neutrons, the researchers said they measured tritium, a by-product of the fusion of deuterium.

—April 11, 1989

Physicists Debunk Claim
of a New Kind of Fusion

By MALCOLM W. BROWNE

Hopes that a new kind of nuclear fusion might give the world an unlimited source of cheap energy appear to have been dealt a devastating blow by scientific evidence presented in Baltimore.

In two days of meetings lasting until midnight, members of the American Physical Society heard fresh experimental evidence from many researchers that nuclear fusion in a jar of water does not exist.

Physicists seemed generally persuaded, as the sessions ended, that assertions of "cold fusion" were based on nothing more than experimental errors by scientists in Utah.

Furor on Initial Claim

Dr. B. Stanley Pons, professor of chemistry at the University of Utah, and his colleague, Dr. Martin Fleischmann of the University of Southampton in England, touched off a furor by asserting on March 23, 1989, in Salt Lake City that they had achieved nuclear fusion in a jar of water at room temperature.

At a news conference today, nine of the leading speakers were asked if they would now rule the Utah claim as dead. Eight said yes, and one, Dr. Johann Rafelski of the University of Arizona, withheld judgment.

Top physicists directed angry attacks at Dr. Pons and Dr. Fleischmann, calling them incompetent, reciting sarcastic verses about their claims and complaining that they had refused to provide details needed for follow-up experiments. A West European expert said "essentially all" West European attempts to duplicate cold fusion had failed.

Response at Utah University

In a telephone interview, Dr. James Brophy, director of research at the University of Utah, responded, "It is difficult to believe that after five years of experiments Dr. Pons and Dr. Fleischmann could have made some of the errors I've heard have been alleged at the American Physical Society meeting."

The criticism at the regular spring meeting of the society came just before Dr. Pons was scheduled to meet with representatives of President George H. W. Bush and just after the University of Utah asked Congress to provide $25 million to pursue Dr. Pons's research. A university spokesman said Dr. Pons was in Washington and could not be reached to answer questions.

Cold fusion, Dr. Pons and Dr. Fleischmann said, can be initiated in a cell containing heavy water, in whose molecules the heavy form of hydrogen called deuterium is substituted for ordinary hydrogen. When current is passed through the heavy water from a palladium cathode, the Utah team said, the palladium absorbs deuterium atoms, which are forced to fuse, generating heat and neutrons.

Fusion, which powers the sun and hydrogen bombs, normally occurs only at extremely high temperatures. If a means could be found to harness a form of hydrogen fusion as a commercial source of power, some scientists have said, energy shortages could be forestalled.

Some of the new experiments also sought to reproduce the less contentious findings on cold fusion reported independently by Dr. Steven E. Jones and his colleagues at Brigham Young University in Utah. Dr. Jones, who used a device similar to the one in the Pons-Fleischmann experiment, did not claim that any useful energy was produced. But he did report that slightly more neutrons were detected while the cell was operating than could be expected from normal sources. The result suggests at least the possibility of fusion, he said, although it is not likely to be useful as an energy source.

Physicists who have investigated Dr. Jones's report have been fairly restrained in their criticism, acknowledging that Dr. Jones is a careful scientist. But from the outset they have expressed profound skepticism of claims by Dr. Fleischmann and Dr. Pons.

Attempts to Repeat Experiments

Since March 1989, scores of laboratories in the United States and abroad have sought to repeat the cold fusion experiments, and some completed their investigations just hours before the meeting was convened in Baltimore.

The most thoroughgoing of the attempts to validate the Pons-Fleischmann experiment was conducted at the California Institute of Technology. According to Dr. Nathan Lewis, leader of the Caltech team, every possible variant of the Pons-Fleischmann experiment was tried without success.

Using equipment far more sensitive than any available to the Utah group,

Caltech failed to find any symptoms of fusion. The scientists found no emitted neutrons, gamma rays, tritium or helium, although the Utah group reported all these emissions at high levels. And all the cells consumed energy rather than produced it, the Caltech team said.

The Caltech team intentionally reproduced experimental errors leading to the same erroneous conclusions reached by the Utah group, Dr. Lewis said. By failing to install a stirring device in the test cell, temperature differences in the cell led to false estimates of its overall heat, he said. This may have suggested to the Utah group that its cell was producing fusion energy.

Presence of Helium in Test

Noting that Dr. Pons and Dr. Fleischmann had also reported the presence of helium, a fusion product, in the test cell, Dr. Lewis said his group had also found helium. But helium is a trace component of air, and the amount of helium in the cell corresponded to what normally enters from the atmosphere.

"Pons would never answer any of our questions," Dr. Lewis told an audience of 1,800 physicists, "so we asked Los Alamos National Laboratory to put our questions to him instead, since they were in touch with him."

Other scientists said they had tried every possible variation of the Utah experiments.

Dr. Edward F. Redish of the University of Maryland, chairman of the meeting, said that he had telephoned Dr. Fleischmann to invite him to participate in the Baltimore sessions and answer criticism.

"He told me that Dr. Pons would try to come," Dr. Redish said. "But just before the meeting Dr. Pons let us know that he would be too busy discussing cold fusion with a congressional committee to come to Baltimore."

A spokesman for the University of Utah said Dr. Pons was preparing to meet with members of Bush's staff Wednesday.

Failure to Elicit Information

Many speakers at the meeting reported failure in their efforts to elicit information or comments from Dr. Pons. Dr. J. K. Dickens of Oak Ridge National Laboratory in Tennessee said that to duplicate the cell used by the Utah group, his laboratory had been forced to estimate its size.

"One published photograph of the Utah cell showed Pons's hand, and that gave us the scale," he said. Dr. Lewis said his group had also used the photograph showing Dr. Pons's hand as a measure of the cell's size. But Oak Ridge Laboratory, like Caltech, failed to find any evidence of cold fusion after it had built and tested the cell.

Physicists asked Dr. Lewis if he could account for the burst of heat that Dr. Pons reported as having destroyed one of the Utah cells. "My understanding," Dr. Lewis said, "is that Pons's son was there at the time, not Pons himself. I understand that someone turned the current off for a while. When that happens, hydrogen naturally bubbles out of the palladium cathode, and creates a hazard of fire or explosion. It is a simple chemical reaction that has nothing to do with fusion."

Other Reports of Failures

Among other major research groups that gave details of experiments failing to validate the Pons-Fleischmann results were representatives of the Massachusetts Institute of Technology, Lawrence Berkeley Laboratory in California and the University of Rochester. Before the meeting, a joint research group of Brookhaven National Laboratory and Yale University also reported failure to find evidence of the existence of cold fusion.

Dr. Douglas R. O. Morrison, a physicist representing CERN, the European scientific consortium for nuclear research, reported that "essentially all" West European attempts to duplicate the Pons-Fleischmann experiment had failed. The entire episode, he said, was an example of "pathological science," in which an erroneous experiment initially gained some support, then prompted skepticism and finally led to denunciation.

Most of the initial support has eroded. The Georgia Institute of Technology withdrew an early report that it had partly confirmed the Pons-Fleischmann experiment.

At Stanford University, Prof. Robert A. Huggins repeated the Pons-Fleischmann experiment, and obtained results that seemed to suggest fusion. But Dr. Walter E. Meyerhof, professor of physics at Stanford, told scientists at the Baltimore gathering that he had carefully studied his colleagues's apparatus and found that the experiment was flawed because of the system used to measure heat. Nevertheless, Dr. Huggins, a materials scientist, said in a telephone interview that he is "more confident than ever" in his results.

While most critics of the Utah work limited themselves to discussion of experimental results, some directed their ire at Dr. Pons and Dr. Fleischmann themselves.

"Incompetence and Delusion"

Dr. Steven E. Koonin of Caltech called the Utah report a result of "the incompetence and delusion of Pons and Fleischmann." The audience of scientists sat in stunned silence for a moment before bursting into applause.

Referring to a possible error in temperature measurements by the Utah group, Dr. Meyerhof of Stanford University offered this contribution:

> Tens of millions of dollars at stake, Dear Brother, because some
> scientist put a thermometer at one place and not another.

Dr. Brophy of the University of Utah said the Utah team, like all other scientific groups, welcomed criticism by other scientists.

"Any scientist can be proved to be slightly in error or greatly in error," he said. "If Dr. Pons and Dr. Fleischmann have made errors, they will acknowledge them. But so far none of their critics have published their criticisms, and they are conducting science by press conference, as we have been accused of doing."

Dr. Brophy said his group was not disturbed by the vote by eight of nine physicists calling the Utah experiment dead. "Pons and Fleischmann will be speaking themselves at a meeting of the Electrochemical Society in Los Angeles, and the vote there would be likely to be different," he said.

Dr. Jones himself spoke at the meeting, and although participants questioned him sharply about his experiment, the questioning was generally friendly.

He drew cheers and laughter when he concluded his talk by saying, "Is this a shortcut to fusion energy? Read my lips: No!" He defended his own experiment, describing his results as a "fragile flower" that would never grow into a "tree" producing useful energy, but could nevertheless "beautify" science.

Some critics, however, continued to insist that Dr. Jones's results also stem from experimental error, rather than fusion.

Dr. Dickens of Oak Ridge noted that Dr. Jones had used relatively crude neutron-detecting equipment, and had measured only a very small excess of neutrons over what could be expected from natural sources without any fusion.

—*May 3, 1989*

In the Quantum World, Keys to New Codes

By JAMES GLANZ

Seven decades ago, Einstein and his scientific allies imagined ways to prove that quantum mechanics, the strange rules that describe the world of the very small, were just too spooky to be true.

Among other things, Einstein showed that, according to quantum mechanics, measuring one particle could instantly change the properties of another particle, no matter how far apart they were. He considered this apparent action-at-a-distance, called entanglement, too absurd to be found in nature, and he wielded his thought experiments like weapons to expose the strange implications that this process would have if it could happen.

But experiments described in three forthcoming papers in the journal *Physical Review Letters* give a measure of just how badly Einstein has been routed. The experiments show not only that entanglement does happen— which has been known for some time—but that it might be used to create unbreakable codes for the secure transmission of bank transactions and diplomatic communiqués.

"It was first considered as pure philosophy," said Prof. Nicolas Gisin, a physicist at the University of Geneva, who is a coauthor of one of the papers. "Now we discuss that these same strange aspects of quantum mechanics can be of some use in securing the Internet, let's say."

The measured properties of entangled particles are, as the name implies, fatefully entwined. It is as if two coins, flipped simultaneously again and again on opposite sides of the globe, always came out both heads or both tails—even though the sequence was unsurprisingly random. Coins do not act this way, but quantum particles can.

As with many things quantum, the true implications of entanglement are far subtler than they at first appear. The process does not, for example, let scientists send information instantly from one place to another.

But it does allow information to be passed in a secure way, since any measurement of either particle leaves its trace on the other—meaning that an eavesdropper can always be detected, no matter how clever or technically sophisticated.

That guarantee of safe passage would allow people at any two sites to share unbreakable codes using quantum entanglement. The experiments demonstrated several versions of this trick, called quantum cryptography. It differs from all traditional cryptography, in which great pains are taken to transmit codes securely or to make them mathematically hard to crack. But spies, with luck or great pains, can and do secretly intercept or crack them.

"The advantage with quantum cryptography is you don't have to worry about any of that," said Dr. Paul Kwiat, a physicist at Los Alamos National Laboratory and a coauthor of another of the new papers. "If you do it right, you're only limited by the laws of quantum physics."

And Thomas Jennewein of the University of Vienna, lead author on the third paper, said that turning those ideas into an off-the-shelf technology was no longer unthinkable: "We realized a complete quantum cryptography system, almost ready to use."

Though they all relied on entangled particles of light, or photons, to make codes, the experiments each did something a little different to make quantum entanglement into a practical tool for cryptographers.

Mr. Jennewein and his collaborators built a system that generated quantum codes very quickly and used them to encrypt the binary string of zeroes and ones representing a digital color picture. Then they sent the coded picture over an ordinary computer network—the usual technique—and decoded it on the other side. Dr. Kwiat's team actually caught a simulated eavesdropper interposed in a quantum transmission, showing that the idea works, while Professor Gisin and coworkers built a system that might be especially compatible with existing systems of fiber optics and electronics in the real world of telecommunications.

"They all, with some justification, claim to be the first serious quantum cryptography experiments with entangled, two-photon sources," said Dr. Charles Bennett, an IBM fellow, who is one of the field's innovators.

As Einstein discovered, the most problematic aspect of the quantum world is simply inducing oneself to believe that it works the way it does. But once they are accepted, those strange rules are what make quantum cryptography go. According to quantum mechanics, for instance, particles do not have any definite properties until they are measured by a detector or observed in another way. Until they are measured, particles can have a potential existence in two or more places at once. But once they are measured, particles suddenly have an actual existence in only one of those spots—wherever they are detected—and the other possibilities vanish.

The same principle holds for the polarization of a photon, the quantum carrier of light, which can vibrate both horizontally and vertically or at two different diagonal angles, until it is actually measured. Once the photon encounters filters that, like Polaroid sunglasses, allow only one polarization to pass through, the photon's polarization becomes known and definite.

Quantum entanglement operates in this weird world, linking the potential states of two particles. If a laser photon passes through a special "nonlinear" crystal, for example, it can split into two photons each with two possible polarizations.

But the photons may be entangled, so that if one is measured and has a horizontal polarization, then the other will have that polarization, too. If the first one measured has a vertical polarization, the second will also be vertical.

The essence of quantum cryptography involves creating entangled pairs of photons and sending each to one of two people, traditionally called Alice and Bob, who want to generate and share a secret code. Alice and Bob each measure the photons' polarizations using filters set in randomly selected directions.

Only if both happen to choose the same direction for a particular photon, and the entangled photons turn out to be polarized in that direction, are Alice and Bob guaranteed a detection on that particular trial. If the photons are polarized in the other direction on that trial, both Alice and Bob are guaranteed a nondetection.

So Alice and Bob phone each other on an unsecure line after collecting enough photons for their code. They reveal to each other, publicly, their detector settings, but not whether they made a detection, in each case. They then use only the results of the measurements when they happened to have the same settings. (In an actual experiment, Alice and Bob have a pair of detectors each so that the photons in question are always captured, though in different places.)

But because of the quantum entanglement effect, they do not need to reveal the results of the measurements—detection or no. They know that they have both gotten the same answer. Each detection in those instances becomes a binary one, and each nondetection, a binary zero. The sequence, known to Bob and Alice but to no one else, becomes the code.

And what if an eavesdropper, usually called Eve, has tried to intercept the photons? Bob and Alice can discover this by comparing just a few of their measurements. If they are not virtually all the same, the entanglement has been disturbed, and the intruder has been found.

Dr. Artur Ekert, a theorist at the University of Oxford whose work paved the way for the new demonstrations, said, "I think the three papers open a new chapter in experimental quantum cryptography." While making the world of commerce and espionage more secure, they may also bend all but the most resilient of minds.

—May 2, 2000

Computing, One Atom at a Time

By GEORGE JOHNSON

The only hint that anything extraordinary is happening inside the brown stucco building at Los Alamos National Laboratory is a small metal sign posted in front: "Warning! Magnetic Field in Use. Remain on Sidewalk." Come much closer and you risk having the magnetic strips on your credit cards erased.

The powerful field is emanating from the supercooled superconducting magnets inside a tanklike machine called a nuclear magnetic resonance spectrometer.

The device itself is unremarkable. NMR machines are used in chemistry labs across the world to map the architecture of molecules by sensing how their atoms dance to the beat of electromagnetic waves. Hospitals and clinics use the same technology, called magnetic resonance imaging, or MRI, to scan the tissues of the human body.

The machine at Los Alamos has been enlisted on a recent morning for a grander purpose: to carry out an experiment in quantum computing. By using radio waves to manipulate atoms like so many quantum abacus beads, the Los Alamos scientists will coax a molecule called crotonic acid into executing a simple computer program.

Last year they set a record, carrying out a calculation involving seven atoms. This year they are shooting for 10. That may not sound like many.

Each atom can be thought of as a little switch, a register that holds a 1 or a 0, and the latest Pentium chip contains 42 million such devices. But the paradoxical laws of quantum mechanics confer a powerful advantage: A single atom can do two calculations at once. Two atoms can do four, three atoms can do eight.

By the time you reach 10, doubling and doubling and doubling along the way, you have an invisibly tiny computer that can carry out $1,024^{(2^{10})}$ calculations at the same time.

If scientists can find ways to leverage this achievement to embrace 20 atoms, they will be able to execute a million simultaneous calculations. Double that again to 40 atoms, and 10 trillion calculations can be done in tandem.

The goal, still but a distant glimmer, is to harness thousands of atoms, resulting in a machine so powerful that it would easily break codes now considered impenetrable and solve other problems that are impossible for even the fastest supercomputer.

"We are at the border of a new territory," said Dr. Raymond Laflamme, one of the leaders of the Los Alamos project. "All the experiments today are a very small step, but they show that there is not a wall."

"The big question," he added, "is whether we can make the transition from theory to practice."

The program that he and a colleague, Dr. Emanuel Knill, are now running—a procedure for detecting and correcting the errors that inevitably crop up during the exceedingly delicate quantum calculations—is being watched with interest by other theorists.

"Quantum error correction is vitally important for future quantum technologies," said Dr. John Preskill, a physicist at the California Institute of Technology. "Until the idea of quantum error correction was discovered in 1995, there was great skepticism about whether large-scale quantum computers capable of outperforming conventional digital computers would ever be practical."

Now Dr. Knill and Dr. Laflamme are demonstrating that what was shown to be true in theory works in practice as well. Their experiment is also a landmark in another way.

Researchers have recently used NMR to get molecules to execute rudimentary programs, like searching a database using fewer steps than required by an ordinary computer. (As a sign of how primitive the technology remains, the database consisted of a list of only eight numbers.) Dr. Knill and Dr. Laflamme's error-correcting algorithm is still quite simple, compared with, say, Microsoft Word, but it is one of the most complex pieces of quantum software yet run.

Less than a decade ago, quantum computing was just an intellectual parlor game, a way for theorists to test their mettle by imagining absurdly small computers with parts the size of individual atoms. At its root, computation is just a matter of shuffling bits, the 1's and 0's of binary arithmetic. So suppose an atom pointing up means 1 and an atom pointing down means 0. Flip around these bits by zapping the atoms with laser beams or radio waves and the result is an extremely tiny computer.

But that would be just the beginning of its power. Quantum mechanics, the rules governing subatomic particles, dictates that these quantum bits, called qubits (pronounced KYEW-bits), can also be in a "superposition," indicating 1 and 0 at the same time. Two atoms can simultaneously be in four states: 00, 01, 10 and 11. Three atoms can say eight things at once: 000, 001, 010, 011, 100, 101, 110 and 111. For each atom added to the chain, the number of possibilities increases exponentially, by a power of 2. Put together a few dozen atoms, it seemed, and they could perform vast numbers of calculations simultaneously.

All this was of little more than academic interest until 1994, when Dr. Peter Shor, a researcher at AT&T Laboratories in Florham Park, New Jersey, proved that a quantum computer could rapidly find the factors of long numbers, a problem that flummoxes human brains and supercomputers. Since the codes that are used to protect military and financial secrets depend on the near impossibility of this task, government money began pouring into places like Los Alamos, allowing theorists like Dr. Laflamme and Dr. Knill to begin turning the thought experiments into reality.

"There is no fundamental physical barrier that makes quantum computing impossible," Dr. Knill said. "The technology, as it exists, is a long way from meeting the goal. But we see no reason in principle why the goal cannot be met."

For the past few years, laboratories have been using exotic technologies to isolate small numbers of atoms, prodding them into performing simple calculations. Dr. Laflamme and Dr. Knill's group is among those that have been trying a different method: using the off-the-shelf technology of NMR, in which molecules—strings of atoms—are trapped in intense magnetic fields and manipulated with radio waves.

This approach is possible because the cores of some atoms—the nuclei—are endowed with a quality called spin. They act like little tops, rotating in the presence of a magnetic field. If the nucleus is rotating counterclockwise, its axis of spin points upward. Flip it over and it rotates clockwise, a condition called downward spin.

Nudging these nuclei with pulses of high-frequency radio waves causes them to shift between the two positions, up and down. And since the molecules emit feeble electromagnetic signals, the progress of the experiment can be monitored on a computer screen.

This technique is a proven tool for chemists, who use NMR to generate charts called spectra that give clues to the structure of chemical compounds. Several years ago, scientists at Stanford, the Massachusetts Institute of Technology, the IBM Almaden Research Center, the University of Oxford and elsewhere realized that NMR could be used for a very different purpose. Call up "1" and down "0" and you have a tiny molecular switch.

As Dr. Laflamme put it: "People had been doing quantum computing all along. They just didn't know it."

During the recent experiment, he and Dr. Knill sat in front of a computer workstation that was wired and programmed to control the NMR machine. Their goal: to get a string of five nuclei to carry out their error-correcting algorithm.

Errors occur when a bit is accidentally flipped so that it says 1 when it really means 0, or vice versa. Ordinary computers can protect against this by using redundancy. In one scheme, data are sent in triplicate, so 101 becomes 111000111. Simple little programs watch out for corrupted triplets like 010 or 110, restoring the errant bits so they match the other two.

For quantum computing, error correction is more convoluted. A qubit is protected by using an intricate scheme that effectively spreads its value across a cluster of five qubits that are all "entangled" quantum mechanically. That means that if one of the qubits becomes scrambled, its original value can be retrieved by analyzing the other four.

In the experiment, the five qubits will be represented by the nuclei of five atoms in a molecule of crotonic acid. Schematically, this can be thought of as a string of five beads, though the arrangement is slightly more complex. Four of the beads are carbon nuclei—or actually isotopes called carbon 13. (Since ordinary carbon 12 is spinless, the physicists called on a Los Alamos chemist, Dr. Rudy Martinez, to synthesize crotonic acid using carbon 13.) The other bead on the string actually consists of a cluster of three hydrogen nuclei, part of a structure called a methyl group, that is treated as a single processor.

In the scientists' notebooks, the five-qubit sequence is abbreviated like this: $M \ C^1 \ C^2 \ C^3 \ C^4$—a methyl followed by four carbons. These are the tokens that will be used to compute. As the calculation unfolds, the tiny signals emitted by the molecules will be monitored and displayed by the computer as a horizontal zigzag line.

First a small flask of crotonic acid containing about (10^{21}) (a billion trillion) of the molecules is placed inside the core of the NMR spectrometer. A bath of liquid nitrogen and liquid helium cools the machine's superconducting coils, allowing electricity to course through them unimpeded, generating the intense magnetic field that the sign outside warns about.

Tapping on the keyboard, Dr. Knill "shims" the magnets—straightening out the kinks in the electromagnetic field. The effect is not unlike what happens when a carpenter uses little wooden wedges to shim a windowframe so it is perfectly horizontal.

After the machine is calibrated, the experiment can begin. At first, the nuclei in the molecules are pointing every which way, creating a predominantly random soup. But the strong magnetic field causes a fraction of the molecules, about one in a 100 million, to line up so that all their nuclei are pointing up: 11111.

This subset of uniformly aligned molecules—about 10 trillion of them—will be used to carry out the computation. This is possible because the five nuclei within the molecules each resonate at a different frequency. Using pulses of radio waves, an operator sitting at the controls of an NMR machine can choose an individual nucleus—carbon number 2, for example—and strike it like a bell. Throughout the flask, trillions of C^2 nuclei will chime in synchrony.

Apply the pulse for just the right duration and the C^2's can be rotated to point down for 0. Another pulse will cause them to point to 1 again. And a pulse of half the duration will cause them to hover in quantum superposition, potentially saying 1 and 0 at the same time.

What has been described so far is just the quantum version of a light switch. The reason a molecule can be used to calculate is that its nuclei, like the tiny switches inside a computer chip, interact with one another: a radio pulse will cause a certain nucleus to change from 1 to 0—but only if the nucleus to its left is 1. In an ordinary computer these kinds of arrangements are called logic gates, the building blocks of computation. String enough of them together and any calculation can be performed.

Dr. Knill starts the quantum algorithm by directing the machine to emit a short burst of pulses. This causes the first carbon nucleus, C^1, to point down while the other four qubits remain up. Throughout the soup, trillions of molecules now say 10111, a pattern that can be displayed on the computer screen as a horizontal line with a peak—the spectrum. A single qubit of quantum information has been stored.

Another series of pulses then protects this information by encoding it among the cluster of qubits, flipping them up or down according to the rules of the correction algorithm.

Then an error is deliberately introduced. A radio pulse is used to flip one of the qubits—just as might accidentally happen during a computation. But the original message has not been lost. A final barrage of pulses is unleashed, which analyzes the cluster, ferrets out the error and fixes the mistake. On the computer screen the peak of the spectrum shifts, showing that the experiment has been a success.

There are no whoops of joy or uncorking of champagne. Since last fall, the experiment has been repeated many times. But it is always satisfying for the theorists to see that it really works.

—March 27, 2001

Brain Surgery, Without Knife or Blood, Gains Favor

By LAURIE TARKAN

Without so much as a nick to her scalp, Cheryl Hogarth had major brain surgery on a tumor that had grown to nearly the size of a ping-pong ball deep within her brain.

Ms. Hogarth, who at 37 was told her malignant tumor was inoperable and given a prognosis of six months to live, went under the Gamma Knife, no-knife surgery that blasts its target with hundreds of high-intensity radiation beams in a single session. There was none of the cutting, bleeding, general anesthesia, ear-to-ear scar or long recovery associated with traditional craniotomy. She went home that evening.

Two years after the surgery, Ms. Hogarth, a Sacramento mother of two, is a survivor. She takes chemotherapy to supplement the treatment, and the tumor has not grown. "I now have hope that I will be here to watch my children grow up," she said.

Called radiosurgery, this bloodless procedure has grown exponentially in the past several years, accounting for nearly 10 percent of brain operations in 1999.

Along the way, however, questions have arisen about the wisdom of using the device, from doctors who question its long-term effects and effectiveness.

The number of centers in the United States using the Gamma Knife, made by Elekta, increased from 32 in 1997 to 72 today. In addition, 75 centers use other types of radiosurgery tools, known as the X-Knife, CyberKnife and Clinac, machines designed to treat the whole body that have been modified to treat the brain.

Though the Gamma Knife has been used in Europe since the 1960s and was approved by the Food and Drug Administration in 1987, improvements in the technology, software and in imaging technology have made it more precise and easier to use.

Radiosurgery has become routine at some medical centers for a handful of brain disorders, including brain metastasis (when breast, skin, lung or other cancer spreads to the brain); acoustic neuromas (tumors of the hearing nerve); meningiomas (tumors occurring on the protective layers of the brain); trigeminal neuralgia, a functional disorder causing severe facial pain; and arteriovenous mal-formation, a tangle of blood vessels that disrupts blood flow in the brain and is the leading cause of stroke in young people. It is being used experimentally to treat such functional disorders as epilepsy and Parkinson's disease.

But its very popularity has led to a debate among neurosurgeons and oncologists. While studies show that radiosurgery has benefits over open craniotomy for a few conditions, some doctors worry that the research is not solid for other conditions commonly treated with radiosurgery. They express concerns about long-term effects of such intense doses of radiation and about the effectiveness of the procedure in deactivating the cancer and preventing recurrences.

The Gamma Knife is designed to give a precise, high-power shot of radiation to a tumor, while minimizing the amount of radiation exposure to surrounding tissue. The dose is divided among 201 fine beams, homing in on the target from as many different directions. Tissue surrounding each beam receives a minimal dose, and falloff is immediate. This design allows surgeons to deliver a dose 10 to 20 times greater than a standard radiation treatment delivers.

The blast of radiation prevents tumor cells from dividing by damaging the DNA and cuts off the blood supply, essentially starving the tumor. It can, however, take many months to see the effects of the radiation. The tumor is deactivated but may or may not shrink and may never completely disappear, which is disconcerting to some physicians and patients.

Dr. L. Dade Lunsford, a neurosurgeon who is an investor in Elekta and a member of its scientific advisory board, said radiosurgery had created a "real revolution" in the treatment of conditions like brain metastasis and deep-seated tumors.

"As few as four years ago, when a person got cancer that spread to the brain, that was the kiss of death," said Dr. Lunsford, who heads the department of neurosurgery at the University of Pittsburgh, the first center in North America to use a Gamma Knife. "Most metastatic patients died of progression of cancer in the brain."

Open surgery is often not an option in patients with multiple metastases because of the risks of cutting into several areas of the brain, or when a tumor is too deep or the patient too frail to undergo a craniotomy.

Dr. Lunsford said that radiosurgery could be used to control brain tumors and their debilitating symptoms in more than 60 percent of the estimated 150,000 patients with metastatic cancer each year. He estimated that fewer than 10 percent of these patients receive radiosurgery.

Several obstacles to treatment exist, including cost, training bias, resistance to radiosurgery and the belief "that brain disease is hopeless," Dr. Lunsford said.

But Dr. Raymond Sawaya, chairman of the neurosurgery department at the M. D. Anderson Cancer Center at the University of Texas, says the lack of long-term research is why more patients with brain metastasis are not treated. "A lot of people are stating in their papers that radiosurgery is the equivalent to surgery, but the answer is not out there," he said.

Studies are, at best, about 15 years old, not long enough to determine chances of recurrence or long-term damage.

For now, though, only about 10 percent of patients with metastatic brain cancer live longer than two years.

The research on radiosurgically treating acoustic neuromas, which affect 2,700 people a year, is more solid, since this was the first disorder to be treated with the Gamma Knife more than 30 years ago. Studies show that radiosurgery is at least as successful as traditional surgery, and it avoids the risks of open surgery, including a 1 in 200 risk of dying, Dr. Lunsford said.

Otologists, ear surgeons who perform traditional acoustic neuroma surgery, are among the loudest critics of radiosurgery. They claim that because the tumor is left inside with radiosurgery, there will be more recurrences of acoustic neuromas. To date, the best studies show rates of recurrences are similar.

Radiosurgery is showing promise for treating meningiomas, which affect 22,000 patients a year. Surgeons say it is particularly useful to treat small tumors in areas where there is a high concentration of critical cranial nerves and surgery is dangerous. But this use is relatively new and its long-term effectiveness unknown.

Radiosurgery may never replace craniotomy. It cannot be used to treat tumors larger than three centimeters (just over an inch), which excludes many primary brain tumors. In larger tumors, there is an increased risk of radiation damage to surrounding areas and of swelling.

Many large tumors need to be surgically removed immediately to relieve the pressure on surrounding tissue, rather than left to shrink away slowly. Radiosurgery is not an option when a tumor sits directly on an underlying structure, like an optic nerve or pituitary gland, which would be devastated by the radiation.

For all conditions treated by radiosurgery, each case needs to be assessed individually, looking at issues including the size and number of tumors, their location, the patient's prognosis, the age and the condition of the patient.

"There are certain tumors that are better treated with the Gamma Knife and there are tumors better treated with surgery," said Dr. Bernadine Donahue, associate professor of radiation oncology at New York University Medical Center.

Said Dr. Sawaya, "In the real world, we might use various modalities in combination rather than in competition with each other."

The benefits of radiosurgery are clear. It avoids many of the risks of craniotomy, including risks of anesthesia (none is needed), infection and hemorrhage, and loss of brain function from cutting through brain tissue, as well as drawbacks like scarring, hospital stays that range from several days to two weeks and long convalescence of four to six weeks.

Radiosurgery patients are released the same day or the next. It's also less costly than surgery, which may be why insurance companies typically cover it. Some companies limit coverage for newer radiosurgery tools like the CyberKnife because of lack of clinical data.

The most common side effect of radiosurgery is swelling of brain tissue, which can be controlled with steroids. Possible risks include radiation damage to neural structures or damage to nearby cells.

"The issue with radiosurgery is, What's going to happen in 10 to 20 years?" said Dr. Edie E. Zusman, director of adult neurology at Sutter Neuroscience Institute in Sacramento. "We were concerned about using a malignant therapy for benign diseases that could be cured surgically, but long-term studies of 10 to 15 years are coming out that have shown that this side effect all of us feared does not seem to be happening."

The success of radiosurgery also varies depending on the type of tool used. Both the Gamma Knife and the others, which are variations of a linear accelerator, have advantages for different types of problems. The best treatment center is one that has access to both types of machines, where decisions about treatment are made based on what's best for the individual patient, not based on what machine is available for use, said Becky Emerick, director of the International Radiosurgery Support Association, based in Harrisburg, Pennsylvania.

For problems in which open surgery and radiosurgery are both possible, many neurosurgeons still prefer open surgery, especially for young patients. Arteriovenous malformations, for instance, tend to occur in children and teenagers who have many years of life ahead.

"We don't know the risk for someone we expect to have a 50-year life expectancy," Dr. Donahue said.

In some cases, though, the preference is based on what doctors specialize in, what they're familiar with or what tools they have access to, said Dr. Donahue. But some consumer advocates believe this bias isn't fair. Many patients are not being informed about radiosurgery, Ms. Emerick said.

"It's beginning to be almost negligent if you cut open a head when you don't have to," she said.

—April 29, 2003

CHAPTER 3

The Fate of the Universe

The Greatest Telescope in the World

By GARRETT P. SERVISS

The first step of an enormous advance in the size of telescopes has just been taken by the officials of the Carnegie Institution's Solar Observatory on Mount Wilson, California. The sum of $45,000 having been contributed by John D. Hooker of Los Angeles to cover the expense, a glass disk of unexampled magnitude has been ordered made in France, and from this disk George Ritchey will construct a mirror for a new reflecting telescope 100 inches in diameter. Mr. Ritchey has nearly completed a mirror of 60 inches diameter in the shops of the observatory at Pasadena, and hitherto this has been looked upon as an undertaking of extraordinary boldness; but the 100-inch mirror will so far exceed the 60-inch in its power to grasp the light of the heavenly bodies that to compare them is like comparing a full-grown man with a child. To show this it is only necessary to remember that the capacity of a telescopic object-glass, or mirror, to collect light varies directly with the square of its diameter. The square of 60 is 3,600, and the square of 100 is 10,000; therefore, the power of the contemplated telescope will be to that of the telescope nearly completed as 10,000 is to 3,600, or nearly three times greater! The famous Lord Rosse telescope, constructed in Ireland in the middle of the 19th century—and still in existence, and to some extent in use—is 72 inches in diameter; but it is so imperfect in figure, and in reflecting quality, that it cannot be compared with either of these two Californian giants.

In truth, a 100-inch telescope is no less than an innovation, and many astronomers, knowing the immense difficulties which beset the use of large instruments, difficulties both mechanical and atmospheric, will be disposed to doubt the success of this undertaking. But Prof. George E. Hale, the director of the Solar Observatory, is confident that it will not fail, and Mr. Ritchey does not shrink from so critical a test of his skill and knowledge. Even the casting of a disk of optical glass of such enormous proportions is a thing not to be undertaken without misgivings; but here again the fear of failure is not allowed to prevail, and the Royal Plate Glass Company of St. Gobain, France, to which the order for the disk has already been given, promises a satisfactory fulfillment of its contract. The new disk will be 13 inches thick. The 60-inch disk is eight inches thick, and weighs a ton. These facts give an indication of the enormous weight the projected mirror will possess, and suggest to those who are familiar with the mechanical problems involved in the mounting of huge telescopes some of the practical difficulties that will have to be overcome before this new astronomic eye opens its mighty lids to penetrate among the yet-undiscovered marvels of the starry universe.

Of course, things done on this scale are not accomplished quickly, and it is estimated that about four years must elapse before the monster instrument will be mounted for use on the summit of Mount Wilson. In the meantime the 60-inch disc will have been put to its work; and even this instrument will so far exceed any now in use, as far as light-gathering capacity is concerned, that it will beyond question greatly extend the boundaries of astronomical observation, thus opening the field, as it were, for the display of the far mightier powers of the colossus that is to follow.

One or two other comparisons between the projected instrument and certain famous telescopes now in existence may be of interest before we consider the question of what the great new mirror may be expected to accomplish for astronomy. The Lick telescope on Mount Hamilton, for many years the most powerful in the world, is 36 inches in diameter, and its successor in the leadership of giant astronomic instruments, the Yerkes telescope of the University of Chicago, is 40 inches in diameter. Both of these, it is true, are refracting, not reflecting, telescopes—that is to say, they depend for their power to grasp the light of the heavenly bodies, and to bring it to a focus where a magnifiable image of the object looked at is formed, not upon a concave mirror, but upon a convex object glass— but, nevertheless, the amount of light that they can collect depends just as in the case of a reflecting telescope, upon their "aperture" or diameter. Consequently, leaving out of sight as of no material consequence for our purpose the relative capacities of the two kinds of instruments for utilizing the light that enters them, we may, as before, base the comparison upon the squares of their diameters. Then, the square of 36 being 1,296, and the square of 100 being 10,000, we see that the new Mount Wilson telescope will exceed the Lick telescope in light-grasping power almost eight times; and, the square of 40 being 1,600, it will exceed the great Yerkes telescope, in like manner, more than six times. No more is needed to demonstrate the truth of what we said at the beginning, that this ambitious California project marks an enormous step in advance for the size, and, it may now be added, the power of telescopes.

It remains to consider the extremely interesting questions: What will the new telescope be capable of doing? What discoveries in the depths of the heavens may be expected from it? The best authority, in this case, should be Prof. Hale himself, since it is under his direction that the instrument is being constructed, and will be used when it is completed.

To begin with, it may be said that probably the construction of so large a telescope would never have been undertaken by persons as competent to judge the chances of success as are Prof. Hale and his coadjutors but for their previous knowledge of the singular excellence of Mount Wilson as a station for astronomical observations. The local conditions of the atmosphere on the summit of that peak, about 7,000 feet above sea level, have been found to be in some respects ideal. "The remarkable calm of the summer nights on Mount Wilson" has excited the enthusiastic admiration of all astronomers who have personally experienced its effects. This permits the use there of higher powers than can be advantageously employed elsewhere. Steadiness of the air and absence of disturbing currents are more important than mere clearness. But the air is also very clear on Mount Wilson.

In such an atmosphere, then, the new instrument will have every opportunity to demonstrate its power and its usefulness. "Even if it should prove," writes Prof. Hale, "that only a few nights in the course of a year can be utilized to the fullest advantage, the construction of such a telescope would nevertheless be desirable. For under ordinary conditions, which are much finer than those in the eastern part of the United States, results of the highest value can be obtained in many classes of work, such as the photography of stellar spectra, the measurement of the heat radiation of stars, etc. The immense amount of light which this mirror will collect should render it particularly suitable for spectroscopic work of all kinds."

This, somewhat technical general statement, the significance of which will be appreciated by all astronomers, may be supplemented with more specific facts.

To the general public the power of a telescope seems to be summed up in its magnifying power. "How many times does it magnify?" is the question usually asked. Another question is, "How near does it bring the moon?" The astronomer often smiles at these questions, and yet he must feel a certain sympathy with them. He knows—alas!—that the larger his telescope and the greater the quantity of light it collects the less effectively, in proportion, can the images that it gives of the heavenly bodies be magnified. For instance, a four-inch telescope of the best make will bear, when the atmosphere is perfectly steady, a magnifying power of

400 diameters in observations of the planets and the moon. Such a power may be said to "bring the moon" within a visual distance of only 600 miles. But a 40-inch telescope will by no means, in any known condition of the atmosphere, bear a corresponding magnifying power of 4,000 diameters, bringing the moon within an apparent distance of only 60 miles. Aerial currents disturbing the light rays, and imperfections of the image due to many causes that are active with very large object-glasses, or mirrors, prevent the employment with such instruments of magnifying powers proportionately great. If this were not so, it might be expected that the new 100-inch telescope would afford a power of 10,000 diameters, bringing the moon (to employ again the popular illustration) apparently within about 24 miles of the observer's eye.

But no such magnifying power will be employed. Even if 5,000 should occasionally be used, it would be only as an experiment. Yet, even in this respect, the new instrument will offer great advantages, for it has been found that the same magnifying powers, when used on a large telescope, give much better and clearer views of the planetary phenomena—double stars and other objects—than when used with smaller glasses.

But the great advantage of the 100-inch telescope will lie, as already intimated, in its enormous light-grasping capacity. This, independently of any question of magnification of the image, will endow it, to an unexampled degree, with what William Herschel called "penetrating power," i.e., the ability to look deeper into the profundities of space than any eye has ever yet been able to range. This power will manifest itself both by making the faint distant stars that existing telescopes reveal appear plainer and brighter, and by revealing probably millions of other, at present unknown, stars and others objects, such as nebulae, whose light is too faint sensibly to impress either the eye or the photographic plate with the instruments now in use.

Prof. Hale expects that the new telescope will greatly increase the possibility of finding intermediate types of stars, which have hitherto escaped the ken of the largest instruments, showing the transition from the so-called solar stars, which resemble our sun, to stars of the "fourth type," which appear to be suns nearly extinguished. Thus it will give indispensable aid in establishing a knowledge of the course arid steps of stellar evolution.

Another field in which the power of so gigantic an instrument will instantly exhibit their supremacy is that of the spiral nebulae. These strange objects, the discovery of which, in great numbers, has, in the past few years, upset the old

Laplacean cosmogony, exhibit extraordinary phenomena, which seem eloquent with meaning, but which escape satisfactory interpretation largely because more perfect views of them than can at present be obtained are needed. Even the 60-inch will certainly vastly increase our knowledge of these objects, but the 100-inch may confidently be expected to afford an astonishing revelation.

—January 27, 1907

Einstein Expounds His New Theory

Now that the Royal Society, at its meeting in London on Nov. 6, 1919, has put the stamp of its official authority on Dr. Albert Einstein's much-debated new "theory of relativity," man's conception of the universe seems likely to undergo radical changes. Indeed, there are German savants who believe that since the promulgation of Newton's theory of gravitation no discovery of such importance has been made in the world of science.

When *The New York Times* correspondent called at his home to gather from his own lips an interpretation of what to laymen must appear the book with the seven seals, Dr. Einstein himself modestly put aside the suggestion that his theory might have the same revolutionary effect on the human mind as Newton's theses. The doctor lives on the top floor of a fashionable apartment house on one of the few elevated spots in Berlin—so to say, close to the stars which he studies, not with a telescope, but rather with the mental eye, and so far only as they come within the range of his mathematical formulas; for he is not an astronomer but a physicist.

It was from his lofty library, in which this conversation took place, that he observed years ago a man dropping from a neighboring roof—luckily onto a pile of soft rubbish—and escaping almost without injury. This man told Dr. Einstein that in falling he experienced no sensation commonly considered as the effect of gravity, which, according to Newton's theory, would pull him down violently toward the Earth. This incident, followed by further researches along the same line, started in his mind a complicated chain of thoughts leading finally, as he expressed it, "not to a disavowal of Newton's theory of gravitation, but to a sublimation or supplement of it."

When he read in the message from *The Times* requesting the interview a reference to Dr. Einstein's statement to his publishers on the submission of his last book that not more than 12 persons in all the world could understand it, coupled with the editor's request that Dr. Einstein put his theory in terms comprehensible to a larger number than 12, the doctor laughed good-naturedly, but still insisted on the difficulty of making himself understood by laymen.

"However," he said, "I am trying to talk as plainly as possible. To begin with the difference between my conception and Newton's law of gravitation: Please imagine the Earth removed, and in its place suspended a box as big as a room or a whole house, and inside a man naturally floating in the center, there being no force whatever pulling him. Imagine, further, this box being, by a rope or other contrivance, suddenly jerked to one side, which is scientifically termed 'difform motion,' as opposed to 'uniform motion.' The person would then naturally reach bottom on the opposite side. The result would consequently be the same as if he obeyed Newton's law of gravitation, while, in fact, there is no gravitation exerted whatever, which proves that difform motion will in every case produce the same effects as gravitation.

"I have applied this new idea to every kind of difform motion and have thus developed mathematical formulas which I am convinced give more precise results than those based on Newton's theory. Newton's formulas, however, are such close approximations that it was difficult to find by observation any obvious disagreement with experience.

"One such case, however, was presented by the motion of the planet Mercury, which for a long time baffled astronomers. This is now completely cleared up by my formulas, as the Astronomer Royal, Sir Frank Dyson, stated at the meeting of the Royal Society.

"Another case was the deflection of rays of light when passing through the field of gravitation. No such deflections are explicable by Newton's theory of gravitation.

"According to my theory of difform motion, such deflections must take place when rays pass close to any gravitating mass, difform motion then coming into activity.

"The crucial test was supplied by the last total solar eclipse, when observations proved that the rays of fixed stars, having to pass close to the sun to reach the Earth, were deflected the exact amount demanded by my formulas, confirming my idea that what so far has been regarded as the effect of gravitation is really the effect of difform motion. Elaborate apparatus and the closest and most indefatigable attention to the difficult task enabled that English expedition, composed of the most talented scientists, to reach those conclusions."

"Why is your idea termed 'the theory of relativity?'" asked the correspondent.

"The term *relativity* refers to time and space," Dr. Einstein replied. "According to Galileo and Newton, time and space were absolute entities, and the moving systems of the universe were dependent on this absolute time and space. On this conception was built the science of mechanics. The resulting formulas sufficed for all motions of a slow nature; it was found, however, that they would not conform to the rapid motions apparent in electrodynamics.

"This led the Dutch professor, Hendrik Lorentz, and myself to develop the theory of special relativity. Briefly, it discards absolute time and space and makes them in every instance relative to moving systems. By this theory all phenomena in electrodynamics, as well as mechanics, hitherto irreducible by the old formulas—and there are multitudes—were satisfactorily explained.

"Till now it was believed that time and space existed by themselves, even if there was nothing else—no sun, no Earth, no stars—while now we know that time and space are not the vessel for the universe, but could not exist at all if there were no contents, namely, no sun, Earth, and other celestial bodies.

"This special relativity, forming the first part of my theory, relates to all systems moving with uniform motion; that is, moving in a straight line with equal velocity.

"Gradually I was led to the idea, seeming a very paradox in science, that it might apply equally to all moving systems, even of difform motion, and thus I developed the conception of general relativity which forms the second part of my theory.

"It was during the development of the formulas for difform motions that the incident of the man falling from the roof gave me the idea that gravitation might be explained by difform motion."

"If there is no absolute time, or space, supposedly forming the vessel of the universe," the correspondent asked, "what, becomes of the ether?"

"There is no ether, as hitherto conceived by science, which is proved by the well-known experiment of the celebrated American savant, Albert Michelson, showing that no influence by the motion of the Earth on the ether is perceptible through change in velocity of light, such as ought to be produced if the old conception were true."

"Are you yourself absolutely convinced of the correctness of this revolutionary theory of relativity, or are there still any reservations?"

"Yes, I am," Dr. Einstein answered. "My theory is confirmed by the two crucial cases mentioned before. But there is still one test outstanding, namely, the spectroscopic. According to my theory, the lines of the spectra of fixed stars must be slightly shifted through the influence of gravitation exerted by the very stars from which they emanate. So far, however, the results of the examinations have been contradictory; but I have no doubt of final confirmation, even through this test."

Just then an old grandfather's clock in the library chimed the midday hour, reminding Dr. Einstein of some appointment in another part of Berlin, and old-fashioned time and space enforced their wonted absolute tyranny over him who had spoken so contemptuously of their existence, thus terminating the interview.

—*December 3, 1919*

Lights All Askew in the Heavens

Efforts made to put in words intelligible to the non-scientific public the Einstein theory of light proved by the eclipse expedition so far have not been very successful. The new theory was discussed at a recent meeting of the Royal Society and Royal Astronomical Society. Sir Joseph Thomson, president of the Royal Society, declares it is not possible to put Einstein's theory into really intelligible words, yet at the same time Thomson adds:

"The results of the eclipse expedition demonstrating that the rays of light from the stars are bent or deflected from their normal course by other aerial bodies acting upon them and consequently the inference that light has weight form a most important contribution to the laws of gravity given us since Newton laid down his principles."

Thomson states that the difference between theories of Newton and those of Einstein are infinitesimal in a popular sense, and as they are purely mathematical and can only be expressed in strictly scientific terms it is useless to endeavor to detail them for the man in the street.

"What is easily understandable," he continued, "is that Einstein predicted the deflection of the starlight when it passed the sun, and the recent eclipse has provided a demonstration of the correctness of the prediction.

"His second theory as to the anomalous motion of the planet Mercury has also been verified, but his third prediction, which dealt with certain sun lines, is still indefinite."

Asked if recent discoveries meant a reversal of the laws of gravity as defined by Newton, Sir Joseph said they held good for ordinary purposes, but in highly mathematical problems the new conceptions of Einstein, whereby space became warped or curled under certain circumstances, would have to be taken into account.

Vastly different conceptions which are involved in this discovery and the necessity for taking Einstein's theory more into account were voiced by a member of the expedition, who pointed out that it meant, among other things, that two lines normally known as parallel do meet eventually, that a circle is not really circular, that three angles of a triangle do not necessarily make the sum total of two right angles.

"Enough has been said to show the importance of Einstein's theory, even if it cannot be expressed clearly in words," laughed this astronomer.

Dr. W.J.S. Lockyer, another astronomer, said:

"The discoveries, while very important, did not, however, affect anything on this Earth. They do not personally concern ordinary human beings; only astronomers are affected. It has hitherto been understood that light traveled in a straight line. Now we find it travels in a curve. It therefore follows that any object, such as a star, is not necessarily in the direction in which it appears to be astronomically.

"This is very important, of course. For one thing, a star may be a considerable distance farther away than we have hitherto counted it. This will not affect navigation, but it means corrections will have to be made."

One of the speakers at the Royal Society's meeting suggested that Euclid was knocked out. Schoolboys should not rejoice prematurely, for it is pointed out that Euclid laid down the axiom that parallel straight lines, if produced ever so far, would not meet. He said nothing about light lines.

Some cynics suggest that the Einstein theory is only a scientific version of the well-known phenomenon that a coin in a basin of water is not on the spot where it seems to be and ask what is new in the refraction of light.

Albert Einstein is a Swiss citizen, about 50 years of age. After occupying a position as professor of mathematical physics at the Swiss Federal Polytechnic School in Zürich and afterward at Prague University, he was elected a member of Emperor William's Scientific Academy in Berlin at the outbreak of the war. Dr. Einstein protested against the German professors' manifesto approving of Germany's participation in the war, and at its conclusion he welcomed the revolution. He has been living in Berlin for about six years.

When he offered his last important work to the publishers, he warned them there were not more than 12 persons in the whole world who would understand it, but the publishers took the risk.

—*November 10, 1919*

Science Seeks Secret of Life in Star Rays

By WALDEMAR KAEMPFFERT

That the Earth is constantly flooded with X-rays which come from distant heavenly bodies and are at least a thousand times more penetrating than any produced on the Earth, is the most striking discovery that twentieth-century science has thus far made. As a result of the studies of the rays made by the Swiss physicists, Albert Gockel and Victor Francis Hess, the German Werner Kolhörster, and the American Nobel Prize winner, Professor R. A. Millikan, astronomers have been compelled to abandon established theories on the way stars are evolved from glowing masses of gas called nebulae. The question has even been raised whether these cosmic X-rays may not have played their part in the creation of life on the Earth.

That there is some relation between X-rays and the activity of living cells cancer specialists will readily concede. Although the cause of cancer is still to be determined, it is definitely known that, for some reason, living cells suddenly begin to breed wild and to produce cancerous growths. It is also known that X-rays can both stimulate cells into reproductive activity and check their growth. If the comparatively feeble X-rays which puny laboratory machines generate can affect living matter so profoundly, what may not be expected of the enormously more powerful rays generated on a cosmic scale in nebulae? Clearly, there is some relation between life—perhaps the origin of life—and these mysterious X-rays that come to us from space. Science once scoffed at the alchemists because they cherished the illusion that base metals could be transmuted into gold. Now transmutation of one element into another is a proven fact. Must we similarly concede that the astrologers of old who insisted that the stars influence life were not hopelessly blind?

To explain these cosmic X-rays we must begin with electrons, infinitesimal particles of electricity, the stuff of which atoms are composed. In size an electron is to an atom what a baseball is to a cathedral. Every atom is composed of a central nucleus, or sun, around which one or more electrons revolve—a miniature solar system. One atom differs from another solely in the number of electrons revolving around the nucleus. If we could add electrons to a nucleus or snatch them away, it would be easy to transmute base metals into gold.

Free Electrons

In an atom electrons are under restraint—held in their orbits by the powerful attraction that they and the nucleus have for each other. But there are free electrons, too. A dynamo pushes them through a wire, and then we speak of a current of electricity. Hot bodies project them freely into space—the glowing filaments of the vacuum tubes used in radio broadcasting and reception. Indeed radio is now utterly dependent on the control of electrons thus emitted, and so is transatlantic telephoning. In an X-ray tube, electrons are shot in a visible, glowing stream against a piece of metal. The atoms of the metal are violently agitated, and their agitations generate in the ether minute, invisible waves, which we call X-rays and which penetrate wood and flesh. So, the origin of the cosmic X-rays, which are now transforming our conceptions of the stars, must be sought in electrons shot forth at a terrific speed and under terrific pressure from some heavenly body.

It takes a very dense metal, like lead, to stop X-rays, whether they are generated by nature in her radium or by man in his tubes. Hence it is customary for the operators of X-ray machines to protect themselves by standing behind lead-glass screens when they manipulate switches, or by wearing lead-lined gloves and aprons. Radium is always kept in safes thickly lined with lead. In some X-ray and radium laboratories, even the walls, floors and ceilings are lined with lead.

X-rays from tubes or radioactive substances have a curious effect on air. They create a terrific commotion in the atoms of which the air gases are composed. The effect may be compared with that which would be produced if some external force were to tear the solar system apart. Every atom, we must not forget, is a miniature solar system. When the X-rays are directed on the atoms of which the air is composed, they tear away electrons. These homeless electrons wander about trying to find a resting place. They are called "ions"—travelers—and so are atoms which have lost an electron. Air in which ions move freely about is electrically conductive or "ionized." When it is not ionized air is a poor conductor.

What the Electroscope Tells Us

Whether air is thus ionized by X-rays or the gamma rays of radium is easily determined by a very sensitive instrument called an electroscope. It has two delicate gold leaves, which dilate and contract as it is affected by slight electric charges. A very old instrument, the electroscope was long ago used to test the electrical conductivity or ionization of the air.

It was observed that a charged electroscope always leaked. It could not hold a load of electricity indefinitely. After much experimenting, physicists came to the conclusion that the air draws off the charge because air is always ionized more or less. The electrons of the air, if they happen to be of the right kind (positive or negative) pull out the electrons of the electroscope, and this at a rate which can be measured.

But what ionizes the air of the atmosphere—makes it conductive and enables it to discharge an electroscope gradually? The answer was not found until radium and other radioactive substances were discovered. Here were sources of gamma rays—X-rays. Air could be ionized by X-rays. Therefore radioactive substances in the earth must ionize the atmosphere and enable the air to affect electroscopes. Mathematical computations proved the theory. It was found that 130 tons of radium distributed in the Earth's crust were enough to ionize or electrify the air. When it was determined that the Earth's crust actually did contain about that amount of radioactive substances, and in a layer about eight inches thick, the case for the gamma rays was regarded as proved.

It occurred to the Swiss physicist, Gockel, that it would be a scientifically grateful task to measure the amount of this ionization of the air at high altitudes. After all, previous measurements had been made in laboratories, and no one knew how far-reaching the effect of the gamma or X-rays of radioactive substances really was. So Gockel took up electroscopes in balloons to a height of 13,000 feet. He expected to find less ionization at that height than on the ground. A minute quantity of radium in the Earth's crust could hardly send out gamma rays which would have an effect miles above sea level. There must be a gradual decrease in their energy. But Gockel found no decrease at all. Actually, the gamma rays seemed to be just as active at 13,000 feet as they were on the ground.

Another Experiment

Here was a scientific puzzle to be solved. Another Swiss, Hess, and the German, Kolhörster, determined to go still higher. Their measurements were made at an altitude of 5.6 miles. Hess and Kolhörster found that not only was Gockel right, but that he was much too conservative.

Whatever the rays were that ionized or electrified the air, they were eight times more powerful at great heights than at the surface of the Earth. These cannot be gamma rays that affect instruments, Hess and Kolhörster reasoned. Radium buried in rocks surely cannot ionize air at a height of over $5\frac{1}{2}$ miles. Where do the rays come from? They must have a cosmic origin, Kolhörster concluded. They strike the Earth from some unknown point in space.

It was just before the outbreak of the war that Hess and Kolhörster made their announcement of cosmic X-rays more penetrating than man could produce. So startling was their discovery that scientists were not quite prepared to accept it. Besides, might not the atmosphere be radioactive on its own account, and might not these supposedly cosmic rays be thus explained?

After the treaty of peace was signed and physicists and chemists gave up bombs, machine guns and poison gases to return to electrons, X-rays and radium, Professor Millikan, who had won the Nobel Prize for the extraordinary feat of precisely measuring the charge of a single, invisible, infinitesimal electron, determined to settle once and for all whether Hess and Kolhörster's rays came from some heavenly body or from the atmosphere.

First of all Millikan devised an extremely sensitive electroscope of his own and provided it with the means of writing down a record of its impressions of ionization. He decided for the time being not to make any allowances for the atmosphere, but to reach even greater heights than Hess and Kolhörster had attained.

At 10 Miles' Height

Up floated a balloon, mile after mile, mile after mile, until it reached a height of nearly 10 miles. It was a balloon of the kind meteorologists use to explore the upper atmosphere—a little gas bag freighted only with accurate scientific recording instruments. When it came down and Millikan recovered it with its precious load, he read an even more thrilling story than any told by Gockel, Hess or Kolhörster. The higher the altitude the more powerful the rays became.

Now came the crucial test of excluding the atmosphere entirely. Millikan decided to sink his instruments in the water of a mountain lake. He had to be high, so as to cut off the radioactivity of the ground as much as possible. He had to sink his instruments in water, because water would exclude the atmosphere. Not ordinary water, but water which had never touched the ground at the surface of the Earth and never been contaminated by radioactive substances. He must have noncommital water. And so he looked around for a mountain lake fed by melting snows.

He found his lake beneath the brow of Mount Whitney—Lake Muir. Into its icy water, 11,800 feet above the level of the sea, he sank his instruments foot by foot, and foot by foot he took the readings. The rate of discharge of the electroscope decreased as he reached a greater and greater depth. At 50 feet it became constant.

Constancy could mean but one thing: The rays were indeed of cosmic origin. They came from overhead. Why, it is easy enough to grasp. Millikan knew that the atmosphere absorbs X-rays. In fact, its absorbing power at Lake Muir is equal to 23 feet of water. He had reached a depth of 50 feet. That made a total absorbing power of 73 feet of water. And 73 feet of water—that was equivalent to six feet of lead.

Here were rays so powerful that they could actually pierce six feet of lead and still influence an electroscope. A few inches of lead will stop the most penetrating X-rays that can be produced in a laboratory. How puny were man's efforts compared with those of the cosmos! With what enormous energy something in space must be bombarding the Earth! These rays that come from somewhere in space reach us with a vigor seemingly undiminished; and yet they have probably traveled with the speed of light millions, billions, probably quintillions, of years.

Before he announced his triumph Millikan determined to check himself. There must be no doubt about the origin of these rays or about their almost incredible penetrating power. He took his instruments 300 miles farther south to snow-fed Arrowhead Lake, which is 6,675 feet lower than Lake Muir. He figured out the difference in absorbing power of the atmosphere between the two elevations. Six feet was the answer. If he lowered his electroscope into Arrowhead Lake, the reading at a given depth ought to agree with a corresponding reading taken six feet lower in Lake Muir. He made the experiment. The readings tallied. The case for cosmic origin was proved beyond a doubt.

Cosmic origin? That is not specific enough. From precisely what point or body in space do these rays come? Because he could find no measurable variation in the intensity of the rays between midday and midnight, Millikan concluded that the rays shoot through space equally in all directions. Then Kolhörster returned to the scene. He climbed the Jungfraujoch in the Alps, chopped a deep hole in the ice, placed his instruments in the hole and let the hole sweep as much of the heavens as it could while the Earth turned on its axis.

Whence the Strongest Rays Emanate

Kolhörster found that when the Milky Way was most nearly overhead, the rays were most intense. And the greatest intensity was observed when the constellations of Hercules and Andromeda were best placed relative to the hole.

All this does not mean that Millikan was wrong. The rays do come from all points in space, but the chief center of X-ray activity seems to be something in Andromeda.

What is this something? Turn the telescope on Andromeda. Here is a vast, spinning mist of incandescent gas—an island universe in which stars are being born. This is nature's X-ray tube, according to Professor J. H. Jeans, president of the Royal Astronomical Society and one of the foremost astrophysicists of our time.

Professor Jeans recently delivered a lecture in London in which he showed where cosmical physics now stands and the overwhelming import of the X-rays shot forth by the spiral nebulae, particularly that in Andromeda. That lecture presented in vivid form the opinions which are held by most astronomers and seek to explain these mysterious, powerful rays in terms of the stars, the progressive degradation of the decrepit Earth and life itself. For that reason it is summarized in the following paragraphs.

In laboratories on the Earth, physicists have succeeded in stripping a few atoms of their outer electrons and leaving the nuclei more or less bare, but no scientist has yet succeeded in combining electrons to build up matter to suit himself. It is now generally thought that in stars atoms are similarly broken up. But nature works with such enormous pressures and temperatures that she can build matter as well as destroy it. Within a star, which should be conceived as a sun, the temperature must be 31.5 million degrees Centigrade. No atom can survive that heat. Hence Jeans and others regard the core of a star as a mixture of bare nuclei and electrons, trying to unite into an atomic system, but prevented from doing so in the face of the heat and general whirl. But as we pass outward and the temperature falls, nuclei and electrons have a chance to attract one another and to build up atoms. Thus in the crucible of a hot star matter is being created.

Within the star there is radiation—short, hard X-rays more like the gamma rays of radium than any X-rays we can produce. So dense is the structure of a star that this radiation never has a chance to escape as such. These ultra-X-rays are transformed into ordinary light in accordance with principles discovered by Professor Arthur H. Compton. Hence we cannot expect to receive X-rays from such bodies as the sun.

How Radiation Is Produced

Turn now to the nebulae, such as that spiral cloud in Andromeda. They are diaphanous in structure compared with any star. Atoms are spaced so far apart that it is easy for X-rays to slip through them and reach the Earth. Huge as Andromeda is we never seemed to receive from it the amount of radiation expected. Now we know the reason. We looked for light which could be seen, but most of Andromeda's energy reaches us in the form of invisible X-rays. In fact Jeans concludes that only about $1/3,000$th part of the total radiation generated within Andromeda is transformed into visible light; the remaining 2,999 parts are the cosmic rays measured by Millikan. Fully one-half the X-rays that strike the Earth come from Andromeda, according to Jeans. Furthermore, "the primary radiation of the universe is not visible light, but short-wave radiation of a hardness which would have seemed incredible at the beginning of the present century."

Now, it appears, the primary physical process of the universe is the conversion of matter into radiation, most of it of the X-ray type. And how is this radiation produced? Solely by the annihilation of matter, in the eyes of Jeans. We see the process on the Earth in the vestige of the radium still left.

By *annihilation* Jeans means something very different from what occurs when coal is burned. "Whereas the ordinary combustion of a ton of coal provides energy enough to drive an express locomotive for an hour, the annihilation of a ton of coal would provide enough energy for all the heating, lighting, power, and transport in Great Britain for a century." In no other way can the enormous amounts of energy radiated be accounted for.

313

We must therefore start with matter heavier than any we know on this Earth—heavier than uranium. It is matter, which, like uranium and radium, is breaking up into something else in the course of millions of years. Heat and pressure cannot accelerate this process. "What is annihilating the matter of the stars," says Jeans, "is merely the passage of time."

A Supreme Question

So it is possible to arrange stars according to the heaviness of the matter of which they are composed. And the heaviest stars invariably prove to be the youngest. Somewhere in the scale of progressively older bodies we must place the sun and with it the Earth. At once we are confronted with the question: How did the degradation of matter produce life?

"Primeval matter must go on transforming itself into radiation for millions and millions of years to produce an infinitesimal amount of inert ash on which life can exist," Jeans decides. So the Earth is a kind of cinder, the residue of a planetary lump of disintegrating matter endowed with the property of self-annihilation.

Hence, Jean concludes, life may be "the final climax toward which the whole creation moves, for which the millions and millions of years of transformation of matter in uninhabited stars and nebulae, and of waste radiation into desert space have been only an incredibly extravagant preparation." Or perhaps it may be "a mere accidental and possibly quite unimportant by-product of natural processes which have some other and more stupendous end in view." Worse still "is it of the nature of a disease which affects matter in its old age?" Or, finally, "is it the only reality which creates instead of being created by the colossal masses of the stars and nebulae and the almost inconceivably long vistas of astronomical time?"

—January 23, 1927

Giant Telescope of Immense Range to Dwarf All Others

Announcement was made by the California Institute of Technology at Pasadena that funds have been provided for the erection and maintenance of a great telescope of the reflector type, with a mirror 200 inches in diameter, planned to surpass by from five to 10 times the power of the present largest astronomical instrument in the world, the 100-inch Hooker telescope on Mount Wilson, near Pasadena, and for the construction of an astrophysical laboratory to supplement the work of the telescope. The actual construction work may be begun within a few months.

International Education Board Gift

The new telescope and laboratory are a gift from the International Education Board, whose headquarters are in New York, to the California Institute of Technology, which will supply money for their maintenance. Their operation will be directed by the institute and the Mount Wilson Observatory of the Carnegie Institution, acting in close cooperation. Search is now being made for a mountaintop, not too far from Mount Wilson, that will be suitable as a site for the telescope. The astrophysical laboratory will be set up on the campus of the institute. Owing to the labor and time required for casting and grinding the mirror, a few years must elapse before the telescope can be ready for use. The amount of the funds provided for the enterprise is not made public, but it must be large, for the 100-inch telescope on Mount Wilson cost nearly $600,000.

It is expected that the new telescope will penetrate hundreds of millions of light-years into space, according to some estimates even a billion light-years, and bring under observation countless millions of now unseen stars and nebulae, opening up a vast unexplored field of knowledge, besides adding much to present knowledge of the nearer objects visible with the aid of existing instruments. In the astrophysical laboratory, stellar conditions as recorded by the telescope with the aid of the camera, the spectroscope and other instruments will be studied and, when possible, experimentally reproduced for further analysis.

Hundreds of Millions of New Stars

It is estimated that the 60-inch telescope on Mount Wilson, which was for years the largest in active use in the world, has within its photographic range about 1 billion stars. The 100-inch telescope, added approximately 500 million more stars to that number. Although Dr. Walter S. Adams, director of Mount Wilson Observatory, points out that the number of stars within our own system appears to be decreasing as the distance from the Earth is increased, he expects that the 200-inch telescope will add several hundred million more stars to those now visible. It may also solve some of the puzzling mysteries of the nebulae and extend the theory that many of the spiral nebulae are really "island universes."

The measurement of the diameters of stars, a branch of astronomical work recently developed, will also be much extended by means of the new telescope. The 20-foot interferometer, designed by Professor A. A. Michelson of Chicago, and attached to the 100-inch telescope, has enabled observers at Mount Wilson to measure the diameters of a few of the larger stars. The 200-inch telescope will make possible the use of a much larger interferometer and the measuring of more distant and smaller stars. The interferometer will also be useful in the study of very close double stars.

These and other uses of the new instrument and its accessories are fully set forth in the formal announcement published below.

An "Observatory Council" has been appointed by the California Institute of Technology under the chairmanship of Dr. George Ellery Hale, former director of Mount Wilson Observatory, which will have full responsibility for the completion of the project and the expenditure of the funds. Its plans call for a preliminary study by a group of experts of methods of making telescope mirror disks of the size to be used in this case, and a detailed study by a group of astronomers, physicists, opticians, instrument makers and engineers of the various optical and mechanical questions involved in the design of the telescope and all other instruments and buildings to be provided.

Problem of the Great Mirror

The construction of the 200-inch mirror presents the most difficult problem. For the first time in the manufacture of large mirrors an attempt will be made to use fused quartz, which is almost proof against the changes of temperature that affect glass and other materials of which mirrors are made. A glass disk of the size to be used in the new telescope would become so heated after 10 minutes or so of grinding as to require a rest of 24 hours to cool off. There would be no such interruption of the grinding of a quartz disk. This indifference of quartz to the effects of heat accounts for part of the expected high power of the new instrument. It will probably require months to cool the fused quartz in the mold. Experiments with this substance are now being made by Dr. Elihu Thomson, director of the Thomson Research Laboratory of the General Electric Company.

A block of fused quartz of the proportions that would be required for the 200-inch mirror will weigh approximately 30 tons. As cast, this will probably not be free from bubbles, but on its surface will be fused a layer of crystal-clear quartz. This will be ground to a paraboloidal surface, which must be accurate to the order of one 500,000th of an inch. In grinding the mirror an iron disk, the same size as the mirror, will first be used. Later, when the mirror is found to be near the required figure, a wooden disk will be substituted for the iron one and jeweler's rouge for emery. The surface of the wooden disk will be covered with squares of resin coated with beeswax to keep the surface of the mirror from being scratched. The last step is to give the paraboloidal surface a coating of silver to secure the necessary reflecting qualities.

Because of its size and the possible dangers that would be encountered in trucking the huge mirror over mountain roads, the erection of shops at the site of the observatory, in which the mirror could be cast and ground, is under consideration. The numerous smaller auxiliary mirrors that will be used in connection with the 200-inch reflector will be made elsewhere and transported to the observatory site.

It has not been decided whether the dome of the observatory is to be of the hemispherical or the drum type, but in order to keep down its size and cost, and for other reasons, the telescope will be made as short as is consistent with the best work. This will be accomplished by using a focal ratio of probably 3.3 instead of the ratio of 5 used in the 100-inch telescope. In this case the focal length of the telescope, but little short of the actual length, will be 200 inches multiplied by 3.3, or 55 feet.

—*October 29, 1928*

Studies of the Cosmic Ray
Point to Endless Creation

By WILLIAM L. LAURENCE

At a meeting of the National Academy of Science at Pasadena, Dr. Robert Millikan, who finds himself thoroughly at home among infinitesimal entities, told of his latest researches in the far-off realm of the cosmic ray, which he himself had originally discovered in the summer of 1925 and which may well prove to be the greatest scientific achievement of our age—a signpost somehow pointing the way to the very doorstep of Creation's laboratory, in which matter itself is fashioned out of primeval electrons and protons.

It is doubtful if any single scientific discovery announced in the breathtaking age of exploration in the unknown appeals to the imagination of men as much as the cosmic ray. Men of the twentieth century must contemplate these new mysterious entities with an awe similar in its nature to the awe in which primitive man, believing in the existence of animated spirits dwelling invisibly all around him, stood in the presence of thunder, lightning, fire, earthquakes, volcanoes and other elemental forces, which he knew had the power to do him harm or good but over which he had absolutely no control.

Knowledge through Research

For mysterious as these rays still are, the researches of Dr. Millikan have given us an amazing amount of information about them in the short five years since their discovery. We not only know for certain that they exist, we know also their wavelength and their power of penetration. Most interesting of all, from the human standpoint, we know we are exposed to their action 24 hours a day no matter when we are.

Now, by analogy, we know that short, high-frequency waves, from ultraviolet down, have some effect upon our health. Ultraviolet rays are healthful; X-rays and gamma rays of radium, if used in the right quantities, are beneficial in the treatment of tumors and cancer and very harmful to health when not properly used. But these cosmic rays are hundreds of times greater in their penetration than any of these. The strongest of them penetrate 16 feet of lead.

Yet for the present we do not know what the effect of continual exposure to these smallest and most powerful rays so far discovered may be. As someone said, this radiation may be necessary to life, or, again, it may be the very thing which causes our gradual disintegration and leads to our final destruction at the end of 70 years or so. But what we do know is that these cosmic rays break up millions of atoms every second in the body of each one of us; whether for the better or for worse we do not know.

The energy of this cosmic ray is so great that it would take a 60-million-volt X-ray tube to produce some of its milder forms.

Studies of the Discoverer

Since the announcement of his original discovery in 1925, there has hardly been a year in which Dr. Millikan has not added some new knowledge on the latest among the wonders of science. In the first years after definitely establishing their existence, Dr. Millikan devoted himself to studies as to their source and origin. These researches are still going on and still continue to be the center of his investigations. However, there have been many colorful by-products developed by him through his studies in this field, some of which are likely to prove revolutionary in our concepts about the end of the universe and the creation of matter.

Dr. Millikan recently returned from Churchill on the west shore of Hudson Bay, about 700 miles south of the magnetic pole. To that remote point he had carried a perfected electroscope, protected by lead shields weighing 600 pounds, and watched night and day for a week to determine the effect of the magnetic currents on the intensity of the cosmic rays. His latest observations, he told the scientists at Pasadena, have led him to three important findings, one of these pointing to a means of putting the cosmic ray to work—finding a practical use for it for the first time.

It is upon a humble task that this giant among cosmic forces is employed in his first harnessing by man. This power that some day may supply all the world's energy can be used now at man's bidding only to forecast the weather. Undoubtedly, much greater and more important uses will be found for it in the near future, for it must be remembered that radio waves are not yet 50 years old. Yet even this first humble task is likely to be of vast practical importance to science.

Rays and Earth's Atmosphere

Daily fluctuations in the intensity of the rays observed at Churchill were found by Dr. Millikan to be connected with the changes in the Earth's atmosphere; it was determined that the constancy of the cosmic rays could be put to work to detect them.

"The air," Dr. Millikan told his colleagues, "is simply an absorbing blanket interposed between us and a constant source of radiation coming into the Earth uniformly from all directions. Every eruption or wave or ripple in that blanket is accurately reflected by the cosmic ray electroscope of the type used in the investigation."

Important as this new discovery may be in its immediate application, and from the point of view of the layman, Dr. Millikan made one other announcement which is of even greater interest to the scientist.

There were some physicists who held, and some who still hold, that these cosmic rays might be high-speed electrons. Dr. Millikan, who disagrees with this theory, offered proof of his own theory that the rays come to the Earth as ether waves of extremely short length. Theoretical calculations by Dr. P. S. Epstein, one of Dr. Millikan's colleagues, showed that if the high-speed electron theory of cosmic radiation were correct, there should be a larger concentration near the magnetic fields of the Earth because of the influence of the Earth's field upon the speeding electrons. But Dr. Millikan reported he had not found the slightest difference in the intensity of cosmic rays at Churchill and at Pasadena.

The Shortest Waves Known

These "ether waves of extremely short wavelength" are by far the shortest ether waves that science has yet discovered. Light is the most familiar variety of ether wave. Heat waves that come to us from the sun are also ether waves, differing from light only in the wavelengths. Radio waves are merely elongated members of the same family. Ultraviolet waves are shorter than light waves but a thousand times longer than X-rays, which are only .000,001 millimeters long. Gamma rays from radium are midgets by comparison with X-rays, being .000,000,002 millimeters in length, but they assume dimensions that are staggering when put alongside the cosmic rays, the longest in the series being .000,000,000,067 and the shortest .000,000,000,040 millimeters.

In penetrating power, these cosmic rays are so effective that ordinary matter scarcely stops them at all. The power is by far the most penetrating so far discovered by man. In its weakest form it penetrates six feet of lead and in its strongest it goes through as much as 18 feet of lead. This is made possible by the extreme shortness of the rays. The atoms of matter may be exceedingly small, so small that many billions of them can dance on the point of a needle with room to spare. Yet cosmic rays are smaller still. The longest have wavelengths less than one-thousandth the diameter of an atom. The shortest are probably even smaller than the electrons, the stuff of which atoms are made. Hence they can sift through the chink between the electrons like sand through a sieve.

Vibrations at Light Speed

In this connection, however, it must be remembered that no matter how high or how low the frequency of the vibration of any of the varieties of waves may be, they all travel at the same speed, the speed of light, or 186,300 miles per second in a vacuum.

The story of the discovery of the cosmic ray is one of the romances of modern science. Like all scientific romances it began with a tiny little clue, avidly snatched at and pursued relentlessly to its ultimate source. It began in 1901, when a British scientist, Professor C.T.R. Wilson of Cambridge, found that an electric charge in an electroscope, even when tightly sealed up in a case, somehow leaked out.

Professor Wilson tried to find an explanation. The only reason he could think of was that the leakage was due to the presence of radium or some other radioactive substance. A little radium is present almost everywhere, even in the air itself. The smallest amount of radium would send out gamma rays, which in turn would create ions. These ions would discharge the gold leaves in an electroscope.

This clue was taken up in 1902 by Sir Ernest Rutherford, then working in Canada, and by Dr. J. C. McLennan of the University of Toronto. They covered their apparatus with five tons of solid lead to keep all radioactive substances out of it. They carried it in the winter to the middle of the frozen surface of Lake Ontario, far from shore, to eliminate any possible effect of radium in the soil. In spite of all this, the electricity still kept leaking out of their electroscope. Some mysterious "electricity thief" somehow still managed to work through five tons of lead, no matter where it was placed.

Research Broken by War

It was not until 1915 that Dr. Millikan took up the hunt for the elusive mischief-maker. He no sooner got started, however, than the war intervened. Dr. Millikan became a member of the scientific corps of the army. Seven years more passed. In 1922, Dr. Millikan, with Ira Sprague Bowen, sent up captive balloons at Kelly Field, near San Antonio. Each one of these carried a specially constructed recording machine. It was a marvel of ingenuity. Although it held 300 cubic centimeters of air at 150 pounds pressure, a barometer, thermometer, electroscope, three sets of motion-picture films and a driving mechanism, it weighed only seven ounces. It took Millikan months to design and build it.

This apparatus ascended to a height of almost 10 miles, above nine-tenths of the Earth's atmosphere. The mysterious forces at work were stronger than ever. Could these be gamma rays of radium after all? No, said Millikan. The power of these gamma rays to penetrate air was by that time known. They are greatly weakened after passing a relatively thin layer of air. Where the air is very rare, their power should increase at a much greater rate than the results of the experiment showed. Hence Dr. Millikan reached two conclusions—these rays are much stronger than gamma rays, and they come from the stars in outer space.

The next step was to find out how penetrating these rays were. This time, instead of experimenting on the heights, he made his laboratory the depths of a snow-fed mountain lake, to see how deeply the rays penetrated the water before being stopped. The scene was Muir Lake, at 11,800 feet elevation, just under the summit of Mount Whitney, the highest peak in the United States. There, in August 1925, he sank his instruments to various depths down to 60 feet.

It was here for the first time since the hunt began in 1901 that a spot was found where the "electricity burglar" could not penetrate. At a depth of 60 feet below the surface of the lake the readings on the electroscope stopped, showing that the new rays could penetrate no further. But other experiments, conducted at Lake Constance, in Switzerland, showed that these cosmic rays vary in length, and that some of them, the shortest ones, are capable of penetrating 200 feet of water, an equivalent of 18 feet of lead, and that they are continually showered upon the Earth from all points of space.

Since then Dr. Millikan has gone far afield in the development of a new cosmology based on the results of his further experiments with the cosmic ray. This new cosmology has as its central point the hypothesis that the creation of worlds and of matter is a never-ending process that goes on today and will go on forever. Out of the cosmic ray he fashioned for science a new weapon with which to give battle to the dreadful Second Law of Thermodynamics, according to which the universe must come inevitably to destruction.

"Energy cannot go downhill forever," says Sir James Jeans, who, with Eddington, is firmly convinced of the doomsday of the universe, "and, like a clock weight, it must touch bottom at last."

Against this seemingly deadly weapon of the Second Law of Thermodynamics, Dr. Millikan goes forth in the shining armor of his cosmic ray.

Two Universes Pictured

In his vision, one of the truly noble visions of modern science, Dr. Millikan paints the existence of two universes. One is the vast, "finite though boundless" universe already created, the stellar universe, which may be styled the "universe of being." The other is an equally vast, perhaps even vaster, universe-in-the-making, the interstellar universe, that vast void outside the fringe of the remotest stars, which we may call the extrastellar universe, or the "universe of becoming."

It is only in this stellar "universe of being" that Dr. Millikan admits the complete sway of the Second Law of Thermodynamics. In this universe, energy, as far as we know, is running downhill. It is here that energy is obtained by transformation from higher levels of availability to lower levels, by "the self-destruction in suicide pacts" of electrons and protons, liberating their atomic energy in a cosmic "cry of agony."

But there is this other "universe of becoming," where "the Creator is continually on the job," a universe endlessly in the making, where the Second Law of Thermodynamics is a complete outsider. In this universe, instead of the death agony of annihilated atoms, there is the incessant "birth-cry" of atoms coming into existence. And, says Dr. Millikan, you can hear this "birth-cry" filling the vastness of space, coming to us on the waves of the cosmic ray.

For it is not a mere poetic vision that Dr. Millikan paints for us. "The physical sciences," he says, "far from destroying the right to believe in the continuous formation of the heavy elements out of the lighter, have, by inductive reasoning, based on experimental fact, apparently traced the work of a creative force to its very source and can report to humanity that such creation is probably even now occurring throughout the vast regions of interstellar space."

What is this "inductive reasoning, based on experimental fact"? Before the discovery of the cosmic ray, and the further study of its nature by Dr. Millikan himself within the past year or so, such experimental facts did not exist. These researches, along with further studies along the same lines, promise to furnish keys to doors of nature hitherto securely locked against the scrutiny of man. With these Dr. Millikan hopes to prove that the Second Law of Thermodynamics is no more infallible than was the Law of Conservation of Matter, which Einstein knocked for a row of pins.

Sources of the Rays

There are three possible sources, Dr. Millikan reasons, for the emission of these all-powerful cosmic rays. They may come from the energy released by the annihilation of atoms in the interior of the stars, or from the building of atoms in the stellar regions. Lastly, they may be the result of the birth of heavier atoms from the lighter ones in the interstellar spaces.

By applying the theories of Einstein and Aston to his experiments he concluded that the last source is the only possible one, according to our present knowledge. These experiments tend to produce evidence that the cosmic radiations are due to the building up in the interstellar spaces of the commoner heavy elements, such as iron, out of hydrogen atoms. More than that. From the penetrating power of the ray, Dr. Millikan was able to determine, he says, just which of the elements is being born in space.

Whenever a new atom is born, he tells us, a special kind of cosmic ray is emitted at its birth. There is a close relationship between the various types of cosmic rays and the elements composing the universe, each ray corresponding to the birth of an atom at some time in the evolution of creation. So far he has been able to identify the "birth-rays" of helium, oxygen, silicon and iron atoms.

While the heat of the interior of suns and stars is maintained by the atom-annihilating process, or the "death-cry," this birth of the atoms takes place in the interstellar regions. Why? Because, Millikan reasons, if these rays emanated from the stars they would have their energy cut down by absorption before they reached the Earth's surface. On the other hand, rays originating in the space between the stars would be unimpeded in their progress to the Earth's atmosphere.

This brings him to the final step of his inductive reasoning. Replying pointedly to Eddington and Jeans, he declares: "While this says nothing of the

Second Law of Thermodynamics, or the 'heat-death,' it might seem to contain a bare suggestion that if heavier atom formation from hydrogen is taking place in space as indicated, this hydrogen is somehow being replenished there from the only form of energy that we know to be leaking out constantly from the stars—namely, radiant energy.

"This has been speculatively suggested many times before. Here is, perhaps, a little bit of experimental finger-pointing in that direction."

—September 28, 1930

New Radio Waves Traced
to Center of the Milky Way

Discovery of mysterious radio waves which appear to come from the center of the Milky Way galaxy was announced by the Bell Telephone Laboratories. The discovery was made during research studies on static by Karl G. Jansky of the radio research department at Holmdel, New Jersey, and was described by him in a paper delivered before the International Scientific Radio Union in Washington.

The galactic radio waves, Mr. Jansky said, differ from the cosmic rays and also from the phenomenon of cosmic radiation, described recently before the American Philosophical Society at Philadelphia by Dr. Vesto M. Slipher, director of the Lowell Observatory at Flagstaff, Arizona.

Unlike the cosmic ray, which comes from all directions in space, does not vary with either the time of day or the time of the year, and may be either a photon or an electron, the galactic waves, Mr. Jansky pointed out, seem to come from a definite source in space, vary in intensity with the time of day and time of the year, and are distinctly electro-magnetic waves that can be picked up by a radio set.

New Waves Have High Frequency

The cosmic radiation discovered by Dr. Slipher is a mysterious form of light apparently radiated independently of starlight, originating, Dr. Slipher concluded, at some distance above the Earth's surface, and possibly produced by the Earth's atmosphere.

The galactic radio waves, the announcement says, are short waves, 14.6 meters, at a frequency of about 20 million cycles a second. The intensity of these waves is very low, so that a delicate apparatus is required for their detection.

Unlike most forms of radio disturbances, the report says, these newly found waves do not appear to be due to any terrestrial phenomena, but rather to come from some point far off in space—probably far beyond our solar system.

If these waves came from a terrestrial origin, it was reasoned, then they should have the same intensity all the year around. But their intensity varies regularly with the time of day and with the seasons, and they get much weaker when the Earth, moving in its orbit, interposes itself between the radio receiver and the source.

A preliminary report, published in the *Proceedings of the Institute of Radio Engineers* in December 1932, described studies which showed the presence of three separate groups of static: static from local thunderstorms, static from distant thunderstorms, and a "steady hiss type static of unknown origin." Further studies in 1933 determine the unknown origin of this third type to be from the direction of the center of the Milky Way, the Earth's own home galaxy.

Direction of Arrival Fixed

The direction from which these waves arrive, the announcement asserts, has been determined by investigations carried on over a considerable period. Measurements of the horizontal component of the waves were taken on several days of each month for an entire year, and by an analysis of these readings at the end of the year their direction of arrival was disclosed.

"The position indicated," it was explained, "is very near to the point where the plane in which the Earth revolves around the sun crosses the center of the Milky Way, and also to that point toward which the solar system is moving with respect to the other stars.

"Further verification of this direction is required, but the discovery, like that of the cosmic rays and of cosmic radiation, raises many cosmological questions of extreme interest."

There is no indication of any kind, Mr. Jansky replied to a question, that these galactic radio waves constitute some kind of interstellar signaling, or that they are the result of some form of intelligence striving for intragalactic communication.

—May 5, 1933

New Dimensions Given to Universe

By WILLIAM L. LAURENCE

The physical universe, according to the latest cosmological calculations based on astronomical observation, is a finite sphere 6 billion light-years in diameter. It is composed of 500 trillion nebulae, each stellar unit being 80 million times as bright as the sun and about 800 million times as massive.

These and other new dimensions of space and its content were presented in Washington, D.C. before scientists attending the annual meeting of the National Academy of Sciences by Dr. Edwin Hubble of Mount Wilson Observatory, who is said to have "seen and probed more of the universe than any other person."

A little over a year ago Dr. Hubble and his colleagues found that nebulae 150 million light-years distant were rushing away from the Earth and from one another at the rate of 15,000 miles a second. This speed of recession, however, became slow indeed by comparison with the speeds of even more distant nebulae presented at the National Academy of Sciences meeting by Dr. Hubble.

Speed Rising with Distance

Nebulae are now observed by the 100-inch telescope at Mount Wilson, Dr. Hubble stated, at distances as great as 300 million to 450 million light-years and their speed of recession corresponds to velocities of 30,000 to 40,000 miles a second, or about one-fifth the velocity of light.

But this, he said, is only the frontier of the observable region. Observations so far show that the greater the distance of the nebulae the greater is the speed with which they are receding.

The speed had been found to increase at a definite rate, which is about 100 miles a second for each million light-years of distance.

As the latest cosmological figures show the physical universe to stretch 6 billion light-years in diameter, the nebulae at the outer rim of the "cosmic bubble" are rushing away at the speed of the velocity of light, 186,300 miles a second.

As the speed of light, according to the theory of relativity, is the maximum possible, and as the uttermost boundary of the physical universe coincides with a point where the outermost nebulae are traveling at the speed of light, this may be taken as another corroboration of relativity.

Beyond this boundary there is literally nothing, for space cannot exist without matter, according to relativity, and the latest cosmogony shows that beyond a point of 6 billion light-years there is no more matter, not even infinitesimal cosmic dust.

View of Expanding Universe

Dr. Hubble first dealt with space within the reach of the 100-inch telescope, finding a remarkable uniformity in this observable region, and then took his audience beyond its rim on an exploratory tour more daring than any hitherto undertaken by a cosmologist.

The observable region, vast as it seems, is less than a drop in a lake as compared with what as yet lies beyond it.

At present, Dr. Hubble stated, the observable region was an insignificant corner of space, no more than 600 million light-years in diameter, containing only 100 million nebulae, a mere handful alongside the entire physical universe with 500,000,000,000,000 times as many nebulae. Thus the observable region constitutes much less than one-tenth of 1 percent of the whole.

This picture of the physical universe, Dr. Hubble explained, was not in the dimensions of the present universe or as it was a few years ago. What is observed today, he pointed out, was the light which started on its journey from the distant nebulae 200 million to 300 million years ago. In these millions of years, he added, the universe had been expanding every second at the speeds he had given.

Hence, when the universe is termed 6 billion light-years in diameter, the dimensions are those of one instant of time 200 million or 300 million years ago.

Since then the universe has increased inconceivably in its dimensions and has decreased correspondingly in its density, though its mass and the amount of matter it contains have remained the same, and it has become a much emptier universe.

Minute Proportion of Matter

Presenting the latest figures on the density of the universe, in terms of that instant of time 200 million or 300 million years ago, Dr. Hubble said:

"The uniform distribution of nebulae means that, on the grand scale, the density of matter in space is uniform, and we can calculate the density.

"There is, on the average, one nebulae, 800 million times as massive as the sun, for every billion billion cubic light-years, or, roughly, one sun per each thousand million cubic light-years.

"In ordinary units this is equivalent to one gram per 10 to the power of 30 cubic centimeters, or one gram to a number of cubic centimeters represented by the numeral 1 followed by thirty zeros.

"This may be visualized as corresponding to a grain of sand in each volume of space equal to the volume of the Earth. The nebulae are scattered very thinly and space is mostly empty.

"In this calculation, we consider only the matter concentrated in the nebulae. There is doubtless matter scattered between the nebulae which has been ignored. How much, we do not know. We can only say that there is not sufficient to be detected— not enough to dim appreciably the most distant nebulae that can be observed.

"On the grand scale we may picture the stellar system [the system our sun belongs to] drifting through the universe as a swarm of bees drifting through the air. From our position somewhere within the system, we look out through the swarm of stars, past the borders, into the universe beyond.

It is empty for the most part—vast stretches of empty space. But here and there, at immense intervals, we find other stellar systems, comparable to our own. They are so distant that in general we do not see the individual stars. They appear as faint patches of light and hence are called nebulae, that is, clouds.

"The nebulae are great beacons scattered through the depths of space. We see a few that appear large and bright. These are the nearer nebulae.

"Then we find them smaller and fainter in constantly increasing numbers and we know we are reaching out into space farther and even farther until, with the faintest nebulae that can be detected with the greatest telescope, we have reached the frontiers of the known universe. The last horizon defines the observable region."

Dr. Hubble then described the principal features of the observable region, how the nebulae are distributed and how their distances are estimated and what the spectroscope reveals about their physical characteristics, especially their apparent extraordinarily rapid recession.

The spectra of nebulae, he said, exhibited a peculiar feature in that the absorption lines were not in their usual position but were displaced toward the red end of the spectrum and that the fainter the nebula, or the farther away it was, the greater was the shift in the direction of the red.

This relation between distance and degree of shift toward the red, Dr. Hubble stated, has been confirmed and extended to the limit at which spectra can be recorded with existing instruments. That is to say, out to 150 million light-years in space, the red shifts increase at a uniform rate.

Nebular Velocity and Distance

In discussing the significance of this characteristic, Dr. Hubble said:

"Red shifts are due either to actual motion or to some hitherto unrecognized principle of physics. On this interpretation, the nebulae are running away from us and the farther away they are the faster they are traveling. The velocities increase by roughly 100 miles per second for each million light-years of distance.

"The present distribution of nebulae can be represented on the assumption that they were once jammed together in our particular region of space, and at a particular instant about 2 billion years ago they started rushing away in all directions at various velocities.

"The slower nebulae, on this assumption, are still in our neighborhood, but the faster nebulae are now far away. The faster they are traveling the farther they have gone.

"The largest red shift actually recorded represents a velocity of about 15,000 miles per second at a distance of roughly 150 million light-years. But nebulae can be photographed out to distances twice or thrice the distances to which their spectra can be recorded.

"Hence, if the observed relation holds to the very frontiers of the observable region, we should encounter red shifts corresponding to velocities of 30,000 or 40,000 miles per second, say, one-fifth the velocity of light.

"Such red shifts are so enormous that we may expect appreciable indirect effects on colors and apparent luminosities. These effects are now under investigation.

"The field is new, but it offers very definite prospects not only of testing the form of the velocity-distance relation beyond the reach of the spectrograph but even of critically testing the interpretation of red shifts as actual motion.

"With this possibility in view, the cautious observer refrains from committing himself to the present interpretation and employs the colorless term 'apparent velocity.'"

The Cosmological Equation

In concluding his lecture, Dr. Hubble turned to consideration of the universe which might be inferred to exist from the sample that is known.

"Mathematics deals with possible worlds, that is, logically consistent systems," he said. "Science attempts to determine the actual world in which we live. So, in cosmology, mathematics presents us with an infinite array of possible universes.

"Considerations of the laws of nature led to the theory of relativity, developed by the genius of Einstein and now generally accepted. According to relativity, the large-scale geometry of space is determined by the contents of space. This dependence is expressed in Einstein's famous cosmological equation and modern cosmology is largely a series of attempts to solve the equation.

"The expanding universe, with its momentary dimensions as previously described, is the latest widely accepted development in cosmology.

Looks to 200-inch Telescope

"Further radical advances in cosmology will probably await the accumulation of more observational data—the elimination of more types of possible worlds. The data will come either from more detailed investigations of the present observable region or from a significant enlargement of the region itself.

"The latter alternative will be achieved with the 200-inch reflector under construction for the California Institute of Technology with the assistance and cooperation of the Carnegie Institution of Washington. This great telescope is expected to enlarge the available sample of the universe some 10 times in a single step.

"I believe the 200-inch reflector will definitely answer the question of the interpretation of red shifts, whether or not they represent actual motion; and if they do represent motions—if the universe is expanding—may indicate the particular type of expansion. This prospect is the climax of the story."

—*April 24, 1934*

Finds Galaxy Goes 100 Miles a Second

By LAWRENCE E. DAVIES

The first definite answer to the question concerning how fast our own Milky Way galactic system of a thousand million giant suns is speeding through the vastness of space was supplied in Philadelphia by Dr. Edwin Hubble of the Mount Wilson Observatory.

On the basis of a mass of data collected through the painstaking efforts of observers at Mount Wilson and elsewhere, he ventured the statement that our galaxy is headed, at the rate of about 100 miles a second, in the general direction of the constellation of Draco, the "Dragon's Head," with a long body that winds between the Great Bear and the Little Bear in the northern sky.

Although the fact that an astronomer finally had estimated the velocity of our galaxy, on the basis of actual observation, was in itself of extraordinary interest to science, the data announced by Dr. Hubble also supplied one possible conclusion regarded as in accord with the theory that the universe is expanding.

So far, however, he said, the tests had not been conclusive in that direction, and, further, more exhaustive tests would have to await the scanning of even more distant outposts in the extragalactic system by the 200-inch telescope being installed on Mount Palomar.

Honored for Study of Nebulae

Dr. Hubble disclosed his findings in an address on "The Motion of the Stellar System among the Nebulae" before an audience of scientists at the Franklin Institute of the State of Pennsylvania, where he received the Franklin Medal, the highest scientific award of the institute, "in recognition of his extensive study of the nebulae, particularly those outside our galaxy, as a result of which the dimensions of observed space have been greatly increased."

Dr. Hubble has, in his own words, "kept an eye on" the problem of the motion of the galactic system for the last 10 years. It has been only recently, however, that data became sufficiently extensive to enable him to venture a conclusion. The Earth itself is moving around the sun at the rate of about 18 miles a second and the sun at 170 miles a second, is traveling through a 200 million-year orbit about the center of our galaxy, located behind a mask of dust clouds in the middle of the Milky Way in the general direction of the southern Constellation Sagittarius.

Basic Problem Is Stated

A problem confronting astronomers has been the determination of the motion of the galaxy itself, as, like other nebulae in the celestial "swarm of bees" comprising the universe, it travels toward the north galactic pole.

Since the motion of the Earth with respect to the sun was relatively small and was accurately known, Dr. Hubble said, the observed motions of stars and of nebulae generally were referred to the sun as origin. The problem thus could be restated, he went on, "as that of the motion of the sun with respect to the nebulae."

"This motion is evidently a combination of the sun's motion within the stellar system and the motion of the stellar system through the realm of the nebulae," he explained.

The velocity of the combined motion was roughly 200 miles a second. The final step, however, was the removal from the solar motion with respect to the nebulae of that part representing the sun's motion within the galactic system. This left a velocity of around 100 miles per second as the motion of the galactic system among the nebulae.

"A relatively small velocity and a high inclination to the plane of the Milky Way seem to be definitely indicated," Dr. Hubble said, "although precise numerical values cannot be derived from the data now available.

"An inspection of the data used in the investigation and a study of the possible ways of assembling additional data suggest that the situation will not be much improved until distances of nebulae have been determined with an accuracy much greater than that of the estimates now available."

"Nearby Nebulae" Studied

Involved in the investigation were not only the comparatively "nearby nebulae" but also those even nearer neighbors, about a dozen in number—from 100,000 to 1.5 million light-years distant—which form with the galactic system what astronomers call a "local group."

In figuring the motion of the galaxy on the basis of data concerning the "nearby nebulae," allowance had to be made for the "red shift," that minute displacement toward the red lines of the spectrum. Dr. Hubble pointed out, however, that for some time astronomers had believed that the law of red shifts might not operate within the "local group." For this reason, the nebulae in this group were not included in the general solutions but were examined separately.

Observed velocities uncorrected for red shifts led to a solar motion of about 165 miles a second, agreeing with galactic rotation, and when they were corrected for red shifts the direction of the solar motion was not materially altered, but the velocity was almost doubled to 300 miles a second.

"The explanation," Dr. Hubble said, "appears to be as follows: With the exception of the magellanic clouds which, as close satellites of the galactic system, are relatively unaffected by red shifts, the members of the local group are seen in the general direction of the apex of the solar motion as determined from the uncorrected velocities.

Data on the "Red Shifts"

"This determination indicates that the galactic system is at rest with respect to the other members in respect to radial motion. Consequently, when corrections for red shifts are applied, the residual velocities are systematically negative (the galactic system and the other members as a unit are approaching one another).

"The assumption that the galactic system, situated at one end of the local group, is approaching the other members (or the center of the group) seems reasonable but is inconsistent with the results derived in the first section of the present investigation. With respect to nebulae in the general field, the galactic system is moving at right angles to the direction of the other members of the local group. Consequently, with respect to the fundamental reference frame, the other members of the local group, as a unit, are moving toward the galactic system.

"This situation seems to indicate either that the galactic system, with its satellites, is a triple nebulae in the general field but in the immediate vicinity of a small group of nebulae which is moving toward the galactic system, or that the law of red shifts does not operate within a physically connected system such as the local group."

In Dr. Hubble's opinion the latter view seems the more likely.

Ideas on "Expanding Universe"

"The conclusion," he said, "is consistent with current theories of an expanding universe—the cosmic repulsion which within a physically connected group of nebulae is overcome by gravitational attraction between the group members.

"But the result does not establish the expansion as the only possible explanation. Other data are available which at the moment point in another direction. The present investigation merely adds another item of information concerning the solution of the problem.

"Only if such items continue to accumulate may we hope to formulate a consistent picture of the universe, derived from the observed characteristics from the sample available for inspection."

The "sample" available so far is that observable by the present 100-inch Mount Wilson mirror.

Dr. Hubble confessed that his idea of an "expanding universe" was that "it is a fascinating bit of speculation which can be tested by actual observation," with the tests not yet conclusive.

—May 18, 1938

Palomar Observers Dazzled in First Use of 200-inch Lens

By WILLIAM L. LAURENCE

Man took his first blinking glances at the heavens last night through his 200-inch eye in the Hale telescope atop Palomar Mountain and was dazzled by a new radiance from the light of distant stars, four times brighter than any similar star-image he had ever seen before.

Newspapermen and guests attending the dedication ceremonies were invited to take a peek at the heavens through the big eye, equivalent in light-gathering power to that of a million human eyes.

This followed an earlier demonstration of the highly intricate mechanism and motions of the telescope and its 137-foot diameter dome, one of the great engineering achievements of modern times.

Under conditions of poor visibility we gazed, through a small eyepiece, at the planet Saturn with its three-ring, 42,000-mile-wide "circus," and its satellites. Its image was magnified 700 times, that being the limit set by the atmospheric conditions. Magnification, however, was not the purpose for which the big eye was built.

What was unique about our glimpse of Saturn was the brightness of its image. It was four times as bright as any image of Saturn ever seen before by human eyes with the most powerful telescope, brighter than a bright moon on a clear night. We could see three of nine satellites, shimmering about her, and at least two of its three concentric rings.

The effulgent light made one blink at first in the darkness, producing the effect of a strong flashlight suddenly brought close to the eye.

Saturn is about 900 million miles distant from the sun, and has a mass 95 times that of the Earth. It is a gaseous planet with a density only 0.13 that of the Earth, by far the least dense of all the planets. Its surface markings usually show a yellowish zone at the Equator and a greenish cap at the Poles.

Of Saturn's nine satellites, the five inner ones move in circular orbits, and nearly in the plane of the rings. The largest, Titan, has a diameter of about 2,600 miles and a brightness one 1,000th that of the planet.

The outer of the three flat concentric rings (A, B, and C) has a diameter of 171,000 miles, and is 10,000 miles wide. The division between rings A and B, known as the Cassini division, is about 3,000 miles wide. Ring B is about 16,000 miles wide and brighter than A. Ring C is a crepe ring, separated from ring B by 1,000 miles. The width of the whole ring system is about 42,000 miles.

There is a clear space of 7,000 miles between the inner edge of the ring system and the planet. The thickness of the rings is about 10 miles. Their structure is a flock, or swarm, of separate particles.

This was one of the few times when the Hale telescope will be used for viewing any stellar body visually. For, in reality, the giant telescope is a huge camera. The 200-inch mirror is an instrument for gathering light that will enable astronomers to photograph objects a billion light-years from the Earth, twice as far as is possible with the 100-inch Mount Wilson telescope.

A light-year is the distance light travels in one year at a speed of 186,300 miles per second, or about six trillion miles.

While the telescope has the light-gathering power of a million human eyes, that does not mean that a human eye with a million times its normal visual power would be able to see as far in space as this telescope will, for the human eye cannot accumulate light as does a photographic plate.

Astronomers use photography for a number of reasons. First, the eye is very limited to what it can actually see, and what it can see is recorded only in the mind of one person. Second, photographic plates can record light that the eye will miss, for they can be exposed to light from an object for any given length of time—hours if necessary. Third, a photograph becomes a permanent measurable record that can be made available and studied by all astronomers.

Telescope's Movement Shown

Equally as impressive and awe-inspiring, producing the feeling of sudden transportation to a strange never-never land where all things move at the same time in all directions, was the operation of the giant telescope itself, which was put through its paces for the benefit of newspapermen and guests.

First we saw the metal covering of the 200-inch mirror open up in the manner of petals of a gigantic flower.

We stood in wonder in the presence of man's biggest artificial eye with which man will be able to see a universe eight times the volume of the universe he could observe until now. It took 180,000 man-hours of labor to give it its parabolic (concave) surface, accurate to within two-millionths of an inch. It weighs $14^{3/4}$ tons. It is about 24 inches thick at the edges and $20^{1/2}$ inches at the center.

By the pressure of a button at the control desk, the giant telescope, weighing a million pounds, started moving, first from east to west (right ascension) and then from zenith to horizon (declination).

Then the giant dome, 137 feet in diameter, started slowly moving about, in the manner of a merry-go-round, and one was gradually transported to a fourth-dimensional environment, in which east and west, north and south, merged into one another and then vanished in a strange loss of awareness of space and time.

Controls Highly Elaborate

The telescope has the most elaborate control system ever used by an astronomer. It has an electric remote indicating system of right ascension (moving the yoke from east to west) and declination (moving the tube from zenith to horizon).

At a control desk located beneath the north-bearing pillars, an assistant can control the huge instrument. He can dial any star position desired, press a button and the telescope will move to that position automatically and begin following the object across the skies. This automatic setting system is accurate to less than one second of arc.

In addition to controlling right ascension and declination movements, it also has numerous switches for energizing other devices and also indicators giving the zenith angle of the telescope, position of the windscreen, rates of motion of right ascension and declination, the focus position, sidereal and Pacific standard time. Both talk box and telephone communication with the control desk is provided.

Dummy Telescope Used

By means of a small dummy telescope, rotation of the dome is controlled and synchronized with movement of the telescope, so that the dome slit is always in the proper position.

So friction-free and delicately balanced is the telescope that only a one-twelfth horsepower electric motor is required to move it at celestial rate. For faster movement a two-horsepower motor is used.

Variations in driving rate, caused by atmospheric refraction, and other effects, such as slight deformation of the telescope structure, are calculated by a mechanical computer which automatically adjusts the frequency of the time standard to the proper tracking rate.

The observatory dome is 137 feet in diameter and 135 feet high. It moves on a circular track and has split shutters riding on horizontal rails. The revolving part of the dome is of butt-welded steel construction. Its total weight is 1,000 tons. The entire building is insulated to keep temperature rise to a minimum, so that there will be little difference between inside and outside temperature when the dome is opened for observation.

—June 5, 1948

Studies Reported in Star Evolution

By WILLIAM L. LAURENCE

S tudies of the light of distant galaxies that started its journey to the Earth 200 million years ago, traveling at a steady speed of 186,300 miles per second, suggest that we may be seeing for the first time the evolution of stars as the progress from birth to maturity and death.

These studies, which promise to make it possible for man on his insignificant speck of dust to roll back the eons of time and view the cosmos at large as it appeared hundreds of millions of years ago, were outlined in Pasadena at the joint meeting of the American Astronomical Society and the Astronomical Society of the Pacific at the California Institute of Technology.

Leading astronomers, such as Dr. Edwin Hubble of the Mount Wilson Observatory, described them as a "revolutionary development in astronomy."

The investigations, made possible by a new highly sensitive type of photoelectric cell developed by the Radio Corporation of America during World War II, were made at the Mount Wilson Observatory by Drs. Joel Stebbins and Albert E. Whitford of the Washburn Observatory of the University of Wisconsin.

The new electric eye, known also as a photomultiplier, makes possible the accurate measurement of light from distant nebulae too faint to be measured by means heretofore available.

Comparisons Were Made

With this new light detector, Drs. Stebbins and Whitford measured for the first time the faint light from distant nebulae at a distance of 200 million light-years away, and compared them with the light of nebulae (galaxies composed of hundreds of millions of stars) nearer to us.

These measurements revealed a phenomenon that startled the astronomers. The farther out a nebula is in space, the redder is the light emanating from it. In each case it was found that the excess amount of red light was two and a half times greater than could be accounted for in the well-known and still puzzling phenomenon of the "red shift" in the spectra of nebulae.

The red shift effect is observed through the spectra of the light of the distant nebulae, whereas the extra-reddening effect observed by Drs. Stebbins and

Whitford has been discovered through actual photographs of the nebulae, taken first in white light and then through filters that eliminate all the colors except red.

The spectrum red shift, namely, the displacement of all the lines in the spectra of the nebulae toward the red end, has been interpreted as meaning that the nebulae are rushing away from us and from each other at the explosive speed of 100 miles per second for each million light-years, so that the nebulae at a distance of 200 million light-years, for example, would be receding from us at the rate of 20,000 miles per second.

Universe Expands Rapidly

This red shift effect is thus generally interpreted as meaning that the universe is expanding at an explosive rate.

Comparing the amount of the red shift in the spectra of distant nebulae with the color of the nebulae as revealed by the filtered light photographs, Drs. Stebbins and Whitford found to their surprise that the redness of the nebulae was two and a half times greater all along the line than could be accounted for by the red shift effect as revealed by the spectrum.

One explanation for the new cosmic mystery is that the intergalactic spaces are filled with cosmic dust that absorb the light and thus produce an effect similar to the reddening of the sun at twilight.

What is regarded as the most logical explanation has been offered by Dr. Martin Schwarzschild of the Princeton University Observatory. According to Dr. Schwarzschild's hypothesis, the distant galaxies are redder than the nearer galaxies because we are seeing them as they were 200 million years ago, and hence at a younger stage in their evolution than the nearer galaxies.

Now in our galaxy of the Milky Way we observe a few super-giant red stars, such as Antares and Betelgeus. These super-giant red stars are known to lead a very fast life, burning up their energy at a much faster pace than stars such as our sun, and thus growing old much faster.

Explanation of the Light

The reason we find so few of these super-giant red stars in our own immediate vicinity in the cosmos, it is believed, is because it takes much less time for the light of the galaxies in which these stars are found to reach us (since they are nearer to us) and hence we see the galaxies at a stage in their evolution when most of their red stars had been burned up as a result of their fast living. Therefore, the total light of the near galaxies is less red than the light of the distant ones.

The light of the distant galaxies, on the other hand, such as that measured by Drs. Stebbins and Whitford, has taken 200 million years to reach us. This means that we are now seeing these galaxies as they were 200 million years ago, when they still contained a very large number of the fast-burning super-giant red stars. If further studies corroborate this hypothesis, man will be able to study his cosmos at its various stages of evolution, just as he can study trees in a forest at their various stages of growth.

The farther he reaches out in space, the younger will be the stage in the evolution of the universe. If he reaches out far enough, he may conceivably see the very beginning of time when his universe was delivered newborn in the depths of space, emitting its first "cosmic cry."

—*June 29, 1948*

Universe Growing, Dr. Hubble Thinks

Evidence from the new 200-inch telescope at Mount Palomar Observatory near San Diego, California, supporting the theory that the universe is expanding, was presented in London by Dr. Edwin Hubble, astronomer of Mount Wilson Observatory at Pasadena, California, to members of the Royal Institution.

Thirty years ago Dr. Hubble said the region of space that could be observed was restricted to Earth's system of stars or galaxy. "Since then the great modern telescopes have enormously enlarged the observable region and have enabled astronomers to identify vast numbers of independent stellar systems known as extragalactic nebulae, each like our own galaxy. It is a plausible assumption that the region of space that can be explored is now so large that any general features associated with it can be taken as applying to the universe as a whole. Two such features have so far been established: The first is the fact that roughly the same number of nebulae is found in each region of the sky, and the second is what is known as the law of the red shifts.

"The great light-gathering power of modern telescopes not only enables more distant objects to be observed but also allows light received from them to be analyzed by means of the spectroscope. When this is done, a surprising thing is found; namely, that in the spectra of all nebulae characteristic lines are displaced from their normal positions toward the red end of the spectrum and this displacement is greater the more distant the nebula. The law of the red shifts states that displacement of a given spectrum line is proportional to the distance of the nebula.

"It is a simple physical fact that a displacement of the spectrum lines toward the red occurs whenever the source of light is moving away from the observer. The nebula shifts may therefore be interpreted as meaning that all the nebulae are receding from us and the more distant the nebula the greater is its recessional velocity, in other words, that the universe is expanding.

"Furthermore from the estimated distances of nebulae and the velocities corresponding to the observed red shifts it is possible to deduce that at some time in the past all the nebulae were quite close together and that from this time the duration of the present expansion is roughly equal to the known age of the Earth."

As observed evidence, Dr. Hubble presented slides showing spectra of nebulae ranging out to the greatest distance at which they can be analyzed with the new 200-inch reflecting telescope.

—*May 16, 1953*

Birth of Universe Traced to Blast

By WILLIAM L. LAURENCE

Observations of 800 galaxies show that the material universe was born in a gigantic cosmic explosion some five and a half billion years ago.

Since that original event, data discovered with the world's two largest telescopes indicate that the fragments of the explosion, countless galaxies each made up of billions of giant stars, have been receding from us and from each other at speeds directly proportional to their distance from us—the farther the distance the greater the speed. The observations were described at the annual meeting of the American Association for the Advancement of Science in Berkeley, California.

Studies over the last 20 years have shown that the speed of the recession is 180 kilometers (112 miles) for each million parsecs. The star system farthest in space so far measured was the hydra cluster at a distance of 1.1 billion light-years, roughly 333 million parsecs.

This cluster, it was found, recedes from us at 60,000 kilometers (37, 283 miles) per second, or one-fifth the speed of light. There is hope, it was reported, that the present equipment will make it possible to extend measurements, to objects receding with speeds of 100,000 kilometers (62, 137 miles) per second, one-third the speed of light, or distances of 555 million parsecs.

Studies Began in 1929

Astronomical distances are measured in light-years and parsecs. A light-year is the distance light travels in one year and a parsec is 3.3 light-years. The speed of light is 300,000 kilometers (186,300miles) a second, so that a light-year is nearly six trillion miles.

The studies were outlined today by Dr. Allan R. Sandage, staff member of the Mount Wilson and Mount Palomar Observatories. There the world's two largest telescopes, the 100- and 200-inch, respectively, are located.

Dr. Sandage collaborated in the 20-year project with Dr. M. L. Humason of Mount Wilson and Mount Palomar and with Dr. N. U. Mayall of the Lick Observatory of the University of California.

The original observations indicating that the universe was expanding were made in 1929 by the late Dr. Edwin P. Hubble. He observed that the light of distant star systems shifted toward the red end of the spectrum and that the red shift increased with the distance of the light source.

The only interpretation to explain the red shift was that the galaxies were receding from us and from each other at a speed directly proportional to the distance of the galaxy.

Observations Extended

In 1929, however, Dr. Hubble had data on the distances and recessional speeds of only 24 galaxies. By 1936 Drs. Hubble and Humason had accumulated data on 106 isolated galaxies and on 10 giant clusters in an effort to probe into space as far as possible and to place the observational aspects of the expansion of the universe on a definitive basis.

More data were needed, however. Consequently, an extensive program of observation was begun in 1936 by Dr. Humason at Mount Wilson and Dr. Mayall of Lick to obtain data on the red shifts of many galaxies.

The latest studies have shown, it was reported, that the speed of the recession was about one-third the speed calculated by Dr. Hubble. Rather than receding at the rate of 536 kilometers (333 miles) per million parsecs, as Dr. Hubble had estimated, the new observations indicate that the speed at which the galaxies recede from us and from each other is only 180 kilometers (112 miles) per million parsecs.

Since the speed of light is the maximum speed possible in the universe, this means that the maximum speed at which the galaxies could recede would be reached at about 1.666 million parsecs, or a distance of five and a half billion light-years. This corresponds to the present figures about the age of the crust of the Earth and the oldest stars in our nearby galaxy.

This would mean also that the radius of the expanding universe was three times greater than it had been believed to be in 1936 and 50 percent greater than the current figure.

—December 28, 1954

Radio Telescope to Expose Space

By JOHN W. FINNEY

In an isolated valley in the mountains of West Virginia the Navy is building a massive radio telescope that should be able to peer out to the edges of space. It is expected to answer age-old questions about the creation and nature of the universe.

In size and complexity, the telescope will dwarf anything yet constructed for study of the universe. Its construction, by 1962, will rank as a major engineering feat.

With a saucer-shaped antenna 600 feet in diameter, the telescope will be the largest ever built to tune in on the radio signals created in the stars and planets. These signals are weakly received on Earth.

Theoretically, the telescope should be able to look out into space 38 billion light-years—or 19 times farther than the 200-inch optical telescope on Mount Palomar, California. The range of the instrument is virtually incomprehensible when it is realized that one light-year—the distance traveled in one year by light moving at 186,300 miles a second—is about 6,000,000,000,000 miles.

The theoretical range of the telescope may exceed the size of the universe, for no one is quite sure how big the universe is. There is considerable doubt among astronomers and cosmologists that the universe's dimensions approach 38 billion light-years.

Major Solutions Hoped For

Because of its tremendous range, the telescope is expected to provide new clues—and perhaps conclusive answers—to such cosmological questions as: How big is the universe? When did it start? What is its shape? How was it created?

The $79 million facility, known officially as the Naval Radio Research Station, is being built on a knoll surrounded by hills in the Allegheny Mountain village of Sugar Grove, W. Virginia. The plans were developed by the Naval Research Laboratory.

The technical director for the project is James H. Trexler, a radio communications engineer at the Naval Laboratory. He first conceived of the telescope some 10 years ago and has carried the project along through the developmental stages.

Since its inception, the project has been carried out with a secrecy comparable to the World War II development of radar at the Research Laboratory.

347

This has been so because the telescope, aside from its usefulness as a scientific research tool, will have certain military applications of high priority.

Because of the immensity of the facility, however, it has been impossible to keep the project completely secret. Hence, it is now possible to discuss the size of the telescope and some of the scientific missions it will perform.

Construction of the telescope poses extraordinary engineering problems. As one Navy official put it, the project "will be an engineering feat as unusual as was the building of the Brooklyn Bridge a century ago."

The engineering difficulties become apparent with a description of the size of the telescope and the fine precision that must be maintained if it is to receive the radio signals from space.

The saucer-shaped dish, which serves as the reflecting mirror of the telescope to capture the radio signals, will be wider across than the 555-foot Washington Monument is high. Into the dish will go 20,000 tons of steel and aluminum—or more than used in a modern destroyer.

Protruding from the center of the dish will be a long, straight steel structure more than 100 feet high. This will be the antenna proper to receive the radio signals reflected from the dish.

To receive the signals accurately, the dish and the antenna must maintain their designed shape and alignment to within a fraction of an inch. The structure, however, will be subjected to tremendous distorting stresses, both from the rotation of the dish in its cradle and from the elements.

The heat of the sun's rays and the pressures of the winds, which will not be uniform across the dish, will tend to twist the telescope out of alignment.

To keep the telescope in the desired shape, the Navy has devised an intricate system of servo motors that will position the sections of the reflector dish and the antenna and keep them in alignment as they are subjected to distorting pressures.

Like a Ferris Wheel

The 20,000-ton dish and antenna will be mounted on what is expected to look like a huge Ferris wheel. This wheel will rotate nearly 180 degrees so that the dish can be elevated in any direction between the horizons.

In turn, the Ferris wheel structure will be mounted on huge rollers mounted on a circular track on the ground. This will permit the telescope to turn as though on a great turntable. Millions of watts of energy will be needed to elevate and rotate the telescope.

In some respects the telescope will "sail" in space much like a ballistic missile. Mounted at the base of the antenna will be an inertial guidance system, like those used in missiles. This system will determine the direction in which the telescope is pointed in reference to a line in space. Directions from the inertial guidance system in turn will tell the motors on the ground in which direction to point the telescope.

Preliminary construction work on the foundation, which began early in 1959, is now well advanced. To provide the huge turntable track, more than 30 feet of Earth, amounting to some 1.5 million cubic yards, was scraped off the top of a knoll. No dynamite blasting was permitted lest the underlying rock formation be disturbed.

Even the selection of the site posed special problems. For several years Navy teams toured the nation looking for a site free from radio interference and not subject to hurricanes, tornadoes or earthquakes. The Navy finally settled on a West Virginia mountain site only 30 miles from the National Radio Observatory, which is being built by the National Science Foundation at Green Bank, West Virginia.

A 140-foot radio telescope is now being built at the Green Bank Observatory for scientific research. The largest radio telescope now in operation is the 250-foot dish at Jodrell Bank in England. The Soviet Union is reported to be planning a 350-foot radio telescope.

Interference Limited

Once the Navy telescope goes into operation, there will be further problems in preventing manmade signals from interfering with reception of the celestial signals. In 1956 the West Virginia Legislature passed laws providing strict controls of electrical interferences within the vicinity of the facility, including interference from electric razors and television sets.

The Federal Communications Commission has established a "radio quiet" zone more than 1,000 square miles in area around the two radio telescope observatories. The Federal Aviation Agency is rerouting airplanes and transferring electric ground equipment.

There are also sociological problems in preparing the mountain village of

Sugar Grove, with its 21 inhabitants, to become a scientific outpost in the universe. By 1962 more than 500 scientific, civilian and military personnel are expected to be stationed at the facility.

One of the leading questions that Navy scientists hope the telescope will help answer is the cosmological debate between an "evolutionary" and "steady-state" universe—a debate that goes to the very creation of the universe and its shape and size.

According to the evolutionary theory, the universe was created some 5 million years ago when a super-hot, super-dense glob containing all matter exploded. Subsequently, the matter collected and condensed to form the stars and the galaxies. According to the steady-state theory, the universe had no finite beginning and matter is continually being created throughout space.

The telescope, which will be particularly tuned in to receive the characteristic "radio song" given off by neutral hydrogen, should also be able to determine whether hydrogen, the basic building block of all matter, exists in intergalactic space. If the hydrogen is found in sufficient quantity, this will support the steady-state theory.

Another critical experiment will check on the shifts in radio signals and thus measure the velocity at which the universe is expanding. A finding that the more distant galaxies are slowing down would tend to confirm the evolutionary theory.

And once the density and velocity of the galaxies have been determined, it should be possible to take the next theoretical step. That would be to determine the geometry of the universe, such as whether it is flat and infinite, or curved like the surface of a sphere and finite, or perhaps "negatively" curved like a saddle and infinite.

Ultimately, the telescope should be able to serve as a receiving station to pick up signals from manmade rockets exploring the outer stretches of the solar system. The Navy estimates that the telescope will be able to pick up television signals from as far away as the planet Saturn, which never comes closer than 745 million miles from Earth.

—June 19, 1959

Rival Cosmologies

By WALTER SULLIVAN

Is the universe growing older? Or is it ageless, with new galaxies, new stars, new planets—perhaps new life—forever being continuously created?

This question, marking the difference between the "steady-state" and the "big bang" cosmologic theories, has held the attention of astronomers from many lands. They discussed it first at a small conference in Santa Barbara, California, and then at the General Assembly of the International Astronomical Union at the University of California at Berkeley.

In attendance, at one or both meetings, were almost all of the leading figures in the cosmological debate that has been carried on in recent years with increasing fervor. Both sides pointed to recent discoveries and observations which, they feel, support their point of view.

Heart of Controversy

In summing up the controversy Dr. Hermann Bondi of Kings College at the University of London set forth the points on which virtually all are agreed: .

(1) First, all of the distant galaxies are moving away from us. A galaxy is an assembly of stars, the Milky Way galaxy has a population of stars reckoned in the billions. The galaxies themselves, within range of optical or radio observation, also probably number in the billions. Expansion of the universe was postulated before it was observed. The receding motion of the galaxies, indicating expansion, has been observed in the stretching, or "reddening" of their light waves, the so-called "red shift."

(2) The dimmer the galaxy, the greater is the red shift. This indicates that galaxies farthest away are receding fastest.

(3) The distribution of matter throughout the universe is uniform. Optical observations suggesting this have recently received strong confirmation from the observations of radio-astronomy.

(4) The farther we look, the deeper we penetrate into the past.
Thus, as Dr. Bondi puts it, "Geography turns into history," as
astronomers explore the most remote sky areas.

In 1927 Abbé Georges Lemaître, a Belgian astronomer-cleric, published the "big bang" hypothesis to account for the expanding motion of the galaxies. If you run this motion backwards five billion years, he said, it seems evident that all of the galaxies originated at one point.

In explaining this idea to newsmen at the conference of the Astronomical Union he described this starting point as "a kind of bottom in space and time." What existed at the beginning was one single entity with no structure. There followed the "beginning of multiplicity" and the flying apart of the components.

Abbé Lemaître denies the allegation of his opponents that his views are tailored to fit the concept of a divine creation. His cosmology differs from the "steady-state" view (the idea that the universe is being continuously created) for one thing in that it permits the existence of superlatives, such as a "largest" galaxy and a "hottest" star, whereas the other cosmology does not.

One of the chief problems in present-day astronomy and cosmology arises from the puzzling recent discovery that some stars seem to be roughly twice as old as the "universe." This finding was cited in two of the talks given to the 1,000 astronomers who attended general sessions of the Astronomical Union Conference. One was by Dr. Jan H. Oort of the Netherlands, president of the organization, and the other by Dr. Martin Schwarzschild of Princeton University.

Dr. Schwarzschild noted that, although the beginning of the expansion of the universe is put at 13 billion years ago, the ages of some star clusters in our galaxy are now estimated at 25 billion years. The latter ages are based on theoretical stellar life histories.

To account for this discrepancy, Abbé Lemaître has postulated that the presently observed rate of expansion between the galaxies (or clusters of galaxies) may apply only to the present time. The rate may have been greater in the past. Therefore the true age of the universe may be considerably more than it appears to be from the observed present rate.

The great difficulty in the opposite view—the "steady-state" concept of the universe —is that it theoretically requires the manufacture of new matter from nothing. If the universe is forever expanding, yet forever uniform in density, new matter must form. This would violate the law of the conservation of matter and energy that scientists have long considered established.

Protagonists of the steady-state view, however, point out that relativity has enforced a revision of other seemingly irrevocable laws. Furthermore, according to Dr. Fred Hoyle of University of Cambridge, England, who was at Berkeley, it is necessary to produce only one atom of hydrogen in a bucketful of space every 10 million years to make up for the expansion.

Looking at the Past

The most widely discussed tests of the rival cosmologies at the California meetings were those based on the assumption that if the universe is aging, we should be able to see it in a more youthful form by looking far enough into the past.

Some, for example, believe the elliptical galaxies are a senile form. And galaxies of other configurations, such as ours, are believed to be a typical spiral galaxy which must be younger. But observations are difficult for the most distant galaxies observable visually are mere pinpoints and their structures cannot be determined.

A new approach that has produced results, which some regard as damaging to the "steady-state" theory, is the analysis of radio signals from distant objects.

Unfortunately, the radio spectrum is not divided into emission lines whose shift toward the red can be used as a gauge of distance. To get around this Dr. Martin Ryle of University of Cambridge has for several years been analyzing, in terms of relative strengths, the sources of radio "noise" that dot the sky.

He assumes that the weaker a source the farther away it is. Furthermore the weaker sources should be more numerous, because the field of view expands with distance.

If the universe is thinning out, as required by the "Big Bang" concept, one would expect to see a still greater density of radio sources at great distance (far in the past). This is what Dr. Ryle reports he has found.

The Other View

Backers of a "steady-state" universe, however, question whether the weak sources observed by Dr. Ryle are really very far away. Dr. Hoyle notes that only 20 or 30 radio sources have been identified optically so that their distances can be measured. And he cites recent observations made at the University of Manchester with antennas some 100 miles apart.

By a closed television circuit between the two points it has been possible to make phase comparisons of incoming radio waves and thus determine the width of the radio sources. Three sources were found to be very narrow and therefore presumably far away. A large percentage, however, were large and irregularly shaped, indicating that they are comparatively near.

Despite this challenge to Dr. Ryle's results, the "steady-state" team admits that it is fighting an uphill battle. And, Dr. Bondi says, they are going to be fighting it for a long time to come.

—August 27, 1961

Satellite Challenges Theory of Universe

By JOHN W. FINNEY

The first astronomical satellite launched by the United States has sent back word challenging the theory that the universe is in a steady state.

There has been a running cosmological debate over how the universe was created. Now *Explorer XI* has supplied indirect evidence supporting the theory that the universe all started with a great primeval explosion billions of years ago.

The evidence indicates that antimatter is not being created in the distant reaches of the universe. And if antimatter is not being created, then probably matter is not being created, as postulated by the steady-state theory on the origin of the universe.

The evidence is based on the lack of extremely powerful gamma rays such as would be created when matter and antimatter meet and annihilate each other.

The *Explorer XI* satellite was sent aloft on April 27 to measure the intensity of paths of gamma rays—a form of electromagnetic radiation similar to X-rays—streaking through space with the speed of light.

Inside the satellite was a "gamma ray telescope" developed by two Massachusetts Institute of Technology physicists, Drs. William Kraushaar and George Clark.

A report on the first and still preliminary findings made by the telescope was made by Dr. Robert Jastrow, chief of the theoretical division of the National Aeronautics and Space Administration's Goddard Space Flight Center.

Preliminary results from the satellite, Dr. Jastrow said in a speech before the National Rocket Club, are that the level of gamma radiation is so low as to "rule out one version of the steady state of cosmology which holds that matter and anti-matter are being created simultaneously."

Continuous Creation

According to the steady-state theory, matter is continuously being created throughout the universe at the rate of about one hydrogen atom per second per 10,000 cubic miles. Thus, according to this theory, as the universe expands, new matter forms new stars and galaxies, leaving the universe in a steady state.

355

If matter is being created, however, it has been suggested by the cosmologists that particles of antimatter must also be created simultaneously. Particles of antimatter are the same atomically as particles of matter except that they have the opposite electrical charge or opposite spin.

After a considerable length of time, perhaps on an average of 10 million years, this newly created antimatter could be expected to combine with matter in the galaxies.

Matter Turns to Energy

In the annihilating process of combination, the atoms of matter and antimatter would leave behind only energy in the form of extremely strong gamma rays.

The intensity of these gamma rays would be about 100 times as great as that of those produced by the interaction of cosmic rays with gas in the galaxies.

Thus far, Dr. Jastrow said, the intensity of the gamma rays observed by the satellite was far below the level that would be expected from the annihilation of matter and antimatter.

In a negative, indirect way, therefore, the findings of the satellite tend to support the opposing cosmological explanation of the origin of the universe, known as the evolutionary or "big bang" theory.

This theory, which is more widely endorsed by astronomers and cosmologists, hypothesizes that the present expanding universe is the aftermath of a gigantic explosion more than 10 billion years ago of an extremely dense glob of matter.

—December 20, 1961

Signals Imply a "Big Bang" Universe

By WALTER SULLIVAN

Scientists at the Bell Telephone Laboratories have observed what a group at Princeton University believes may be remnants of an explosion that gave birth to the universe.

These remnants are thought to have originated in the burst of light from that cataclysmic event.

Such a primordial explosion is embodied in the "big bang" theory of the universe. It seeks to explain the observation that virtually all distant galaxies are flying away from the Earth. Their motion implies that they all originated at a single point 10 or 15 billion years ago.

The Bell observations, made by Drs. Arno A. Penzias and Robert W. Wilson from a hilltop in Holmdel, New Jersey, were of radio waves that appear to be flying in all directions through the universe. Since radio waves and light waves are identical, except for their wavelength, these are thought to be remnants of light waves from the primordial flash.

The waves were stretched into radio waves by the vast expansion of the universe that has occurred since the explosion and release of the waves from the expanding gas cloud born of the fireball.

In what may prove to be one of the most remarkable coincidences in scientific history, the existence of such waves was predicted at Princeton University at the same time that the scientists at the Bell Laboratories were puzzling over an observation of almost identical waves that they could not explain.

The Princeton group, led by Dr. Robert H. Dicke, professor of physics, was unaware of the Bell observation. Those at Bell had not heard of the Princeton prediction.

Like the recent discovery of objects, known as quasars, that lie near the fringes of the observable universe, the new observations may enable scientists to choose the correct picture of the universe: Is it infinite or limited in extent? Is it eternal and unchanging? Was it born in a single "big bang," or is it oscillating?

It is clear that Dr. Dicke and others would like to see an oscillating universe come out triumphant. The idea of a universe born "from nothing" in a single explosion raises philosophical as well as scientific problems.

An oscillating universe gets around the problem of origin. The galaxies fly apart in the manner currently observed. Then, at a certain point, they begin to fall back together again.

Finally, the night sky becomes brilliant with the light of converging galaxies. In a frightful cataclysm they fall together into a mass of fragmented atoms, then burst forth as a new fireball. This scatters hydrogen in all directions, from which new elements and new galaxies are formed.

Quasar Observations

The observations of quasars from Mount Palomar in California have persuaded Dr. Allan R. Sandage of the observatory there that the universe may be oscillating at a rate of one "bang" every 82 billion years. However, further observations are needed for a clear-cut answer, he said in a recent telephone interview.

The study of the Bell Laboratories' observation at Princeton likewise leaves open the question of whether there has been but one explosion or the universe oscillates. However, both Dr. Sandage and Dr. Dicke clearly doubt the steady-state theory in which there is no explosion at all.

Since, in this concept, the age of the universe is infinite, constant expansion would long since have carried all galaxies beyond our range of vision. Hence the theory demands the constant creation of new matter between galaxies to fill the gap.

The Holmdel observations were made with a hornlike antenna designed for experiments in space communications. The antenna played a key role in the development of the Telstar satellite system.

It stands on a hill not far from the field where another Bell Laboratories researcher, Karl G. Jansky, made the discovery in the 1930s that gave birth to a new science: radio astronomy.

Research with the horn antenna in recent years has been directed toward cleaning the receiver system of noise inherent in such systems—the hum in a typical receiver, for example.

A black body at a certain temperature emits a certain pattern of radio waves and hence such "system noise" is expressed in terms of temperature.

New devices, in particular the traveling wave maser, have greatly reduced the system noise in radio receivers and have played a major role in making satellite communications a reality.

In the research at Holmdel it was evident that some of the observed noise came from the system; some came from the warm envelope of air around the Earth and a small amount came from the Milky Way galaxy and other galaxies.

Unexplained Noise Remained

Yet there was always an unexplained residue. As long as the operational requirements of Telstar were paramount, little attention was paid to this residue, but more recently Drs. Penzias and Wilson decided to try for a "cleaner" system.

Working on the Telstar frequency of 4,000 megacycles (a wavelength of 7.3 centimeters) they still were unable to eliminate the leftover noise. They took the huge antenna apart, machined its moving parts and subjected its electric circuits to scrutiny comparable to that used in preparing a manned spacecraft. Perhaps, they thought, radio emissions were leaking into the antenna from something behind it.

Hence they carted a radio transmitter around nearby fields, testing to see if its emissions entered the system. They did not. Finally, they reassembled the whole contraption and found the emissions were still there.

They anticipated 2.3 degrees of noise from the air, and one degree from the antenna, apart from 20 degrees from the receiver, which they had effectively canceled out. The noise to be expected from the Milky Way and other galaxies on that frequency was negligible.

The researchers were baffled and reported as much to their colleagues. It was then that someone told them of the proposal by the Princeton group. The latter predicted noise equivalent to that from a black body at 10 degrees Kelvin (that is 10 degrees above absolute zero, measured on the Centigrade scale).

The residual noise observed by those at Bell was 3.5 degrees, which was considered quite close to the prediction. Dr. Dicke and his colleagues went to Holmdel to inspect the array and apparently went away convinced.

They themselves have built an antenna to observe on a three-centimeter wavelength and expect to begin observations with it shortly.

Could Support Theory

If the effect is detected on that wavelength, too, the chances will be greatly increased that the primordial fireball has, in fact, been detected. It may also be easier to assess whether the universe is "open," expanding into infinite space, or "closed" and oscillating.

The temperature of the fireball is estimated to have been at least 10 billion degrees Kelvin, to begin with. However, the expanding cloud originally was not transparent. Only when it had cooled to about 5,000 degrees Kelvin (-269 degrees Celsius) did the light begin to move freely through space. The continuing expansion of the universe has further stretched the waves until they appear in the radio part of the spectrum, according to the calculations.

The observations at Holmdel were conducted for an entire year. The horn antenna was always aimed at the zenith, across which marched many regions of the universe.

The only specific direction in which no observations were made was toward the Milky Way—the dense portion of our own galaxy—which is too noisy, in radio waves, to permit a valid assessment of other noise sources.

Participate in Study

Those at Princeton who have taken part in the study include in addition to Dr. Dicke, Drs. P.J.E. Peebles, P. G. Roll and D. T. Wilkinson. Because their study has not yet been published, they were reluctant to discuss it. The same was true of Drs. Penzias and Wilson at Bell Laboratories.

Both groups have submitted papers on the subject to *The Astrophysical Journal*.

The parallel with the Jansky observations of 1931–33 is striking because he, too, was seeking to track down radio noise. In this case his attention was on static interfering with long-range radio communications, particularly that originating in thunderstorms.

As in the recent observation, he tracked down all the obvious sources and was left with a perplexing residue. At first he suspected it came from the sun, but colleagues then suggested that it might be from a point among the stars. The source proved to be the core of the Milky Way galaxy.

His discovery led others to aim antennas at the sky and the thriving science of radio astronomy was born.

—*May 21, 1965*

An X-Ray Scanning Satellite May Have Discovered a "Black Hole" in Space

By WALTER SULLIVAN

Observations with an Earth satellite named *Uhuru*, the first vehicle placed in orbit primarily to scan the heavens at X-ray wavelengths, have revealed what, it is suspected, are one or more of the long-postulated "black holes."

A black hole is the hypothetical remnant of a star that has collapsed into so small and dense an object that light cannot escape from it. The reason for its invisibility would be its extremely intense gravity.

The gravity of a massive object like the sun increases the wavelengths of light emanating from it. In terms of visible light, the stronger the gravity, the redder the light becomes. In a black hole, gravity is so strong that the wavelengths are infinitely lengthened before they can escape.

Hence the object would be invisible at all wavelengths, including those of radio, visible light and X-rays. The X-ray pulses that have been reported, it was suspected, were generated perhaps by the spinning of the central object outside its most intense gravitational field.

On Dec. 12, 1970, the *Uhuru* satellite was launched for the National Aeronautics and Space Administration from a movable platform operated by Italian personnel in the Indian Ocean off East Africa. The word *uhuru* means "freedom" in Swahili. The satellite is No. 42 in the *Explorer* series.

Equatorial Orbit

Because the platform was on the Equator it was possible to place the satellite in an equatorial orbit by launching it due east. New discoveries are emerging from analysis of its data by a team at American Science and Engineering, Inc., in Cambridge, Massachusetts. The discovery of a possible black hole was reported yesterday to the American Astronomical Society in Baton Rouge, Louisiana, by Dr. Riccardo Giacconi of the Cambridge team and was elaborated on in telephone interviews.

The object is one of a number of previously known sources of X-ray emissions, and is known as Cygnus X-l. The designation means it is X-ray source No. 1 in the constellation Cygnus, the Swan. The new finding is that it is pulsing in a highly rhythmic manner at a rate of at least 15 times a second.

The rate could be double that, but this would not be evident because of the limited sampling rate of the satellite. The X-ray pulses are reminiscent of those from pulsars—objects that emit very rhythmic radio pulses (and, in one case, light and X-ray pulses).

Radio Pulses Absent

However, as noted by Dr. Herbert Gursky of the Cambridge team, what is extraordinary about this object is the absence of any observable radio pulses. Furthermore, intense efforts over the years to identify a peculiar-looking star in its vicinity have been unsuccessful.

Currently, Dr. Jerome Kristian of the Hale Observatories in California is seeking to find an object in that part of the sky whose light is flickering at 15 times a second. His results so far have been negative, according to Dr. Gursky, but a more thorough search will be possible when the sky is extremely dark, during the next new moon.

It is the apparent absence of any radio or light pulses associated with those at X-ray wavelengths that has led to the suspicion that a black hole is under observation. Pulsars, pulsing generally at slower rates and at radio wavelengths, are believed to be neutron stars.

It is assumed that their pulse rates represent their spin rates, averaging about one second. Neutron stars were presumably formed by the collapse of a burned-out star and differ from the hypothetical black holes in that their density and gravity are not sufficient to make them invisible.

Another Spot Observed

This was not, however, true of all of them. One source, in the southern constellation Circinus, seems to be varying in the same manner as Cygnus X-l, although no precise pulse rate has been identified.

Another pulsing source of X-rays, Dr. Gursky said, may be the object in Centaurus reported, early in March 1971, to be "flaring" on a time scale as short as 15 seconds. The variations were reported by Dr. Walter H. G. Lewin of the Massachusetts Institute of Technology. They were based on observations with a high-flying balloon.

Only by more detailed analysis of the incoming *Uhuru* data will it be possible to determine whether these other objects are pulsing rhythmically—and whether a number of other such otherwise invisible objects are adrift in space.

Finally, it will be up to the theorists to assess the likelihood that they are, in fact, black holes and, if so, why they are signaling their presence with such powerful X-ray pulses.

—April 1, 1971

End of Universe in "Black Hole" Foreseen

By WALTER SULLIVAN

A veteran explorer of gravitational theory has told members of the National Academy of Sciences that the collapse of the universe into a single, great "black hole" with the annihilation of all matter and all physical laws seemed ultimately inevitable.

Dr. John A. Wheeler of Princeton University said the prospect of such a collapse "confronts the physics of our day with the greatest crisis in all the history of science."

It casts doubt on such seemingly basic laws as the indestructibility of the large atomic particles, or "baryons."

Dr. Wheeler, the final speaker in a series of meetings to honor the 500th anniversary of the birth of Copernicus, said the evidence is "incontrovertible" that the present expansion of the universe was slowing.

Three "Black Holes" Postulated

The Copernicus celebration was organized jointly by the National Academy of Sciences and the Smithsonian Institution.

Dr. Wheeler noted that in the last year or so, three objects in the heavens have been identified as possible or probable "black holes." Such super-dense objects had been predicted as the end products of the collapse of large stars when they no longer produced enough outward flowing heat to counter the enormous weight of their own material.

So great would the inward pressure be that the material of the star would be squeezed out of existence, as matter, but would still generate gravity so strong that no light—or anything else—could escape from it, thus producing a "black hole" in the sky.

Dr. Wheeler said black holes could serve as examples of what would happen if the entire universe ceased its expansion and began to fall back upon itself. Since its expansion apparently does not have sufficient momentum to continue indefinitely, the universe could reach its maximum size in about 40 billion more years.

The observations on which these estimates have been made assume that the most distant visible galaxies were just as bright when their light began its journey to Earth, billions of years ago, as galaxies seen at closer distances, which are far younger.

Dr. Allan Sandage of the Hale Observatories in California pointed out that a decrease of only 5 percent in the brightness of galaxies for each billion years of their lifetime would alter the calculations and point to an unending expansion.

It is Dr. Sandage's observations of the apparent decrease in the rate of expansion that lie at the basis of current predictions of ultimate collapse. Dr. Sandage recently reported that astronomers have seemingly seen the "edge" of the universe.

He pointed out that the objects believed to be the most brilliant in the heavens, known as quasars, can only be seen out to a limited distance. Looking beyond them is thought to constitute looking back in time to the earliest stage in evolution of the universe and the "edge" that is seen, within a framework of space and time, is the one that existed at this early epoch.

At today's meeting, Dr. Maarten Schmidt of the California Institute of Technology, a co-discoverer of the quasars, reviewed this evidence. He quipped that astronomers must now face the prospect of "cosmic claustrophobia" in their apparent inability to see any further into the past.

Dr. Wheeler pointed out that Einstein calculated, in his studies of gravity in 1915, that the universe should expand, slow down and then fall back together, much as a stone, thrown into the air, slows down and then falls. At the time he adjusted the formulation to avoid this seeming absurdity.

When the expansion was later confirmed, Dr. Wheeler said, Einstein told him this doctoring of his formulation was "the biggest blunder of my life."

—April 26, 1973

First Photo Taken by New Telescope

By WALTER SULLIVAN

The largest telescopic eye ever opened on the southern heavens has produced its first photograph as a prelude to observations that are expected to broaden understanding of the universe.

The telescope, with a mirror 13 feet 2 inches in diameter, is at the Cerro Tololo Inter-American Observatory along the spine of the Andes in Chile. Its cost of $10 million was supported equally by the Ford Foundation and the National Science Foundation.

It is operated by the Association of Universities for Research in Astronomy, which also operates the Kitt Peak National Observatory in Arizona. The new telescope is matched by a similar instrument atop Kitt Peak. An almost identical one is nearing completion in Australia.

Although many of the most significant astronomical objects hang in southern skies, which cannot be observed from northern observatories, almost all of the world's large telescopes have been built where most astronomers live— in the North. This was true of the 16-foot-8-inch reflector on Mount Palomar in California, the biggest in operation, and the 19-foot-9-inch instrument nearing completion in the Soviet Union.

Ethyl Alcohol in Space

Another discovery has been reported from Kitt Peak, made with a radio antenna operated there by the National Radio Astronomy Observatory, which has headquarters in Charlottesville, Virginia. This was the detection of radio emissions indicating the presence far out in space of ethyl alcohol—the key characteristic of intoxicating beverages.

Until a decade or two ago it was believed that such substances could be generated only by living organisms, and they were therefore termed "organic."

However, in recent years the detection of telltale radio waves emitted by these chemicals under various forms of stimulation have revealed their spontaneous formation from elemental gases in space. This is particularly true of the swirling clouds of dust and gas from which new stars and planets are apparently forming. Among the many organic molecules previously detected in this manner have been formic acid, formaldehyde, cyanoacetylene and methyl alcohol. The latter, sometimes referred to as "wood" alcohol, is poisonous.

A vast cloud of dust and gas generating such emissions lies toward the core of the Milky Way galaxy, and, Dr. Patrick E. Palmer of the University of Chicago said yesterday, it was there that the ethyl alcohol was detected.

Dr. Palmer belongs to the international team of radio astronomers, led by Dr. Benjamin M. Zuckerman of the University of Maryland, that made the observations. The cloud from which the emissions are coming, he said, appears to be more than 30 light-years wide (one light-year equaling the distance traversed by light in one year). Its total mass is believed equal to that of a million suns.

Hence it is presumably condensing into a cluster of stars and planetary systems, rather than a single object.

Southern Heavens Photo

The new photograph was of the brightest globular cluster of stars in the sky. Known as 47 Tucanae, it is in the southern heavens and cannot be seen from northern observatories. Globular clusters are assemblages of stars that look, in photographs, like a swarm of bees. A cluster may comprise hundreds of thousands of stars.

The Cerro Tololo telescope, 7,200 feet above sea level, is said to have the light-gathering power of more than one million human eyes. The climate there is said to be ideal for year-round observing. The mirror, made of Cer-Vit, a compound that remains relatively stable in spite of day-night temperature fluctuations, was cast by Owens-Illinois Inc.

After 30 months of grinding and polishing at the Kitt Peak shops in Tucson, Arizona, it was hauled to the site and, following its arrival on Sept. 2, 1974, was installed in its 375-ton movable mounting.

The initial photograph, made on Oct. 18, was designed to test the optics. A year of alignment and shakedown is expected before full-time observing begins.

—October 29, 1974

Galaxy's Speed through Universe Found to Exceed a Million MPH

By WALTER SULLIVAN

By measuring the movements of the Earth against the "glow" left from the fireball in which the universe was born, scientists have found that the Milky Way galaxy, the home of this solar system, is hurtling through space at more than a million miles an hour, relative to the universe as a whole.

This finding by University of California scientists who made a series of high-altitude U-2 flights, as well as other recent observations, show the universe to be remarkably mobile. Yet it appears that there are no bright or dim areas in the fireball's residual radiation, indicating that the explosion was extremely uniform. The new reports only compound the mystery of how, from this seemingly uniform, homogeneous explosion, the present universe evolved with all of its diversity, from galaxies to flowers.

Remnants of Stars

Evidence has also been found hinting at what its discoverer calls the "almost unbelievable" possibility that the sun is orbiting an unobserved companion star, possibly one of the hypothetical bodies called "black holes." This proposition is based on the otherwise unexplained behavior of some pulsars lying toward the core of the Milky Way star system.

Pulsars, so called because they emit highly rhythmic radio pulses, are assumed to be remnants of stars that, having burned up their nuclear fuel, collapsed to objects of great density and rapid spin. Their pulse rates match their spin rates and characteristically slow down as the pulsar ages.

However, six pulsars, all lying in generally the same direction, are hardly slowing at all, and one of them is increasing its pulse rate. Astronomers have sought in vain to explain this satisfactorily.

In a forthcoming article in the British journal *Nature,* Dr. Edward R. Harrison of the University of Massachusetts in Amherst proposes the explanation that an acceleration of the sun and its planets in that direction might be masking the slowdown in pulses.

Such acceleration could be explained, according to Dr. Harrison, if the sun was circling a companion of comparable mass at a distance about 1,000 times greater than that between the Earth and sun.

While, he says, it is hard to believe that a star so close could have gone unobserved, it might be one of very low luminosity, such as a "black dwarf" that has never ignited its hydrogen, a neutron star (some of which manifest themselves as pulsars) or a black hole.

The last named is a hypothetical object formed by the collapse of a star so massive that no force could halt its collapse, until it became so dense that no light, or anything else, could escape its gravity.

Dr. Harrison believes a neutron star or black hole, having been formed by the explosive collapse of a star, would probably be in a hyperbolic orbit, like that of a comet. This could periodically bring it close enough, as Dr. Harrison believes it is now, to have an observable effect on the solar system—and perhaps influence climate on Earth.

At the other end of its orbit it would come close to the "sea" of comets thought to envelop the solar system at a great distance, pulling them out of that region to swing close to the sun.

For a number of years experimenters have sought to use the glow remaining from the assumed "big bang" explosion, the origin of the universe, as an "ether," an all-pervasive medium through which all bodies are flying.

In the last century such an invisible ether was postulated as the medium through which light waves are propagated and through which the Earth flies in its orbit. The famous experiments of Albert A. Michelson and Edward W. Morley in the 1880s, however, showed the speed of light "upwind" and "downwind" in the Earth's orbital flight to be identical, disproving the existence of any "ether wind."

In 1965, however, Drs. Arno Penzias and Robert Wilson detected radiation coming from all parts of the heavens that appeared to be the residue of the big bang and were presumably emitted billions of years ago from a fireball that was essentially at rest. Expansion and "cooling" have lengthened the wavelengths of this flash until they occur primarily at the shortwave (microwave) end of the radio spectrum.

Soon after this discovery, P.J.E. Peebles of Princeton University, one of those who had predicted the nature of this radiation before it was observed, pointed out that it provided a counterpart of the ether as a reference frame against which absolute motion with respect to the whole universe could be determined. A number of efforts have been made to record such motion, but observers on the ground have been hampered by changing transparency of the atmosphere and emissions from the Milky Way.

Drs. Brian E. Corey and David T. Wilkinson of Princeton reported last year on more successful measurements made from balloons. Now three scientists from the University of California at Berkeley have described an extensive survey, spread incompletely over two-thirds of the northern celestial hemisphere, that was made on almost a dozen high-altitude flights.

The aircraft was a U-2, of the type designed for reconnaissance, operated by the Ames Research Center of the National Aeronautics and Space Administration in Mountain View, California. At the flight altitude of 65,000 feet, above 90 percent of the atmosphere, it was possible to observe the "glow" at a wavelength of 0.9 centimeters, where Milky Way noise is minimal.

It was found that the emissions became systematically "warmer" in the direction of the constellation Leo, implying motion of close to 400 kilometers a second in that direction. A kilometer is 0.6 miles.

Since the sun is orbiting the Milky Way galaxy at that speed, and in the opposite direction, the galaxy as a whole, they concluded, must be flying in the direction of Leo twice that fast—at more than 600 kilometers a second.

The three Berkeley experimenters were Dr. Richard A. Muller, Dr. George Smoot and Marc V. Gorenstein, a graduate student.

In a report published in October 1977 in *Physical Review Letters,* they noted that earlier attempts to determine the motion of the Milky Way relative to other nearby galaxies were done by examining the rates at which those galaxies are flying away from this one, as indicated by the lengthening of wavelengths in their light (the so-called "red shift").

In this way, any motions other than those related to uniform expansion of the universe in all directions should become evident. The studies have shown that the Milky Way is moving at about 100 kilometers a second relative to the center of mass of the local group of galaxies.

The local group, in turn, is moving at comparable velocity relative to the super-cluster to which it belongs. The motion deduced from the U-2 observations, however, is relative to the universe as a whole.

In all cases the velocities are far greater than those inferred by Dr. Harrison for the movement of the sun around its hypothetical companion. That, it appears, would require 10,000 years for a single revolution.

The U-2 observations were financed by NASA and the Department of

Energy. Such measurements, which require extreme sensitivity, are being conducted by several groups and are expected to intensify, now it has been shown that they offer a means of learning more about the early phases of the "big bang."

Recording velocities relative to the universe as a whole, Dr. Harrison said in a telephone interview last week, "contradicts what we have been telling young physics students for years—namely that there is no such thing within the universe as absolute motion."

—November 14, 1977

Antennas Sharpen Radio "View" of Heavens

By WALTER SULLIVAN

It is as though, having always been severely myopic, radio astronomers are acquiring glasses giving them 20/20 vision of the heavens.

In what is by far the most ambitious undertaking of its kind, an observatory is taking shape on the Plains of San Augustin 50 miles west of Socorro, New Mexico.

Its 27 dish antennas, each 85 feet in diameter, will produce images of the sky, as "seen" at radio wavelengths, as sharp as those photographed in visible light through the 200-inch telescope on Mount Palomar. The latter, on a California summit, was until recently the most powerful optical instrument on Earth.

The radio antennas will generate images through a process known as aperture synthesis. Otherwise, to match photographs obtained at the far shorter wavelengths observed with the Palomar instrument, a single antenna 17 miles in diameter would be necessary.

In recent years observations at radio wavelengths have led to such discoveries as quasars and pulsars, but the fuzziness of the images obtained has limited interpretation of the observations.

The antennas were used in one major discovery already, and it is widely anticipated that other significant findings will accrue as the new observatory comes into full operation.

Ten antennas are operating now, and were used in the discovery of an object far beyond the Milky Way that is radiating vast amounts of energy across the entire spectrum, from radio waves through visible light to X-rays and possibly gamma rays.

An 11th antenna has been built and is becoming operational. Others are being assembled at a rate of one every six or eight weeks.

Mounted on Railroad Track

Each antenna will be mounted along 38 miles of double railroad track, forming a giant Y with arms 12 to 13 miles long. When the entire array is completed late in 1980, there will be 72 tripod foundations on which the antennas can be mounted.

A 65-ton transporter already operating on the double rails can roll under an antenna, lift its 214-ton assembly off its three piers and carry it to any of the other tripods.

In this way a variety of antenna patterns can be formed to map the sky at radio wavelengths or to scan some small feature of it in detail. To make possible quick "scene changes" there will eventually be two transporters.

The observatory is known as the Very Large Array, or VLA, and will be the chief facility of the National Radio Astronomy Observatory. Until now the latter's primary facility was at Green Bank, West Virginia. The cost of the VLA will be about $78 million. Economies have been achieved by mass-producing the antennas and using surplus rail from government installations.

Cooperative Effort

The discovery of the powerfully radiating object beyond the Milky Way came in a joint effort by several institutions. In October 1977, a European group reported the detection of 13 sources of gamma rays, using the European satellite COS-B. All the sources were close to the plane of the Milky Way galaxy, suggesting that they might be within the galaxy.

It was suggested by Dr. Henry F. Helmken of the Harvard-Smithsonian Center for Astrophysics and by astronomers visiting from India and Italy that some of these sources be scanned by X-ray detectors of the Massachusetts Institute of Technology aboard the satellite SAS-3.

The observations were made from Nov. 24 to 29, and the position of a previously known X-ray source in the constellation Cassiopeia was determined with great accuracy. Dr. Hale V. Bradt of MIT arranged with Dr. Bruce Margon of the University of California at Los Angeles to search the area with the 120-inch telescope at the university's Lick observatory.

Charting the Stars

It was necessary, first, to produce a chart of stars near the X-ray position. This was done with the use of reference stars in the catalogue of the Smithsonian Astrophysical Observatory and an MIT computer. The chart, with the candidate stars numbered, was air-freighted to Dr. Margon.

He obtained spectra from the light of each star and excitedly phoned Dr. Bradt. The spectral lines from one were shifted strongly toward the red, indicating that it was receding far too rapidly to be a star. In terms of uniform expansion of the universe it would be 850 million light-years away.

Dr. Bradt then phoned Dr. Robert M. Hjellming at the VLA site. The antennas, under central computer control, were focused on the same position, and the object was found to be a strong emitter of radio waves despite its enormous distance. It could be either a quasar or a Seyfert galaxy, which is periodically subject to violent eruptions.

Some believe that quasars are the cores of very distant Seyfert galaxies, but no quasar has previously been identified so close to the Milky Way galaxy, which includes the sun and its planets. Because the newly identified object is in the plane of the galaxy, its light has been considerably dimmed by intervening material.

Whether it is the source of gamma rays observed in that region is not certain, as the gamma ray position is not very precise. It would be the first object beyond the Milky Way found to be emitting such extremely high energy radiation with sufficient intensity to be detected over the vast distances between galaxies.

Even without such emissions, the energy source of quasars remains one of the major problems in astronomy.

Overcoming a Limitation

The VLA is being built under the direction of John H. Lancaster. The aperture synthesis method that is the key to its operation was developed in England by Sir Martin Ryle, and won for him a Nobel Prize. It gets around a limitation on the "resolution" of an observing system, or the amount of detail that can be recorded, dependent on the wavelength observed and the size of the observing system.

With the help of a computer, the antennas synthesize an image of the sky comparable in resolution to one obtained from one huge antenna. The method, known as interferometry, depends on recording waves through pairs of antennas.

When all are installed, 351 pair combinations will be used to reconstruct an image, sweeping the sky at a succession of angles as the Earth rotates.

The closest rival to the VLA is at Westerbork in the Netherlands, which has only a dozen antennas mounted on a single line. Both stems use the Earth's rotation as an essential element of the scanning. The combined collecting surfaces of the VLA will equal that of a single antenna 430 feet in diameter. The largest such instrument is the 1,000-foot dish at Arecibo, Puerto Rico.

—December 29, 1977

Island in Hawaii Is Becoming a World Astronomy Center

By WALTER SULLIVAN

In Mount Mauna Kea, Hawaii, atop the most massive protuberance from the Earth's surface—the highest island mountain—astronomers from four nations are developing one of the world's most important observation sites.

Its telescopes, some already in operation, will probe the farthest observable reaches of the universe, and, closer to home, study such targets as the atmospheres of moons orbiting Jupiter and Saturn in ways previously beyond reach.

After tests at sites around the world, many astronomers have become convinced that conditions for observation are better at Mount Mauna Kea than anywhere else in the Northern Hemisphere.

The summit of this volcano, 13,796 feet above the surrounding sea, is so high that clouds rarely pass overhead. Even more important, the air above is extremely dry, making the site ideal for infrared observations—one of the fastest-growing fields of astronomy.

Key to Many Mysteries

Infrared wavelengths—those beyond the red end of the spectrum—are strongly absorbed by water vapor, but their observation is the key to many astronomical mysteries. The infrared telescope being built at Mount Mauna Kea at a cost of $6 million provided by the National Aeronautics and Space Administration has a mirror 120 inches in diameter.

Its design is much like that of an ordinary optical telescope except that it must avoid exposure of the optics to any structural surfaces that might radiate heat—that is, energy at infrared wavelengths.

The site's disadvantages include winds that sometimes exceed 125 miles an hour and a location bristling with cinder cones left by past eruptions, suggesting the possibility of another. There has, however, been no volcanic activity on this mountain for thousands of years.

The atmosphere is so thin that the astronomers do not live at Mount Mauna Kea and some find it difficult to do profound thinking there. They commute from a base station at 9,200 feet elevation, where they can do much of their homework.

Preparation for Earthquakes

Finally, there are occasional earthquakes, one of which has already done slight damage. The telescope mounts have been built with the possibility of temblors in mind.

The advantages, however, are overwhelming. The site's nearness to the Equator means that most of the southern and all of the northern sky can be observed. Astronomers sometimes have rare conditions when clearness and stability of the atmosphere make it possible to observe features only one second of arc in angular width.

Such nights occur fairly often at the Kitt Peak National Observatory in Arizona, considered an excellent site and perhaps the most extensive observing facility in the world. At Mount Mauna Kea, it has been estimated that they occur three times more frequently.

While remoteness of the site is a problem for some sponsoring institutions, notably in France, almost halfway around the world, it also means there are virtually no city lights to contend with.

Normal Streetlamp Glow

Even the most remote mainland American sites are near enough to a community with sodium vapor streetlamps for their glow to be scattered back from overhead. Thus every spectrum recorded from the light of the most distant object shows an overwhelmingly strong sodium line.

At Mount Mauna Kea the nearest large city is Hilo, but like other shoreline communities, it is usually covered by low-lying clouds.

Mauna Kea and its twin summit, Mauna Loa, crown the largest volcanic massif on Earth, rising in a gentle slope more than 30,000 feet from the Pacific floor. While Mount Everest stands 29,141 feet higher above the distant sea, its height above the surrounding land is far less.

The summit of Mauna Kea is a cluster of barren humps of volcanic debris reminiscent of a lunar landscape, but sometimes in winter it is white with snow, providing the Hawaiians with the rare experience of skiing while their fellow islanders are swimming off the beaches far below. A two-stage rope tow is operated, as well as jeep service that shuttles skiers back to the top.

The crown jewels of the summit are three great telescopes all nearing completion. NASA's infrared telescope, when finished early next year, will be operated by the University of Hawaii as a national facility.

France-Canada Project

The chief interest of the space agency is to use the instrument in conjunction with spacecraft making observations of such targets as the moons of the outer planets to determine the temperatures and compositions of their atmospheres. Infrared observations are also opening new vistas on cosmology as well as the formation of stars and planets.

The primary optical telescope, in terms of expected performance, is one with a 144-inch mirror being built as a joint project of Canada, France and the University of Hawaii. It was budgeted in 1973 at $18 million, to be provided equally by France and Canada.

Each of those countries, when the telescope goes into operation some time in 1979, will be entitled to 42.5 percent of the observing time. The University of Hawaii, in return for providing support facilities, will use the remaining 15 percent. The largest telescope on the summit is Britain's 150-inch reflector, to be used for both optical and infrared work. It is a "bargain basement" instrument without the precisely shaped mirror needed for distinguishing tiny features. The mirror reportedly became available after the manufacturer had made it as an experiment. It should be in operation before the end of 1978.

The University of Hawaii has been operating an 88-inch reflector at Mount Mauna Kea since 1970.

—*August 1, 1978*

Gravity "Lens" Is Found in Space

By WALTER SULLIVAN

After more than a year of elaborate observations with optical and radio telescopes in Britain, Germany, the Netherlands and the United States, as well as from an Earth satellite, not only has the discovery of a gravitational "lens" far out in space apparently been confirmed, but another has been reported.

A gravitational lens is one of the more bizarre implications of Einstein's general theory of relativity, which predicted that strong gravity fields could perceptibly bend the paths of light waves and, acting as a lens, could split distant images.

The newly found "lens" (possibly a double lens) appears to have split the image of a quasar on the fringes of the observable universe into three separate images. Quasars, which shine very brightly, are widely believed to be galaxies in an early stage of development.

The original discovery, reported in early 1979, was of "twin" quasars so much alike that some scientists proposed they might be a single quasar whose image had been split by the gravity of an object of enormous mass. Further observations appear to have confirmed this hypothesis.

The light-bending effect was first observed in 1919, when stars seen past the edge of the eclipsed sun appeared out of place because the path of their light had been bent as it passed through the strong gravity near the sun. This implied that a source of powerful gravity between the Earth and a very distant light source, such as a quasar, could act as a lens; that is, light passing on any side of the gravity source would be bent.

In typical glass lenses the bending, or refraction, of light becomes greater at increasing distances from the center of the lens, bringing the light to a single point of focus. In a gravitational lens the greatest bending would occur near the central region, where the gravity is strongest.

This effect of gravity would make the light from the distant source appear like a ring if the gravity source were exactly in line with the light source—an extremely unlikely possibility, considering the thin distribution of objects in the universe. If the gravity source were moved slightly to one side of a direct line, the ring would dissolve into two crescent-shaped images.

This appears to be the case with the "double" quasar discovered last year. It is so distant that its light has taken an estimated 10 billion years to reach Earth. A massive galaxy with several smaller companion galaxies has been detected at half that distance, in a position that would enable its gravity to split the quasar's image in two.

The "triple" quasar, detected recently by Dr. Ray Weismann and others at the University of Arizona, is even more distant and the sources of gravity splitting its image into three may themselves be so far away they cannot be observed.

An effort was made, in the twin quasars first discovered, to detect crescent shapes as predicted by the lens hypothesis, but it was foiled by observing difficulties.

Other tests of the past year have explored whether the two are identical in their emission intensities at numerous optical and radio wavelengths. Observatories on both sides of the Atlantic Ocean made radio maps of each object and the region surrounding it, by using a technique known as interferometry, in which multiple antennas are required.

Observations were made by antennas at University of Cambridge in England, by a trio of antennas at Jodrell Bank in England, by 14 antennas of the Very Large Array of the National Radio Astronomy Observatory in New Mexico and by the European Network, consisting of the radio telescopes at Jodrell Bank, Dwingeloo in the Netherlands and Effelsberg in West Germany.

A radio map produced by the Very Large Array showed what appeared to be twin "jets" emanating from the north quasar, but not from the southern one. This mismatch led scientists at the Massachusetts Institute of Technology who analyzed the results to question the lens hypothesis.

They suggested that the quasars were true twins that "had a common origin, are similar in their basic physical parameters and are evolving in similar fashions."

Finally, scanning with electronic intensification devices mounted on the 200-inch Mount Palomar telescope in California, second largest in the world, revealed an intervening galaxy slightly north of the southern quasar of the pair.

It is now proposed that this galaxy produces the lens effect, with "aberrations" contributed by the gravity of its neighbors. Further evidence has been reported in a recent issue of the British journal *Nature*, based on observations with the *International Ultraviolet Explorer*, a satellite that can observe at ultraviolet wavelengths unable to penetrate the Earth's atmosphere.

Scientists from the Rutherford and Appleton Laboratory in Slough, near London, and University College London have found the ultraviolet spectra of the two quasars virtually identical.

This established that across almost the whole spectrum, from radio waves to ultraviolet light, the twins seem identical. Only in the infrared part of the spectrum do they differ, possibly because of a contribution, along one of the two light paths, from the intervening galaxy.

—June 24, 1980

"Big Bang" Has a Revival in New View of Universe

By WALTER SULLIVAN

A revolution is taking place in the study of the cosmos. Theorists now believe it is possible that the entire observable universe evolved from a concentration of energy no more massive than a single apple.

They have now concluded, too, that this entire universe may be embedded in another universe that is far larger but totally unobservable.

Such concepts, although still regarded by some experts as improbable, appear to be replacing the "big bang" theory long favored by mainstream cosmologists.

The new view proposes, essentially, that the universe was born from virtually nothing. Its substance and energy formed spontaneously during an initial period of expansion when, in the tiniest imaginable fraction of a second, the universe repeatedly doubled its size.

Because the concept envisions an initial stage of almost incredibly rapid expansion, it is called the "inflationary universe" model. In the big bang model, all the stuff of the universe was there from the beginning, although in an incredible concentration of energy that exploded and partially condensed into matter, forming galaxies, stars, planets, and people.

Both theories account for the observation that the galaxies forming the universe are flying apart in all directions. Both also provide an explanation for the "glow" that fills the heavens (in the microwave part of the radio spectrum). That glow is seen as evidence of some earlier formative stage of the universe's evolution.

But several mysteries that have cast doubt on the big bang model seem to be better explained by the inflationary universe, including:

> The extraordinary uniformity of the universe. Its elements all seem so much alike, their atomic structures, their shapes, their relationship to each other. But the farthest corners of the universe are separating from each other so fast that in the big bang model they could never, ever have been in contact, not even in the first moment of time. Why then should everything be so homogeneous, as though conforming to the same blueprint?

Uniform expansion. The constant expansion of the universe is also everywhere uniform and smooth. The wild turbulence to be expected from an explosion is found nowhere.

Rate of expansion. It has seemed to many astronomers just too lucky for belief that the rate of expansion after the big bang matched the requirements of universe formation. Had the explosion been slightly more powerful, the universe would have expanded too fast for galaxies to form. Had it been slightly less powerful, the universe would quickly have collapsed back together.

As Paul C. W. Davies of the University of Newcastle-upon-Tyne in England put it in a recent issue of *The Sciences,* "For some mysterious reason, the vigor of the explosion was matched so delicately to the gravitating power of the universe that the galaxies had just enough speed to escape each other's gravity, yet not so much as to rapidly disperse."

These inconsistencies slip away under the explanatory power of the inflationary universe idea, which is based on the hypothesis that mass and energy appeared from nowhere.

The notion that "something can be made from nothing" derives from unpredictable fluctuations known to take place on the subatomic level. From bits of gravitational or electrical energy, particles form spontaneously, then recombine and vanish almost immediately.

Ordinarily, this would seem to violate the law of conservation, which holds that matter cannot be created from nothing. But nature seems to have a way of "cheating," rapidly creating matter, then making it disappear. With enough energy, however, the particles of matter may persist.

And that is what happened, according to this new theory of the expanding universe. At the end of the universe's inflationary stage, enough energy and matter had appeared to produce the universe we now observe.

In December 1982, Alan H. Guth of the Massachusetts Institute of Technology, who first proposed the theory several years ago, described its revised version to an international astrophysics meeting in Austin, Texas. The "seed" that became the observable universe, he said, may once have been far smaller than an atomic particle, with a mass less than that of an apple. All the rest of its mass (in the form of energy) was created during an initial period of runaway inflation.

If, as many now suspect, the conservation laws can be violated, Dr. Guth told the meeting, "it becomes very plausible that our observed universe emerged from nothing or from almost nothing."

When his theory was first formulated, he said, he recognized that it was incomplete. Then A. D. Linde at the Lebedev Institute in Moscow proposed a way to avoid its drawbacks, as did Paul Steinhardt and Andreas Albrecht at the University of Pennsylvania. The result is the now widely discussed "new inflationary universe."

An essential ingredient of the theory is a fundamental transformation of the universe that occurred at the end of its inflationary phase, when it was still far less than a second old.

The transformation is consistent with the so-called Grand Unification Theories. If the inflationary universe concept is widely accepted, it should become an integral part of these theories, which are at the heart of speculation in contemporary physics.

The theories propose that, in the extremely high temperature environment very soon after the universe was born, the diversity of the particles of matter now known did not yet exist. All particles were identical, and so were the forces controlling them.

Only when the universe cooled off to a mere 20 billion billion billion degrees Fahrenheit did the particles and forces begin segregating into the diverse forms seen today. This "symmetry breaking" is seen as a "phase change" in the nature of matter comparable to the phase change of water to ice.

As Dr. Guth explains it, before that cosmic "phase change" the universe was expanding with an energy content that grew as fast as its volume. The phase change that followed was delayed long enough for the universe to become "supercooled," just as water may remain liquid below its freezing temperature, then freeze almost instantly. Ordinary water, freezing normally, gives up latent heat gradually. But when supercooled water freezes, it gives up latent heat suddenly.

Just as freezing water sheds its latent heat, the phase changes at the end of the universe's inflationary phase released energy, but on a catastrophic scale, like the big bang but later in the evolution of the universe. Part of the energy condensed into matter. Part of it, in less than a second, blew up the universe to a volume billions of times larger than everything that can now be observed. Within this great volume the universe we know, no larger at first than a softball, began expanding to its present dimensions.

A Universe Beyond

Before this new explosion began, all parts of our observed universe were still inside the softball, a small enough volume to put everything in contact with everything else, thus ensuring uniformity throughout. The superlarge universe beyond our vision may be highly irregular, but in it, as Dr. Davies puts it, "would lie substantial oases of uniformity and quiescence, such as the region we now inhabit."

After expanding for about 100,000 years, the universe thinned out enough to become transparent. Its temperature then was about 8,000 degrees Fahrenheit, producing the universal glow now visible at radio wavelengths. The utter uniformity of this glow in all directions (apart from effects of the Earth's motion) testifies to the absence of turbulence when our universe was young.

—March 29, 1983

Cosmic Powerhouse Finally Seen in Detail

By WALTER SULLIVAN

An object considered by many astronomers to be the most remarkable in the universe has been "photographed" in great detail for the first time.

Although the object, Cygnus A, is 600 million light-years away from Earth, it is second only to a relatively nearby supernova remnant in the constellation Cassiopeia as a source of radio emissions in the sky. In one second, Cygnus A produces more energy than the sun generates in 30,000 years. How it produces such enormous amounts of energy is one of the basic problems in astrophysics.

The new image, produced by coordinated radio observations using the multiple antennas of the Very Large Array in New Mexico, is a major improvement over any obtained previously. It shows for the first time what appear to be jets radiating in opposite directions from a central galaxy. They culminate in two strikingly symmetrical clouds of swirling filaments. On the outer edge of each such cloud, as viewed at radio wavelengths, is a compact lobe of exceptional brilliance.

So enormous is the scale of this two-jet structure that it would take light 500,000 years to travel from one side of the structure to the other, moving at 186,300 miles a second.

The image was obtained by Dr. John Dreher of the Massachusetts Institute of Technology, Dr. Rick Perley of the National Radio Astronomy Observatory, which operates the Very Large Array, and Dr. John Cowan of the University of Oklahoma. Its fine detail was brought out by computer processing at the Jet Propulsion Laboratory in Pasadena, California.

The observing scientists liken the clouds at the ends of the jets to cotton candy, "an intricate tracery of filaments, wisps and arcs." Even the smallest of the wisps, they note, is more than a million times larger than the entire solar system.

Because of these newly revealed details, they say, present theories to explain such objects, known as radio galaxies, may have to be "significantly altered."

A number of radio galaxies have been observed in other parts of the universe. Jets have been observed radiating from more than 125 of them, as well as from more distant quasars, but none deliver to the Earth the intense radio power of Cygnus A.

As has now been demonstrated anew for Cygnus A, twin jets typically culminate in lobes of powerful radio emission on opposite sides of a central galaxy, which appears very small relative to the scale of the jets and lobes. In at least one case, a jet has been traced to the core of the galaxy.

Theory on Energy Source

Some mysterious energy source, or "engine," in the core of each galaxy must be responsible for the jets. The leading theory is that the core contains a fast-rotating superdense object, or black hole, with a mass comparable to millions or billions of stars.

A black hole, in theory, is a concentration of mass so dense that gravity prevents the escape of light or anything else. Thus, it could only be observed through the effects of its gravity on nearby matter. The jets, according to this hypothesis, are ejected along opposite ends of the hole's spin axis, racing outward at close to the speed of light.

According to present theory, as these jets encounter the extremely thin gas adrift between the galaxies, the gas is violently disrupted and sheds energy at radio wavelengths. The gas is also pushed outward by the jet, forming the lobes at the end of each jet.

At the outer extremity of each lobe is a "hot spot," or brilliant source of radio energy that, the astronomers believe, was blasted out of the galaxy and channeled along the jet. The wispy lobes, they proposed in the Oct. 1 issue of *Astrophysical Journal Letters*, "are the regions behind the advancing hot spots where 'waste particles' are deposited."

As the astronomers have put it, the central galaxy is, in effect, "blowing up two giant balloons" in space. It is the balloons that are the chief source of radio energy emitted by Cygnus A.

Study Began Four Decades Ago

The history of efforts to observe and understand Cygnus A, so-called because it lies beyond the constellation Cygnus, the Swan, date to the earliest days of radio astronomy, immediately after World War II, when J. Stanley Hey in Britain turned a modified antiaircraft radar on the heavens.

The strongest radio source, in the constellation Cassiopeia, was relatively diffuse and proved to be the nearby remnant of a great star explosion, or supernova. The emissions from Cygnus A, on the other hand, appeared to originate in a tiny point. When radio astronomers in 1951 obtained a precise position for the source, Walter Baade in California aimed the giant telescope on Mount Palomar, then the world's largest, at the spot.

"I knew something was unusual the moment I examined the negatives," he said afterward. "I had never seen anything like it before. It was so much on my mind that while I was driving home for supper, I had to stop the car and think."

He concluded that it was two galaxies in collision. Only recently, when large arrays of radio telescopes became available, has it become possible to map the source in detail, showing its remarkable structure and vast size. Over a three-year period the Very Large Array made four runs of observations at two radio frequencies and with the antennas deployed into various configurations.

The emissions from Cygnus A, recorded for a total of 18 hours, were stored on magnetic tape for computer analysis. The most recent result can be likened to photographing a scene at many exposures to record in detail both its bright and dim areas. The pictures would then be combined to show both bright and dim areas in comparable detail.

—January 1, 1985

New View of Universe Shows
Sea of Bubbles to Which Stars Cling

By WALTER SULLIVAN

A new three-dimensional map of part of the universe shows that it is composed of gigantic "bubbles," with the stars and galaxies, probably including our Milky Way, gathered on the surfaces.

This new view could require basic revisions in theories on the evolution of the universe, scientists say. If the bubble structure is borne out by further mapping, they add, it could support a theory that a multitude of vast, powerful explosions, not the forces of gravity, were the primary shapers of the cosmos.

The new survey, reaching out twice as far as previous ones, has shown a universe that appears to be made up of giant bubblelike voids with galaxies distributed on their surfaces.

"If we are right, these bubbles fill the universe just like suds filling the kitchen sink," said John P. Huchra of the Harvard-Smithsonian Center for Astrophysics in Cambridge, Mass. Professor Huchra is one of the authors of a report on the survey that is to be published in the March 1 issue of *Astrophysical Journal Letters*. His coauthors are Margaret J. Geller and Valerie de Lapparent, also of the Harvard-Smithsonian center.

The observations, say the authors, "pose serious challenges for current models for the formation of a large-scale structure."

Need to Re-weigh Theories

Some scientists say the theory of gigantic explosions in the formation of the universe could account for the bubbles. Other theorists suspect that the bubble structure is related to transitions in the nature of matter that occurred in the earliest history of the universe, which scientists believe is 10 billion to 20 billion years old.

But both schools agree that conventional explanations for the evolution of large-scale structure in which gravity played a dominant role may have to be modified or abandoned.

For the past decade the Harvard-Smithsonian group has been trying to chart the structure of the universe in one region of the sky, using the rate at which the galaxies are receding from the Earth to indicate their distance.

389

Because of the seemingly uniform expansion of the universe, the rate at which each galaxy or cluster of galaxies is receding from the Earth, as indicated by a shift of its light toward the red end of the spectrum, is used as a yardstick of distance.

Most, if not all, stars are organized into the immense assemblages known as galaxies, including the Milky Way galaxy to which our solar system belongs. The galaxies are themselves organized into clusters and superclusters, some of which, earlier surveys have shown, are linked into filaments spanning large parts of the observable universe. It also appears that there are voids containing almost no galaxies, including a large one in the constellation Boötes.

The theory that the formation of the galaxies may have been triggered by shock waves from great explosions was proposed in 1981 by Jeremiah P. Ostriker of Princeton University and Lennox L. Cowie of the Johns Hopkins University. Observational support of the concept, Dr. Ostriker said Friday, "is very good news for me."

Their concept was that early in the history of the universe, new, very massive stars burned out, collapsed and then exploded in supernovas. Closely spaced explosions, initiating one another in a chain reaction, would generate a shock wave that, like a great expanding bubble, would compress material in its path, stimulating the formation of stars and galaxies.

The observation of "starburst" galaxies in which such chain reactions appear to be taking place made the concept seem more plausible.

That a similar process, on a smaller scale, has occurred within the Milky Way is evident in photographs showing filamentary structures formed by expanding shock waves from past supernova explosions. Because of this, parts of the heavens have a striking "bubble bath" appearance.

Bigger Than Envisioned

The Ostriker-Cowie concept envisioned bubbles far smaller than the larger ones found in the new survey, one of which appears to be more than 160 million light-years wide. The Milky Way galaxy, by contrast, is less than 100,000 light-years wide.

That giant bubble found in the survey is nearly circular and its edge, the authors of the report say, "is remarkably sharp." One possibility, Dr. Ostriker said, is that the largest bubbles are consolidations of smaller ones.

Dr. David N. Schramm of the University of Chicago said in a telephone interview that nonexplosive processes could account for the bubbles, such as phase transitions in the nature of matter in the formative stages of the universe. He likened such transitions, generating new structures, to those in which dissolved material forms into needlelike crystals.

Dr. Vera C. Rubin of the Carnegie Institution of Washington, an authority on galactic structure, noted in an interview that the mapping had been based on velocity rather than absolute determinations of distance. Hence, the structures have been charted in what she called "velocity space" rather than "position space." If the velocity of the stars' retreat is nonuniform, she suggested, this could distort the red shift and alter the three-dimensional picture put together by the Harvard-Smithsonian astrophysicists.

The new map shows 1,100 galaxies within a slice of the sky 6 degrees wide and 117 degrees long in the general direction of the Coma Cluster of galaxies. The moon, seen from the Earth, is half a degree wide. The survey has used 327 red shifts obtained by others, 186 obtained by the survey itself and 584 recorded from the Smithsonian observatory on Mount Hopkins, Arizona.

It appears that clusters of galaxies lie at the intersections of large bubbles. The local supercluster of galaxies to which the Milky Way belongs may "sit on the surface of a shell," according to the astrophysicists in Cambridge.

The group hopes to extend their survey to a region 10 times larger than the one covered so far.

—January 5, 1986

Powerful Source of Gravity Detected Deep in the Universe

By WALTER SULLIVAN

Far beyond the Milky Way, astronomers have detected an invisible but incredibly powerful source of gravity that appears to defy conventional explanation.

Its gravitational field, acting as a lens, has split light from a distant quasar in a way that some theorists believe is possible only if the field is the result of fragments from an early stage of the universe, before matter existed, that are still coursing through space.

Such fragments, existing as loops or infinitely long one-dimensional "cosmic strings," would create bizarre gravitational and optical effects along their paths, including the splitting of celestial images.

Six other gravitational lenses have been discovered since 1979, confirming a prediction made by Einstein's General Theory of Relativity. However, the lens discovered recently is so powerful—the split quasar images are 20 times farther apart than any others yet discovered—that it cannot readily be explained in terms of any existing theories.

Probably not since the discovery of quasars in 1963 has an astronomical discovery generated so much excitement and perplexity among theorists. A detailed report on the discovery, written by a team of eight astronomers led by Dr. Edwin L. Turner of Princeton University, will appear in an upcoming issue of the British journal *Nature*. It was widely discussed in the corridors of the spring meeting of the American Physical Society in Washington.

Other than cosmic strings, possible explanations for the power of the lens include a black hole of extremely large mass—equal to one million billion times the mass of the sun—or an extremely dense cluster of galaxies that, for some reason, cannot be seen. So far, efforts to test these hypotheses have not provided clear support for any of them.

Several of those who took part in identifying the double quasar were directly or indirectly involved in the discovery of the first quasars 23 years ago. It was then that one of them, Maarten Schmidt of the California Institute of Technology, found from analysis of light from a starlike source of radio waves that it appeared far more distant than any known object.

Since then several thousand such "quasi-stellar radio sources," or quasars, have been discovered. Assuming the yardstick used to measure cosmic distances—the so-called red shift—is valid, many quasars are so far away their light has taken billions of years to reach the Earth. They are therefore seen as they were during the infancy of the universe and may be galaxies in an early, extremely brilliant stage of development.

Astronomers studying these quasars in 1979 were startled to discover that two such objects appeared to have identical spectral signatures and were at identical distances from the Earth, although they were ever so slightly separated in the sky. These twin quasars were close enough together to be explained by the presence along the light path of a massive concentration of material, such as a cluster of galaxies, whose extreme gravitational force was bending the light.

Einstein's theory that gravity bends light was first demonstrated in a 1919 eclipse of the sun. Stars seen past the rim appeared out of place because their light had been bent by the strong gravitational field near the sun.

The effect on light of a very intense gravitational field can be demonstrated by looking at a small, bright light source through the base of a wineglass whose thickness tapers toward its rim. If alignment is perfect, most of the light will be deflected into a ring. If the alignment is not perfect—if the center of the lens is not directly between the light source and the observer—the light may form two irregular spots with a faint light showing through the center.

Relics of Primordial Vacuum

In the previously discovered "twin" quasars, the images were a few seconds of arc apart. In the newly discovered example the images are 157 arc seconds apart, more than 20 times farther than in any other case.

Proponents of the cosmic string hypothesis have argued for some time that the discovery of such widely split objects would indicate the presence of something far more powerful than the most massive clusters of galaxies yet found—cosmic strings.

The concept derives from the efforts of Alexander Vilenkin of Tufts University, the Soviet theorist Yakov B. Zeldovich and others to account for progression of the exploding universe from a primordial condition of extreme concentration and uniformity to one of great diversity, with galaxies, stars, planets and people.

In particular, the theorists wanted to explain what first provided the gravitational suction that began drawing together material to form what became great clusters of galaxies. Very early in the life of the universe, according to the now widely accepted "inflationary" view of its origin, the universe was without atomic particles and lacked the diverse laws that now govern their behavior.

It then underwent a radical transformation—a phase change comparable to that when liquid water turns to solid ice. Matter and the physical laws took on their present form. The change, however, was not universal and instantaneous. Parts of the earlier universe survived, according to the hypothesis, and now form cosmic "strings" with only one dimension—length—and no end. They must either form closed loops, some reaching across much of the universe, or be infinitely long. They are unrelated to the tiny "superstrings" that, according to some theories, form the basic particles of matter, replacing the conventional concept that such building blocks are pointlike.

Cosmic strings are relics of the primordial vacuum that existed before there was matter, and some may be traveling at close to the speed of light. Negative pressure along their lengths produces enormous tension. Since the strings contain no matter, this tension produces their strange gravitational effects.

Loops would shrink as they radiated gravitational energy and, as they shrank, would form galaxies and, finally, the extreme concentrations of matter in their cores. In 1985, Richard Gott of Princeton proposed in the *Astrophysical Journal* that a string could split light from a quasar into identical images up to 360 seconds of arc apart, twice what has now been observed.

Bohdan Paczynski of Princeton then drew attention to several twin quasars as possible examples of this effect, including the newly identified one. At a recent Physical Society meeting Dr. Vilenkin did likewise.

Testing for Cosmic Glow

One test of splitting by an unseen cluster of galaxies, said Jeremiah P. Ostriker of Princeton, would be the effect of hot gas within the cluster on the backdrop "glow" remaining from birth throes of the universe.

This "cosmic background radiation" in the microwave part of the radio spectrum is normally uniform in all directions, equivalent to that radiated by a black surface warmed six degrees Fahrenheit above a total absence of heat.

It is assumed that this background radiation would be altered by any of the proposed explanations for the latest split quasar images. If the gravity came from a massive cluster of galaxies, their churning clouds of hot gas would knock the cosmic glow to higher, unobserved energy levels, making the glow appear cooler in that direction. A cosmic string would create an abrupt, steplike change in the glow's temperature. A black hole would "suck in" the glow, producing an empty spot.

To test the theories, a communications antenna at Crawford Hill, New Jersey, was trained on the images in such a way that the rotation of the Earth swept its aim across the hypothetical string, black hole or galaxy cluster. Nothing unusual was observed.

Theorists do not believe this rules out any of the hypotheses, since the antenna beam may have been too broad or the receiver too insensitive to detect the effect. According to Dr. Paczynski, the Very Large Array of radio telescopes in New Mexico may be the only facility capable of detecting the black hole effect.

The participants in the observation included Anthony Stark of AT&T Bell Laboratories; Dr. Robert W. Wilson, who shared a Nobel Prize with Dr. Arno Penzias for the original "cosmic glow" observation in 1965; Mark Dragovan, also of Bell Labs, and Dr. Goss.

The authors of the discovery paper, in addition to Dr. Turner and Dr. Schmidt, are Donald P. Schneider of the Institute for Advanced Study at Princeton, Bernard F. Burke, Jacqueline N. Hewitt and Glen I. Langston of the Massachusetts Institute of Technology, James E. Gunn of Princeton University and Charles R. Lawrence of Caltech.

Red Shift Validity Questioned

In their *Nature* paper, they point out that attention was first drawn to this part of the heavens by Halton C. Arp, then with the Hale Observatories in California, and two other astronomers. They found a striking concentration of quasars within a small patch of sky. Dr. Arp has long sought—largely in vain—to persuade astronomers that the red shift is an unreliable measure of quasar distances.

He cited this cluster of quasars, seemingly close to one another but with very different red shifts, as support for his contention. Skeptics argued that the quasars are actually at very different distances, their clustering being happenstance.

Red shift refers to the extent to which light from a distant galaxy has been shifted toward the red, or long-wave, end of the spectrum by the galaxy's rapid motion away from the Earth. The effect is the same as that which lowers the pitch of the horn on a receding vehicle. The galaxies are receding because the universe is expanding like a swelling cloud of gas.

A microbial observer on a grain of dust anywhere within such an expanding cloud would note that a dust grain twice as distant as a nearer one would be receding at twice the velocity. That this effect applies to galaxies seems confirmed by the manner in which those with greater red shifts are also dimmer. This dimming relationship, however, is less clear with quasars.

Dr. Arp and his associates noted that two of the quasars in the cluster, identified as 1146+111 B and C, had identical red shifts, but dismissed the idea that they were split images because of their wide separation.

To test whether the images are identical, detailed spectra of both were obtained on the nights of March 5 and 7, 1986, with the 157-inch reflector of the Kitt Peak National Observatory in Arizona. Exposures as long as 30 minutes were used and the two objects proved alike in so many respects that there is little doubt they are the same.

According to Dr. Ostriker, the possibility remains "that nature is playing a horrible trick on us," but he said he seriously doubts that. On the other hand, Dr. Turner and his colleagues believe the phenomenon responsible for the lens effect may also account for the surprising concentration of quasars in that patch of sky.

Light from the quasars may have been gravitationally amplified and a careful search may reveal more quasar twins. Their distribution might show whether the gravity source is "stringy." Searches for such clues are under way.

Research leading to the discovery was supported by grants from the National Science Foundation and the National Aeronautics and Space Administration.

—*May 6, 1986*

Huge Stellar Explosion Detected Close Enough for Careful Study

By MALCOLM W. BROWNE

Astronomers have reported that the gigantic explosion of a massive star has occurred much closer to the Earth than any observed since 1604. Because such nearby supernovas are so rare, this event is likely to have a dramatic effect on scientific understanding of how stars, galaxies and the entire universe have evolved through the ages.

The cosmic blast occurred 50,000 years ago, but the light it generated is only now reaching Earth.

Dr. Robert E. Williams, director of Cerro Tololo Inter-American Observatory in Chile, said in a telephone interview that astronomers throughout the Southern Hemisphere had deployed their telescopes last night for intense scrutiny of the supernova. He said that if the object lived up to expectations, it was likely to become the brightest star in southern skies for the next 10 days. It would be as bright as the planet Jupiter, he said.

Visible in Southern Hemisphere

The object will not be visible in most of the Northern Hemisphere, although people as far south as Brownsville, Texas, might at some point get a glimpse of it low on the horizon. But for observatories in Chile, Australia and South Africa, the supernova will be within easy view.

Dr. Williams said that because of the importance of the event, his team was televising its observations last night so that astronomers throughout the world could make immediate use of them. His group's main telescope at Cerro Tololo near La Serena, Chile, is one of the world's most powerful and is by far the largest in the Southern Hemisphere. The Inter-American Observatory is one of the major observatories in North and South America operated by National Optical Astronomy Observatories on behalf of the National Science Foundation.

He said that the discovery was made shortly before dawn yesterday by observers at the University of Toronto observatory at Las Campanas, Chile, about 60 miles south of the much larger Cerro Tololo facility.

Small Telescope Used

"Their observer was exposing some routine astronomical photographs" made with "a rather small telescope when he noticed the image of a new star where previous photographs had shown none," Dr. Williams said, adding: "By that time the sun was beginning to rise, and further observations were impossible. But this is an event every astronomer in the world had been hoping would occur in his or her lifetime, and you may be sure we'll make the most of the opportunity. Luckily, the skies over Chile are likely to be clear tonight."

Soon after sunset, observers at the Cerro Tololo telescope quickly confirmed the existence of the new supernova. The exploding star was shining with a brightness of 4th magnitude, comparable to that of a moderately bright star visible to the naked eye. Dr. Williams added that initial measurements of the supernova's light spectrum suggested that it was probably a Type I supernova, that is, one that had formed through the interaction of two stars orbiting each other.

The supernova was observed in the Greater Magellanic Cloud, one of a cluster of small galaxies that are so close to our own Milky Way galaxy as to be considered satellite galaxies. The supernova is 50,000 light-years from Earth, a short distance away in astronomical terms.

The nearest full-size galaxy to ours is the giant spiral known to astronomers as M-31, the Great Nebula in Andromeda, which is more than 2 million light-years away. Most other galaxies are many hundreds of millions of light-years distant, and when supernovas are spotted at such great distances they are comparatively dim and difficult to study.

As director of Cerro Tololo, Dr. Williams supervises a huge instrument with a light-gathering mirror four meters in diameter.

"One of the first things we and other observatories will have to do at this point is to determine whether any star previously photographed could have been the precursor of this supernova," he said. "For that we will have to determine the supernova's position very exactly for comparison with star charts, and we're beginning that work tonight."

He said it was difficult to tell when the supernova might first have become visible on Earth but that it probably began to brighten markedly about one week ago. The supernova will be intensely studied by many observatories in the months and years ahead, he said, and astronomers will be particularly interested in the cloud of debris expected to expand from the explosion.

All supernovas are believed to be the violent deaths of very massive stars that have used up most of their nuclear fuel and are forced by their gigantic gravity to collapse on themselves. The collapse releases so much energy that a final cycle of nuclear fusion occurs, and the resulting explosion briefly produces more light than does an entire galaxy of hundreds of millions of stars.

Among the many reasons astronomers are interested in supernovas is that these cosmic fireworks are thought to create most of the heavy elements in nature, including the carbon from which all living things on Earth is made. An ordinary star made mostly of hydrogen converts most of its fuel by the fusion process into helium and then dies. But stars substantially larger than our sun may continue the fusion cycle, creating elements as heavy as iron before erupting as supernovas, which may generate even heavier elements that are blasted into space.

The brightest part of a supernova explosion flares up in a matter of days and fades rapidly after several weeks. The 1604 explosion, which was observed by the great astronomer Johannes Kepler, left a residue visible through telescopes today as a faint nebula of glowing gas expanding rapidly from the point where the explosion occurred.

Astronomers recognize two main types of supernova. In a Type II supernova, a star begins its life as a particularly massive object and consumes its own substance during its violent death throes. The other type of supernova, Type I, involves a binary star, a system in which two stars orbit each other. In some cases, one of the stars in the pair is so massive and so close to its companion that it draws matter away from the companion, gradually becoming more massive itself. Eventually, such a cannibal star becomes so engorged that it becomes a supernova.

"The spectral measurements we have made of this object tonight," Dr. Williams said, "tell us that it is a Type I supernova created by the interaction of two stars—one of them a degenerate and very massive white dwarf, and the other a normal star. The white dwarf probably drew off matter from its companion until the point that it became a supermassive object doomed to die in a supernova explosion."

A white dwarf is an old, dim star consisting of material so dense that although the star may be no larger than the Earth it has a mass equal to that of the sun.

—February 25, 1987

Elated by Supernova, Astronomers Watch Their Theories Come to Life

By MALCOLM W. BROWNE

The cataclysmic explosion of a nearby star in February 1987 has dramatically strengthened major theories of how such explosions occur, and astrophysicists say the supernova has given them a new sense of confidence in their ability to interpret stellar events.

In particular, scientists believe they were correct in predicting how and in what sequence an exploding star would combine simple atomic nuclei with protons and neutrons to make the heavy atoms essential to complex chemical processes, including the creation of life.

They have evidence that the explosion blasted dense layers of freshly created elemental matter into rings that are flying outward from a small remnant of the former star. The atoms of that remnant have been crushed into a compact, super-dense globe of neutrons.

The supernova explosion that occurred over the Southern Hemisphere in February 1987, some 150,000 light-years from Earth, was the closest such event in four centuries.

"There's an enormous intellectual change taking place because of this object," W. David Arnett of the University of Chicago said in an interview. Dr. Arnett is a leading astrophysicist and is one of the scientists whose theoretical explanations have been confirmed by the recent event.

"Before this thing happened," Dr. Arnett said, "we had a very different way of looking at our work. We'd developed good theories, but we didn't know they were correct, and we didn't act as if they were correct. Now that we're certain we're on the right track, we know what kind of instruments we should build to measure the next supernova in our galaxy. It probably won't be visible like this one, but if it's the same type, it will emit a characteristic pulse of neutrino particles that we can detect."

Supernova SN1987A, as the explosion was designated, flared into view on Feb. 23, 1987, and a Canadian astronomer, Ian Shelton, working at Las Campanas Observatory in Chile, discovered it a few hours after its appearance.

Brightness Reaches Peak

Dr. Robert E. Williams, director of the Cerro Tololo Inter-American Observatory in Chile, said in a telephone interview that the supernova continued to brighten continuously until May 20 when it peaked at magnitude 2.97, the equivalent of a fairly bright star easily visible to the naked eye. Since then it has declined slightly in brightness to about magnitude 3.4.

The supernova has scintillated with cosmic clues that astronomers have had little difficulty fitting into the theoretical models devised over the years to account for the behavior of exploding stars.

Despite the glaring brightness of the supernova as viewed by such giant instruments as Cerro Tololo's telescope, whose light-focusing mirror is four meters in diameter, the explosion is not yielding its secrets easily. The glare is intense light emitted by the outermost shell of the blast—a kind of luminous globe that conceals what is happening within.

The hot shells of gas, analogous to nested balloons, are expanding into space at thousands of kilometers per second, and as they do so, they grow cooler and more rarefied. But for the time being, these expanding gases, mostly in the form of electrically charged atoms called ions, are opaque; neither light nor any other form of electromagnetic radiation can leak out of the depths of the explosion to be observed on Earth.

"We're not really sure exactly when the supernova will become transparent to various wavelengths of radiation," Dr. Williams said, "and this is why it's essential to measure the light spectra with great care every night."

Recently, Dr. Williams's group has begun to detect features of the supernova spectrum (absorption lines) characteristic of the presence of barium and strontium, and this suggests that the observers are already looking deeper into the explosion than was possible before.

"The beautiful results obtained at Cerro Tololo," Dr. Arnett commented, "not only suggest that the gas in the outermost hydrogen shell is becoming transparent. They confirm our prediction that barium and strontium would be created in this kind of process, and would turn up in the underlying helium shell."

Events Leading Up to Explosion

SN1987A is a type of supernova, Dr. Arnett and other theorists believe, in which the progenitor star had been synthesizing heavy elements for a long time before the explosion occurred. In this hypothetical process, called the "S Process" (for slow process), the nuclear fusion that fuels the stellar furnace grafts protons and neutrons on atoms even heavier than iron, creating new elements as heavy even as uranium. The explosion itself creates other elements, and successive layers contain characteristic proportions of various elements.

The explosion starts with a very hot, short-lived star some 20 times the mass of the sun, which has consumed most of the hydrogen of which it was initially composed, fusing hydrogen nuclei together to create helium. As the core cools it can no longer support the crushing gravitational pressure of the star's huge mass, and the core collapses. The inward-rushing matter generates new heat, however, which initiates the fusion of helium. The process of cooling, collapse, reheating and initiation of new stages of fusion can continue up through successively heavier elements up to iron, according to theory.

But iron is the end of the line for a star, because the fusion processes needed to create still heavier elements absorb energy rather than create it. Once a star has burned most of its fuel to iron, it quickly cools and suffers a core collapse so violent that a supernova explosion ensues. In the case of such supernovas as SN1987A, a small, ultradense neutron star is left at the center, a star so dense that atomic nuclei themselves are crushed together into a compacted mass of neutron particles.

At the instant this happens, according to theory, an avalanche of neutrino particles rushes out of the superdense matter in the collapsed star, speeding outward in all directions at the speed of light. Neutrinos have no electrical charge and probably have no mass, and they therefore scarcely interact with matter; a single neutrino speeding through the entire Earth would stand very little chance of colliding with an atom or particle.

But when great floods of neutrinos pass through a substance, the chances become greater that one of them will hit something and register its existence. When the supernova first appeared, it happened that several sensitive neutrino detectors were working on another project involving particle physics.

Two of these detectors, one at Kamioka in Japan and another near Cleveland, independently detected huge surges of neutrinos at the same moment. Subsequent analysis of the data by the group in Japan and by scientists at the University of

California at Irvine, the University of Michigan and Brookhaven National Laboratory on Long Island left no doubt that the neutrino burst had come from the supernova.

Astronomers everywhere exulted over that result. "The neutrinos reached Earth in just the way theory predicted," Dr. Arnett said, "telling us that the core of the supernova had collapsed into a neutron star."

At that point, Dr. Arnett and other theorists calculated the exact characteristics of the progenitor star that would account for such an explosion, and it wasn't long before observers found exactly what was needed. Intensive study of photographic plates made prior to the explosion revealed that a single star, discovered in 1969 by Nicholas Sanduleak of Case Western Reserve University, was at the right spot. Moreover, the star, designated Sk69.202, was a hot supergiant some 20 times the mass of the sun—just what the theorists needed to produce a supernova with the characteristics of SN1987A.

Observations made by the *International Ultraviolet Explorer* satellite suggested initially that Sk69.202 had survived the explosion and could therefore not have been the supernova's progenitor, but reanalysis of the data showed that the progenitor star had definitely disappeared.

In a general way, theorists believe they know how the expanding supernova explosion is structured. An innermost layer of gases rushing away from the neutron star consisted initially of nickel, and is now a mixture of nickel, cobalt and iron. Outside this layer are concentric layers consisting predominantly of silicon, oxygen, neon, carbon, helium and hydrogen. The outermost atoms of the explosion are moving at speeds greater than 25,000 miles per second.

At this point, Dr. Arnett believes, nearly all the light reaching us from the supernova is coming from the outermost layers, where atoms have been excited to produce light by a bombardment of gamma rays coming from within. The gamma rays, produced initially from the radioactive decay of nickel-56 formed in the explosion, are now coming from the decay of cobalt-56, the metal created by the decay of the nickel, he said.

The trouble, astronomers say, is that the gamma rays themselves are not escaping from the opaque shells of gas and therefore cannot be studied directly. The signatures of elements disclosed by gamma-ray spectrums would be invaluable in analyzing the explosion. But the gases are not expected to thin out enough for gamma rays to escape for another year or so, and meanwhile, the production of gamma rays deep inside the explosion may die out in the next four months.

The interval of time in which gamma rays from the supernova reach Earth might be very short, and unless instruments are in place, the opportunity could be missed. Complicating matters, gamma rays do not penetrate Earth's atmosphere and must be observed from space. Since the space shuttle is no longer operating, the United States has lost the chance to orbit its Gamma Ray Observatory, which could have been in place to study the supernova.

Efforts to Detect Gamma Rays

Meanwhile, several scientific groups working in collaboration with the National Aeronautics and Space Administration are launching gamma-ray instruments aboard high-altitude balloons that reach altitudes of 130,000 feet, above most of Earth's atmosphere.

Dr. Thomas A. Prince, an astrophysicist at the California Institute of Technology, recently conducted one such flight from Australia, using his 4,300-pound Gamma Ray Imaging Payload.

"The instrument worked perfectly for several known gamma-ray sources," he said, "notably the Crab Nebula," a luminous cloud of expanding gas left over from a supernova recorded by Chinese astronomers in 1054. "But it was still too early for us to see any gamma radiation from SN1987A."

The supernova has confronted scientists with one major enigma: the appearance of a dim twin of the supernova. Using a device that cancels the twinkling distortion of the supernova image caused by light passing through Earth's turbulent atmosphere, scientists from the Harvard-Smithsonian Center for Astrophysics spotted the peculiar companion in May 1987.

Dr. Arnett says he suspects the seeming companion explosion may be merely a burst of light streaming from a break in the shock wave hurtling out from the supernova. Others have offered different explanations.

"It's an enigma that's going to take serious study," Dr. Arnett said.

He noted that the supernova's behavior has puzzled many observational astronomers, but said, "What has puzzled observers, in many cases, has been perfectly logical for theorists."

"It's rare in astronomy that theory gets as far ahead of observation as we were in this case," Dr. Arnett said, "and it's a very satisfying feeling when observations fall as neatly into place as they have during the past few months. It makes us feel we're doing our job."

—June 16, 1987

Massive Clusters of Galaxies Defy Concepts of the Universe

By JOHN NOBLE WILFORD

As Hamlet would say to an astrophysicist these days, without drawing an argument, there are more things in the heavens "than are dreamt of in your philosophy."

For astrophysicists, who look beyond individual galaxies of stars to clusters of galaxies and, at an even larger scale, to aggregates of clusters called superclusters, are increasingly mystified by what they see. To their bewilderment the universe seems to be organized in even more vast chains of galaxies, which are structures so large and complex that they defy understanding in terms of the current theory of cosmology.

The picture of the universe's structure has become more confused with the discovery, reported recently, of what appears to be an immense conglomeration of galaxies, a feature that is being called a supercluster complex. The complex, which includes our own Milky Way galaxy, is 100 times more massive than any previously known structure and encompasses millions of galaxies, stretching across 10 percent of the observable universe.

The discoverer, R. Brent Tully of the University of Hawaii's Institute of Astronomy, said the observation of such an enormous structure posed "major challenges to conventional theories of galaxy formation."

Dr. Tully said in a telephone interview that other observations suggested the existence of at least four other supercluster complexes "of comparable magnitude," indicating that "this must be a general property of the universe."

Astrophysicists are cautious in their initial reactions, neither rejecting nor accepting Dr. Tully's interpretation of his discovery.

Simon White, a theoretical astrophysicist at the University of Arizona, said, "I think he's got enough here to force people to really try and confirm or deny it with new observations and analysis."

Dr. Tully said he felt "quite comfortable" with his interpretations, calling them "not certain, but highly probable." He based his conclusions on a new supercomputer analysis of the positions of galaxy clusters.

Jeremiah P. Ostriker, a Princeton University astrophysicist, said: "If this is right, it's extremely important because it's hard to produce this kind of structure in current theory. It could mean there's some vital missing ingredient in standard theories of the development of the universe."

Searching for the missing ingredient, should that be necessary, would add to the ferment in cosmology today. Efforts to conceptualize the history and structure of the universe were already running into trouble because of the growing realization that the universe was not as uniform as had been assumed.

Astronomers first glimpsed form and movement in the universe in the 1920s. The familiar constellations are groupings of stars only in the eye of earthly beholders. In reality, Edwin P. Hubble found, billions of stars tend to occur in "island universes," and our Milky Way is only one of a multitude of these galaxies. Analysis of changes in light from the distant stars and galaxies—the displacements toward the red end of the spectrum or red shift—showed that the universe was expanding in all directions.

Is There a Pattern?

Everywhere astronomers looked, the galaxies appeared to be distributed uniformly and without any discernible pattern of larger structure. Fritz Zwicky of the California Institute of Technology saw in the 1930s some evidence for galactic clustering. But even the evidence that solidified support for the big bang, the prevailing theory that the universe exploded into being 10 billion to 20 billion years ago as a superdense fireball, did nothing to change scientists' image of a smooth, uniform universe.

In 1964, radio astronomers at Bell Laboratories detected microwave background radiation throughout the universe, the leftover radio noise of the explosive moment of creation. The smoothness of the background radiation signals indicated that the universe began with a uniform distribution of matter.

Astrophysicists say the new discoveries of a large-scale structure raise questions about developments after the big bang—what events and forces gave such a rich texture to the universe? But they say the findings do not diminish their faith in this concept of how it all began.

According to the big bang theory, the initial expansion was rapid, which would account for the universe's general smoothness.

Wrinkles and Chains

But there must have been some wrinkles in the early distribution of matter. Either some clumping of matter set in motion gravitational forces to form the galaxies and later the clusters of galaxies, or perhaps some large-scale variations in the density of matter survived after the early, rapid expansion of the universe slowed down, and galaxies could have formed around the edges of these vast areas and be strung out in chains.

While cosmologists were still debating these alternative theories, in recent years astronomers with improved telescopes and better computers for building and testing theoretical models began seeing that clusters of galaxies clumped into superclusters. They also observed groups of galaxies occurring in flat sheets and in chains or filaments stretching across millions and millions of light-years. They found some galaxies that appeared to surround a dark void.

Such a variety of observed shapes may be a problem of perspective. Like the blind man and the elephant, astronomers using different observational techniques, X-ray or radio or optical telescopes, may be describing different aspects of the same structures because of their particular points of view. But there was no doubt that they were seeing surprisingly large structures.

By the 1980s, more than ever, cosmologists strongly suspected that there was much more to the universe than met the eye, and this has sorely tested their genius for conceiving cosmic theories. One of their most popular new concepts is that of the missing mass.

Various calculations, based on an assumption that the universe is closed in a balance between energy and matter, suggests that as much as 90 percent of the universe's mass has gone undetected. This dark matter, composed of as-yet-unknown elementary particles, is assumed to exert the dominant gravity in the universe, causing the clumping of galaxies.

Superclusters Cause Problems

"The hypothesis works very well to explain the formation of individual galaxies and groups of galaxies," said Dr. White of the University of Arizona. "But superclusters cause us problems." Dr. Ostriker, who has worked on these problems for years, conceded, "As observers find structure larger and larger, it gets harder and harder to explain things theoretically."

Telescopes with new electronic cameras enable astronomers to examine many galaxies in a single field of view, whereas it used to take them a lifetime to study a few hundred. Large-dish radio antennas, as well as optical telescopes, give them glimpses of the galactic clusters shaped like bubbles. X-ray and optical telescopes provide the raw material for plotting clusters and superclusters on computer-generated maps of the sky.

Dr. Tully was searching for the edge of the Local Supercluster, the region including the Milky Way, when he became aware that the structure was much larger than previously thought. He then looked at the "rich" clusters farther out in space, calculating their motion and plotting their distances from each other.

Using a supercomputer, Dr. Tully constructed maps of the distribution of the clusters of galaxies as they would appear to an observer at various points in outer space. He compared their distance from each other with random spacings between clusters. He concluded that the clusters were related to each other because the probability of their occurring by chance was statistically slight.

As a result, Dr. Tully decided that about 60 of these clusters were concentrated in a single supercluster complex. He calls it the Pisces-Cetus Supercluster Complex, after the constellations in which it is found.

"It's Lumpy"

"It was supposed that on a large scale, things in the universe are smoother and homogeneous," Dr. Tully said. "My findings show that is not the case. It's lumpy. "

Dr. Tully suggested that the theory of cosmic strings "might provide an explanation" for what he is seeing. According to this theory, anomalies in space, called topological defects, were created in the first fraction of a second after the big bang. These defects could provide the focuses for the accumulation of matter. But he acknowledged that if such significant concentrations of matter existed at the moment after the big bang, "they should have given rise to irregularities in the relic background radiation."

The cluster Dr. Tully has identified is about one billion light-years long and 150 million light-years across. Since the universe is thought to be at least 10 billion years old, a distance of one billion light-years represents as much as 10 percent of the expanding universe's reach.

Dr. Tully's research was supported by the National Science Foundation.

—November 10, 1987

Shuttle Soars 381 Miles High, with Telescope and a Dream

By John Noble Wilford

In a thunderous overture to a promised new era in astronomy, the space shuttle *Discovery* rocketed into orbit today with the $1.5 billion Hubble Space Telescope, which scientists believe will give them a commanding view of the universe as it was, is and will be.

Reaching an altitude of 381 miles, the highest ever attained by a shuttle, the five astronauts were in position to deploy the telescope on April 25, 1990. Flight control engineers used the word *perfect* repeatedly in describing the telescope's condition and orbit.

A televised inspection and analysis of engineering data showed that the telescope had survived the launching undamaged and seemed ready to begin a mission of cosmic exploration that could last more than 15 years.

Far Clearer View of Stars

"It's in its element now," said Richard H. Truly, head of the space agency, just after the spaceship settled into orbit with the 43-foot-long telescope tucked in the cargo bay.

High above Earth's murky atmosphere, the 94.5-inch mirror of the Hubble telescope should collect visible and ultraviolet light coming from near the edge of the universe and thus toward the very beginning of time. The largest and most complex scientific instrument ever put in space, the Hubble is expected to observe distant stars and galaxies with a clarity 10 times that ever before achieved.

In their more optimistic moments, astronomers boldly predict that Hubble observations should enable them to learn the age and size of the universe and possibly its fate: whether it will keep expanding forever or eventually collapse on itself.

When liftoff occurred, at 8:34 a.m., astronomers were ecstatic. With the explosive thunderclaps of rocket exhaust beating at their chests, they watched the *Discovery* climb steeply through a lone white cloud and reemerge in the higher blue, a speck of fire trailed by billowing smoke and vapor.

"The few seconds of thrill made the 12 years of effort well worth it," said Dr.

Edward J. Weiler, the chief Hubble scientist for the National Aeronautics and Space Administration.

After 45 years as a dream and 12 years of planning, building and waiting, the often-delayed Hubble telescope was on its way. It had failed to meet its original launching date in 1983 because of serious technical and management troubles. It had been all set to go in 1986, but the *Challenger* disaster that January put the mission on hold again. And on April 10, 1990, the telescope came within four minutes of lifting off. A flawed power unit on the shuttle halted the countdown and had to be replaced, causing a two-week delay.

The launching on April 24 was in doubt for three suspenseful minutes, when countdown clocks stopped at 31 seconds before the planned liftoff. Computers detected that a fuel valve that was supposed to close had remained open.

Reacting swiftly, launching engineers sent a manual command, bypassing the usual computer program. The countdown resumed. Liftoff was only three minutes late, to the relief of anxious officials and scientists.

"The first step is always the hardest," said Mr. Truly, NASA's administrator. "And we're beyond that now."

Plan for Deployment

Indeed, the launching was the first step not only for the long-awaited Hubble mission, but also for a vigorous program of space astronomy planned for this decade.

Scientists tempered their excitement with the recognition that the Hubble telescope was still in the shuttle cargo bay. Release of the telescope into its own orbit is scheduled to occur at 1:57 p.m. April 25.

"It's not the launching but the deployment that will have me biting my nails," remarked Dr. Eric Chaisson, an astrophysicist at the Space Telescope Science Institute in Baltimore.

Michael M. Harrington, director of the procedures to get the telescope ready for full scientific operation, said the spacecraft had responded to all commands, indicating that all its systems were operating satisfactorily.

"We were all elated when we got our first communication from the telescope," Mr. Harrington said at a news briefing this afternoon at the Goddard Space Flight Center in Greenbelt, Md., where operations of the telescope are being directed.

Mechanical Arm Is Tested

Also, Dr. Steven A. Hawley, an astronomer in the *Discovery* crew, unstowed the shuttle's 50-foot-long mechanical arm and tested its performance. The arm was ready.

Early April 25, while Col. Loren J. Shriver of the Air Force and Col. Charles F. Bolden Jr. of the Marine Corps steer the *Discovery*, Dr. Hawley will operate the mechanical arm from controls at the rear of the cabin. Looking out the rear window and on television, he is to grapple the spacecraft housing the telescope. It is roughly the size of a railroad tank car and would weigh 12 tons on Earth.

After the telescope is uncradled, Dr. Hawley is to use the robotic arm to lift it high above the cargo bay. Its door, which opens on the big mirror, must be carefully pointed away from the sun.

At this time, commands will be sent to extend two antennas on the telescope and also extend two arrays of solar cells. With the telescope's batteries limited to about six or seven hours, the astronauts have little time to spare in getting the solar panels out and able to supply electrical energy.

The two other astronauts, Capt. Bruce McCandless 2d of the Navy and Dr. Kathryn D. Sullivan, a scientist, will be standing by in full spacesuits, ready to go outside for a space walk to help release the antennas or solar panels if they should be stuck. The shuttle cabin has already been partly depressurized to shorten the time necessary to get ready for a space walk.

Dr. Charles R. O'Dell, an astronomer at Rice University, who was the chief scientist in the project's formative years, said, "Once we get power out of the solar arrays, then we're basically home-free."

Shortly afterward, if there are no problems, Dr. Hawley is supposed to maneuver the telescope out to the side and release the robotic arm's grip. Deployment should occur on *Discovery*'s 19th orbit of Earth.

Return on Sunday

For the next two days, the *Discovery* is to fly in formation with the telescope, about 40 miles apart, while engineers at the Goddard center continue to check all spacecraft systems. If major trouble arises, the shuttle can bring the Hubble craft back to Earth.

The plan is for the *Discovery* to remain in orbit five days and return Sunday morning, April 29, with a landing at Edwards Air Force Base in California.

The Hubble telescope, named for Edwin P. Hubble, an American astronomer who in the early 20th century discovered that the universe is expanding, will be only beginning its mission. Scientists said full use of the telescope might not begin for six or seven months because extensive testing and focusing were planned.

"It's a complex beast, and takes a lot of time to tune up," Dr. Weiler said.

But the telescope is expected to see its "first light" in a week. The space agency said it would release the Hubble's first test photograph sometime next week.

—April 25, 1990

5,000-Mile Radio Telescope
Set to Probe Depths of Time and Space

By MALCOLM W. BROWNE

After seven years of construction, a radio telescope spread across 5,000 miles—the largest astronomical instrument ever built—is finished.

All 10 of the new telescope's gleaming white antennas, which are scattered across United States territory from the central Pacific Ocean to the Caribbean, are now aimed at a new frontier in deep-space astronomy. Using the Very Long Baseline Array (VLBA), as the new compound telescope is called, scientists have begun to explore a mysterious region of space-time far away and long ago, where speeding galaxies appear to stand still, huge gas clouds create an unsettling illusion that they are moving at 10 times the speed of light, and galactic jets twist and squirm in the grip of powerful but puzzling forces.

The VLBA consists of 10 dish-shaped antennas, each 82 feet in diameter, spread across North America from Mauna Kea in Hawaii to St. Croix in the Virgin Islands. All 10 antennas are exquisitely synchronized by the federally financed National Radio Astronomy Observatory control center in Socorro, New Mexico, to make them function as a single telescope of gigantic size, an arrangement that gives the VLBA far sharper vision than that of any other telescope.

The new instrument's resolving power—the smallest angular distance between two objects that can still be seen as separated from each other—is a fraction of a milliarcsecond: more than 1,000 times better than the best optical telescopes. With that kind of resolving power, an observer in New York City would be able to read a newspaper in San Francisco.

The monster telescope array is as sensitive as it is sharp-eyed. It is so sensitive that any of its antennas could easily detect even the faint radio waves emitted by the warm tissues of a human being. The VLBA cannot see visible light or the X-rays and gamma rays on one side of the electromagnetic spectrum, but the microwave radio waves it does see at the other end of the spectrum are opening a new world to astronomers.

The VLBA, which cost $85 million, will be formally dedicated here on Friday, but even during the past year, before all 10 antennas were on line, the instrument has produced some spectacular images. Some of them reveal details never before seen in very distant objects, even by the most powerful optical telescopes.

For example, astronomers using the unfinished VLBA have discovered a very peculiar jet of radio-emitting particles streaming away from the core of a galaxy named Markarian 501, some 300 million light-years from the Earth. Unlike the straight jets ejected at nearly the speed of light from other active galaxies, this jet is twisted at an angle of nearly 90 degrees, and scientists have not yet found a good explanation. They hope that an understanding of the twisted jet, which has perhaps been warped by some unusual magnetic field, will emerge from further VLBA observations.

Besides revealing details of galaxies as they existed billions of years ago, when the universe was young, the VLBA will serve as a powerful tool for measuring geological changes, and perhaps even predicting earthquakes here on Earth.

Geodesy measurements of distant quasars made by pairs of VLBA radio antennas, for example, will be used to calculate the exact distances between the antennas themselves. These distances of hundreds of miles can be measured to within an accuracy of less than a half-inch of error. Using pairs of astronomical antennas, scientists have already determined that the movement of tectonic plates underlying the Earth's continents is carrying North America away from Europe at a rate of eight-tenths of an inch a year, about the rate at which fingernails grow. The Hawaiian islands, astronomers have determined, are moving westward at a brisk four inches a year.

The United States Geological Survey, the Naval Observatory and the National Aeronautics and Space Administration will be among the agencies using the data collected by the VLBA, and seismologists hope that improved measurement of the movement of tectonic plates in the vicinity of fault lines, rifts and other geological features may eventually lead to better prediction of earthquakes. Geologists and seismologists in Japan, China and California are all in touch with VLBA astronomers and are awaiting more data.

"As with any new and powerful instrument," said Dr. W. Miller Goss, director of the VLBA, "this one will certainly find uses we could not have foreseen. We're sifting through hundreds of research projects proposed by astronomers who hope to use the array, but we'll be able to allot observing time to only about half of the applicants. Some very exciting years lie ahead for this instrument and its users."

The science of radio astronomy began in 1933 with a discovery by Karl Jansky, a scientist working for Bell Telephone Laboratories in New Jersey (renamed AT&T Bell Laboratories), that radio signals were reaching the Earth from outer space. While looking for the source of the static noise that degraded telephone conversations, he accidentally discovered an invisible object called Sagittarius A, something that lies deep within the core of our Milky Way galaxy and emits intense radio waves.

Radio astronomy progressed slowly until the end of World War II, when radar reached maturity. Radar and radio astronomy both exploit the part of the electromagnetic spectrum called microwaves, and their technologies are similar. Like visible light, radar waves can be reflected from circular mirrors shaped as paraboloids, which focus the incoming radiation on small secondary reflectors. A secondary reflector bounces the focused radiation back through a hole in the main mirror, where the radio signal is converted by a receiver into an electrical signal that can be amplified and analyzed.

The Technology from Microwaves to Images of Space

In a modern radio telescope, the focused microwave beam arriving from some remote galaxy or supernova remnant is channeled by a metal horn into the radio receiver, where it impinges on an electronic chip chilled by liquid helium to a temperature only a few degrees above absolute zero, which is some 430 degrees below zero on the Fahrenheit scale. At this temperature, the rapidly moving molecules in the radio receiver chip are slowed almost to a standstill. By eliminating molecular motion, the radio static the chip would otherwise emit is virtually eliminated. The chilled receiver can then listen without interference for the faint radio signals coming from space.

But the real magic in the VLBA and the several smaller arrays that preceded it is a technique called interferometry; in the VLBA, that technique is supported by a computer as powerful as any in the world. Even the mighty Cray computers, used by nuclear weapons laboratories, scarcely rival the Correlator computer that has been designed and built especially for the new radio telescope.

Data from each of the 10 antennas is separately recorded on tape and, at the end of each observing session, the 10 tapes are shipped by the far-flung antenna crews to the Array Operations Center here, where they are processed by the Correlator. The computer must match the tapes' timing and speeds to within a billionth of a second, and it must allow for the differing positions and viewing angles of the antennas, their motions as the Earth rotates and other factors. Finally, the Correlator integrates all this data and, using a mathematical analysis called the Fourier transform, extracts the faint celestial signals buried under the mountains of noise.

The Correlator can perform 750 billion "flops," or simple calculations, per second. At that rate, if the computer were ordered to tally up the checkbooks of 250 million people, each of whom had written 3,000 checks, the calculation would take only one second.

The radio waves coming from a celestial object through a single antenna provide the computer with enough information to construct a rough image of the object, with the help of a phenomenon called interference. Most of the target's radio "light" is focused on a central spot, while some of the radio energy is scattered in concentric bands around the central spot.

These bands of energy are called interference fringes. They are somewhat like the ripples in a pond that spread from the impact of a rock thrown into the water, and, like water waves, they can interfere with each other. If the crests of two fringes happen to coincide, they will combine to produce a crest higher than either one; but if a wave crest happens to coincide with a wave trough, the two will tend to cancel each other out.

A radio image of a celestial object can be greatly improved by combining the signals and fringe patterns produced by two separated antennas; if signals from many widely separated antennas are merged, their combined signals become much stronger and sharper than any produced by a single antenna. The VLBA performs as well in most respects as would a single antenna that extended over 5,000 miles. Converted into computer pixels, the signals from the 10 VLBA antennas can be used to build up detailed images of the objects at which the antennas are aimed.

The VLBA not only reconstructs the contours of very distant objects, but it can also map the speed of objects relative to the Earth, the speeds and directions of their rotations, the rates at which they change shapes, the dynamics of their turbulent gases, and even the distributions of the chemical molecules inside them.

Very Large Array Older Instrument Aids New Effort

The new telescope vastly extends the reach and capabilities of an earlier radio telescope, the Very Large Array (VLA), which was completed in 1980 some 50 miles west of here on the 7,000-foot-high San Agustin Plain. The 27 antennas (plus one spare) that make up the VLA, each one 82 feet in diameter, are mounted on rattlesnake-infested railroad tracks in the shape of a huge Y in which each arm is some 12 miles long. The antennas are periodically moved farther or closer to the center of the Y; when spread out to their maximum extent, the antennas provide the best possible resolution, and when pulled in close to the center the sensitivity of the instrument increases, at the expense of resolution. The VLA antenna system manipulates radio waves the way a camera's zoom lens controls visible light.

The VLA has achieved a series of scientific triumphs which have involved cooperation with other countries, including European and South American nations, Japan and the countries of the former Soviet Union. Far from obsolete, the VLA antennas will often be used in conjunction with those of the VLBA to achieve even further improvements in the quality of their combined images.

Data from radio telescope antennas in Europe, Asia and Russia can feed into this system as well, and astronomers plan to improve the global radio-telescope network even further by extending it into space. Japan hopes to launch an orbiting radio telescope in 1996, and the observations of that instrument will be combined with those of the ground-based network. Russia is working on a similar project.

Allure of Radio Waves

Astronomers who once confined themselves to optical telescopes have become fascinated with radio-emitting objects, partly because radio telescopes can often reach far beyond optical telescopes into space and time. Because the universe is expanding, the more distant an object is, the faster it is receding from Earth, thereby Doppler-shifting the light headed our way toward the red end of the spectrum. Beyond a certain recessional velocity, the light from the object is shifted out of the visible spectrum altogether, becoming "visible" only to instruments sensitive to the infrared or radio regions of the spectrum, which are invisible to the naked eye.

Thus, many of the most interesting galaxies, pulsars, jets and other objects lying at immense distances in space and time from the Earth can be studied only by radio telescopes.

Even as the VLBA observing program begins, its older cousin, the VLA, is embarking on new projects that are also expected to reap impressive scientific harvests. The VLA will soon begin a radio survey of the entire sky, a tally expected to include some two million radio-emitting objects in a new catalogue that will be comparable to the one compiled for visible wavelengths by the Palomar Observatory.

The VLA is also about to turn its 27 antenna eyes on an object much closer to home than galaxies, pulsars and supernova remnants: Titan, a large moon of the planet Saturn. As the VLA antennas focus on Titan, NASA's Goldstone Solar System Radar, a dish antenna 82 feet in diameter at Goldstone, California, will send its 500-kilowatt microwave beam out to Titan. The VLA will listen for the radar echo from Titan, and astronomers expect to use it in measuring the movement of one or more of Titan's geological features, allowing them to calculate its exact period of rotation.

Relativity Playing Tricks with Perception

As radio telescopes observe objects more than about seven billion light-years away—that is, objects as they looked when the universe was about half its current age—relativity begins to play tricks on perception. Normally, any object receding from one's eye appears to grow smaller as it moves away. But beyond about seven billion light-years, objects are receding from Earth at speeds so close to that of light that their images are distorted. Although these images grow fainter and become shifted farther toward the red end of the spectrum, they do not seem, from the perspective of the Earth, to be shrinking.

Radio astronomers have detected another kind of illusion stemming from relativity effects. Blobs of matter ejected by some distant galaxies appear to be moving away from the objects that ejected them at up to 10 times the speed of light, even though the special theory of relativity prohibits anything from exceeding the speed of light. The apparent speed is a "projection effect" that arises when an object is moving almost directly toward or away from the Earth at a speed near that of light, theorists say.

Dr. Kenneth I. Kellerman, one of the astronomers at Socorro, says that a major application of the new telescope will be to test the accuracy of various predictions arising from relativity theory, predictions that could not be verified with any earlier telescopes.

"With the great precision the VLBA gives us," he said, "details of the structure of the nearby universe will be filled in rapidly, and this will lead to improved estimates of the Hubble constant, the rate at which the expansion of the universe increases over distance. The VLBA will also help us determine the rate at which the universal expansion may be slowing down, which is a measure of the amount of matter in the universe.

"In other words, we think the VLBA will shed a lot of light on some of the most important cosmological questions," he said.

Even new solar systems and planets that might one day evolve life may be discovered by the new telescope. Dr. Leonid I. Matveenko of the Institute for Space Research in Moscow, one of the Russian radio astronomers working at Socorro, says that he and a British colleague, Dr. Philip J. Diamond, have already found radio evidence of a protoplanetary system forming around a star in the Orion nebula.

"Mankind is going to discover some wonderful things with these great new radio telescopes," he said.

—August 17, 1993

Big Bang's Defenders Weigh "Fudge Factor," a Blunder of Einstein's, as Fix for New Crisis

By JOHN NOBLE WILFORD

Even though the universe has infinite ways of humbling them, cosmologists are nothing if not resilient and endlessly creative. They have to be, given their daunting task of standing on the shore of a small world and looking beyond the harbor lights of nearby stars to the arching waves of clustered galaxies, seeking on far horizons glimpses of the entire cosmic history, from beginning to probable end.

For more than three decades, their most satisfying reconstructions of that history have rested squarely on the "big bang" theory. In the beginning, according to this model, mass was compressed into a state of infinite density, an initial singularity. Then there was a kind of explosion. Everything—space itself—expanded, thinned out and cooled. At first, all was smooth and virtually uniform. But around some faint wrinkles, called density fluctuations, matter began clumping into stars, galaxies of stars and clusters of gravitationally bound galaxies stretching across the sky.

But the universe seems to keep throwing the cosmologists nasty curves, exposing the woeful limitations of their knowledge about how in the apparently allotted time the cosmos evolved from these beginnings to a present-day structure of such manifest inhomogeneity. There does not seem to have been enough time. And where and what is all the invisible mass, the so-called dark matter, to account for the gravity needed to pull together such vast galactic agglomerations?

Such unsettling questions have left cosmologists shaking their heads and entertaining all manner of modifications in the details of their theories. Perhaps the universe underwent a brief period of accelerated expansion in its very early stages, the widely favored inflationary addendum to the big bang theory. Perhaps most of the universe is composed of invisible exotic particles— WIMPs, for weakly interacting massive particles—that supply much of the gravity-shaping galactic structures. But important aspects of this hypothesis have been attacked recently, and these particles have yet to be discovered. The tinkering continues.

It may be a measure of the current turmoil that some cosmologists are driven to reconsider, tentatively but with increasing interest, an idea that Einstein once proposed but later rejected, saying it was "the greatest blunder of my life." His cosmological constant, as he called it, was a kind of antigravity force, a monumental fudge factor to force the universe he envisaged to conform with some discomfiting implications of his own general theory of relativity.

Recently, theorists got more perplexing news and were talking more openly about the theoretical beauties of the cosmological constant. Einstein, they speculated, might have been right after all, though for the wrong reasons.

Cosmologists were responding to a report of the most accurate measurement yet of the distance to a remote galaxy, made by astronomers using the Hubble Space Telescope. Calculations of the universe's expansion rate from this and other recent observations provide strong evidence that the universe may be much younger than scientists previously estimated. It may be no more than eight billion years old, as compared with other calculations ranging up to 20 billion years.

Since some stars are reliably estimated to be 16 billion years old, the new findings mean that the universe appears to be younger than some of its components, a most disturbing paradox for cosmologists. Calling this "an apparent impossibility," Dr. George H. Jacoby, an astrophysicist at the National Optical Astronomy Observatories in Tucson, Arizonia, said it "will force a reexamination of our universe model and how stellar ages are measured."

Of course, no one is accepting the low age estimate as the last word, not until further measurements are made using other techniques and are extended out to greater distances, beyond disturbing influences of local gravity. But no one is dismissing it, either. This latest age crisis has scientists conceding that the expansion of the universe may not be so simple and pondering what basic physical property they may have overlooked.

"These are exciting times, when push is coming to shove," said Dr. Michael Turner, an astrophysicist at the University of Chicago and the Fermi National Laboratory.

So far, no prominent cosmologists are doubting the essentials of the big bang theory itself. The theory has endured, astrophysicists say, because it logically follows from Einstein's general theory of relativity and the observed outward motions of galaxies in expanding space. It has also passed critical tests.

In 1964, radio telescopes picked up the microwave echoes of the explosion. Other research revealed that the overall chemical makeup of the universe agreed with the theory's predictions. Two years ago, a spacecraft detected the early density fluctuations in the cosmic background radiation out of which presumably emerged the great galactic structures of the present universe.

If any cosmic age estimate based on present expansion rates should be confirmed as being lower than stellar ages, Dr. Alexei V. Filippenko, an astronomer at the University of California at Berkeley, said, "It might lead to a revolution in cosmology," adding: "We may have to consider some new, perhaps wild ideas. Maybe there is another force pushing the universe out or something like that."

Dr. Craig J. Hogan, an astrophysicist at the University of Washington in Seattle, was more explicit. If the age estimates cannot be revised, he said, scientists might be forced to think of introducing "a cosmological constant into the equations of motion of general relativity—the equivalent of self-gravitation of empty space."

Einstein was the first to resort to such a concept, soon after he published his general theory of relativity in 1915. When he applied the equations of his theory to model the universe, he was chagrined to discover that he had constructed a dynamic universe, one that was either expanding or contracting but could not be static. Since there was no evidence at the time that the universe was expanding, Einstein looked for a way to reconcile his theory with a static universe.

Einstein introduced an extra term in his equations, the cosmological constant, which represented a repulsive force of an undetected form of energy that was equal to but opposite that of gravity. There was some mathematical justification for the concept, and it certainly tidied up things, but soon other scientists exposed weaknesses in the equations. In 1929, Edwin P. Hubble, the American astronomer, finally established that the universe was not static, but expanding, a discovery that set the stage for the big bang theory.

Even the great Einstein could make mistakes, and this one cost him the opportunity to make the sensational prediction that the universe is expanding.

"It's not such a crazy idea," Dr. Turner said of the cosmological constant. "There's no known principle that forbids it. If you put it in the equations, it leads to a very good fit to the universe."

For several years, in fact, astrophysicists have been reexamining the concept as a possible solution to their most intractable problems. If the value of the constant is reasonably low, they pointed out, it could solve the problem of the missing mass. It would make up the difference between the observed density of matter in the universe and the critical density that theorists think is the most likely consequence of the big bang and the best fit for subsequent cosmic behavior.

Since the observed matter is 20 percent of the favored critical density, Dr. Turner said, the energy assumed by the constant would need to provide the other 80 percent. This should result in an equilibrium between the forces of gravity and of expansion, meaning the universe would not eventually collapse or keep expanding into a cold, vanishingly thin eternity.

The force represented by the constant would also imply a lower density of matter in the universe, making it easier for various reasons to explain how matter has congregated into exceedingly large structures, like the observed "great wall" of galactic clusters that stretches over more than half a billion light-years. This is something the exotic dark-matter hypotheses have so far failed to do.

And the constant could solve the age problem. Its energy could change the rate at which the universe expands. The early expansion could have been much slower than has been thought, giving the universe more time to develop its large structures, but under the influence of this energy the expansion could have speeded up. The gravity of matter would tend to decelerate expansion, but by acting against that gravity the vacuum energy of the constant might accelerate the expansion by now. Thus, the new low-age estimates may be misleading.

"People would be partying in the streets if the cosmological constant were sensible," Dr. Turner said.

As far as scientists can determine, some laws of physics permit the possibility of such an energy force. Einstein himself established that matter and energy are equivalent, and so energy generates gravity just as matter does. But certain laws of particle physics suggest that the value of the cosmological constant would be too great, which would make the universe fly apart.

"It's a very radical thing, but one that is reasonable," said Dr. Joseph Silk, a cosmologist at the University of California at Berkeley who is the author of *A Short History of the Universe*, to be published by W. H. Freeman.

Cosmologists are extremely reluctant, however, to put too much faith in the constant because of its embarrassing failure the two times before that it was applied. There was Einstein's experience, and also an attempt in the 1940s to use the constant to get out of another age dilemma. At the time, it seemed that the universe, dated at two billion years, was younger than the solar system, dated at four billion to five billion years. It turned out the scientists had wildly overestimated the cosmic expansion rate.

As they learn more, scientists may again find a solution without resorting to the constant. There are reasons to suspect that the new age estimates are much too young. Dr. Silk speculated that the estimates might simply reflect conditions in nearby space, where galaxies are underrepresented and, as a result, there was a more rapid expansion rate.

Dr. Allan R. Sandage of the Carnegie Observatories in Pasadena, California, who staunchly maintains that the universe is closer to 20 billion years old, has promised to report new measurements early next year that would move the estimates toward such an old age. Dr. Silk and Dr. Turner have collaborated on a study, to be reported in January 1995, that shows how a 20-billion-year age would be most compatible with all that is now understood about the universe.

A few scientists suggest that the theoretical models on which stellar ages are estimated may not be complete and may be based on inaccurate assumptions. But most specialists in stellar evolution doubt that the ages of the oldest stars could be reduced to as low as the current low estimates for the cosmic age.

Dr. Pierre R. Demarque, a Yale University astrophysicist who specializes in stellar evolution, said: "We're constantly under pressure from the cosmologists, but we really can't change our numbers that much. They are based on really well-known physical principles."

If nothing else, the crisis in cosmology demonstrates that there is much more to understanding cosmic history than is immediately revealed in the big bang theory, however well established it may have become through rigorous testing.

In *The Origin of the Universe*, published in 1994 by Basic Books, Dr. John D. Barrow, an astronomer at the University of Sussex in England, wrote: "One must understand that the term 'big-bang model' has come to mean a picture of an expanding universe in which the past was hotter and denser than the present. The job of cosmologists is to pin down the expansion history of the universe—to determine how the galaxies formed; why they cluster as they do; why the expansion proceeds at the rate that it does—and to explain the shape of the universe and the balance of matter and radiation existing within it."

A tall order, indeed, but those who take on the universe as a field of study must accept the likelihood that for all their brilliance and ingenuity, many of the ultimate answers will remain beyond their grasp. But they keep trying in the belief that the universe is comprehensible.

Dr. Silk, in his new book, conceded that the creation story told through the big bang theory might in a thousand years be regarded as a 20th-century myth, like the creation stories of antiquity. But he hastened to add, "I am an optimist who finds our current paradigm so compelling that I can only imagine it will eventually be subsumed into a greater theory, without losing its essential features."

—November 1, 1994

Age of Universe Is Now Settled, Astronomer Says

By MALCOLM W. BROWNE

Astronomers working with new measurements by the Hubble Space Telescope believe that they have finally settled a long-standing controversy over the age of the universe. Dr. Allan Sandage of the Carnegie Observatories in Pasadena, California, and his colleagues calculate that the universe is at least 15 billion years old, and they say that recent challenges to the standard theory of how the universe began and evolved stand refuted.

Other astronomers do not agree that the debate is over, but they acknowledge that Dr. Sandage's results, to be reported in a forthcoming issue of *The Astrophysical Journal,* represent a major development for cosmological theory. The new report, the seventh in a series by the same team, was published by Dr. Sandage and colleagues from the University of Basel, Switzerland, and the European Space Agency, who analyzed data from the space telescope and various ground-based observatories.

"We believe that this marks the end of the 'Hubble wars,'" Dr. Sandage said in an interview.

The "Hubble wars" is a debate among astronomers over the value of the Hubble constant, which was named for Edwin P. Hubble, who discovered in 1923 that the wisps of light now known to be galaxies lie far outside Earth's own galaxy, the Milky Way. The Hubble constant is a measure of the rate of expansion of the universe, but astronomers using different ways of estimating this rate have calculated wildly differing values over the years.

One recent estimate suggested that the universe was actually younger than certain globular clusters of stars. If that were true, a large part of the widely accepted big bang theory of how the universe began would have to be reassessed.

Dr. Sandage's investigation has concluded that the seemingly ancient globular star clusters are actually no older than 13 billion years, while their parent universe is at least 15 billion years old—a comfortable margin, consistent with the big bang theory.

But the debate has not ended.

A critic who has frequently challenged Dr. Sandage's estimates of the Hubble constant over the years, Dr. Wendy L. Freedman, who is also a Carnegie

Observatories astrophysicist, remains somewhat skeptical of Dr. Sandage's latest estimate.

"These are excellent observations that represent important progress," she said in an interview, "but they are not final. A measurement is only as good as the ruler used to make the measurement, and it's the reliability of the ruler Dr. Sandage is using that's open to question."

Estimating the expansion rate of the universe is a notoriously difficult problem because of the lack of a single yardstick by which all distances can be measured.

Astronomers are fairly confident that they can measure the speed at which any given object is receding from Earth by measuring its "red shift," the Doppler shift of light from the object toward the red end of the spectrum. Most astronomers also agree that the farther away an object is from Earth (or any other observer position), the faster it is receding; hence, the inference that the entire universe is expanding.

But what astronomers cannot be sure of is the absolute distance from Earth to any very remote object. Only by knowing the distances and the red shifts of many remote galaxies—their recessional velocities—can experts estimate the expansion rate of the universe as a whole: the Hubble constant. Assuming that the big bang theory of the birth of the universe is correct, a low Hubble constant implies that the universe is expanding slowly and that the universe must be old to have reached its current size. By contrast, a high estimate of the Hubble constant implies a rapid expansion and a relatively young universe.

The main approach to calculating the distance to any object is to estimate its true intrinsic brightness and compare that with the brightness of its light reaching Earth; intrinsic brightness is reduced by the square of the distance as light travels outward, and the calculation is simple. If all celestial objects were of the same intrinsic brightness, then their distances could be easily gauged by measuring the amount of their light reaching Earth.

But celestial objects do not shine with uniform brightness.

However, in 1921, an American astronomer, Dr. Henrietta Swan Leavitt, discovered something almost as useful as a celestial object of "standard" brightness. She and Dr. Harlow Shapley found that a certain kind of rhythmically pulsating star, called a Cepheid variable, is remarkably consistent in its behavior: The more rapidly it pulses, the fainter it is. Two Cepheid variable stars with the same pulse rate are assumed to have the same intrinsic brightness.

Because there are Cepheid variables relatively near Earth, their distances can be measured with fair accuracy by other means, so that when a Cepheid variable is identified farther away, its distance can be calculated from its apparent and intrinsic brightness. The latter can be gauged by its pulsation rate.

But although Cepheid variables can be identified and measured in the Milky Way galaxy and a few nearby galaxies, they are much too faint to observe in very distant galaxies. That poses a problem because the measurement of the expansion rate of the universe depends vitally on accurately measuring distances to remote galaxies. So another type of "standard candle"—one vastly brighter than the Cepheid variables—is needed.

Most astronomers believe that a good candidate for a long-range yardstick is a species of supernova known as Type IA, which at its peak produces a million times more light than a Cepheid variable star and therefore can be identified at much greater distances. Type IA stellar explosions are believed to occur when stars some 1.4 times the mass of Earth's sun use up their fuel and collapse, then explode in gigantic fireballs. If Dr. Sandage and like-minded astronomers are right, the peak brilliance of any given Type IA supernova is about the same as the peak brilliance of any other member of this group.

But supernovas near Earth's region of the galaxy are extremely rare, and their peak brightness persists for only a matter of days. Astronomers are therefore forced to cast their nets for Type IA supernovas much farther from our galaxy, and to try to relate their intrinsic brightness to Cepheid variables whose distances from Earth are known. But until recently, the galaxies in which Type IA supernovas have been observed were far too distant to discern Cepheid variables.

However, technical improvements in telescopes, including the invention of the "charge-coupled device," a light-collecting chip, and the Hubble Space Telescope, have simplified the search for a good long-distance standard candle, Dr. Sandage said.

During observations begun in 1992, Dr. Sandage has focused particularly on a galaxy called NGC 4639; the light from this galaxy now reaching Earth began its journey during the heyday of the dinosaurs.

Within this galaxy, a Type IA supernova was detected in 1990, allowing astronomers to measure its peak brightness as viewed from Earth. But the real achievement was that the space telescope was able to see individual stars within this distant galaxy, and 20 of them were Cepheid variables. Because the brightness of Cepheid variables is presumed to be known, the Sandage group was able to calculate the distance of the NGC 4639 galaxy as 82 million light-years; from this, they calculated a new value for the absolute peak brightness of a Type IA supernova.

Applying this value to measurements made over the years of the apparent peak brightness of six other Type IA supernovas, Dr. Sandage recalculated their distances from Earth, reckoned in their recessional speeds and concluded that the Hubble constant is 57. That would mean that there is an increase in the recessional speed of 57 kilometers (35.5 miles) per second for every megaparsec of distance from Earth (or any other observer position). One megaparsec equals 3.26 million light-years.

The debate over the Hubble constant has divided astronomers into two main camps. Dr. Sandage and his partisans have long maintained that this constant is somewhere between 42 and 56, while Dr. Freedman and many other astrophysicists have argued for a value of about 80, which would imply a much younger universe, perhaps as young as eight billion years old.

But the disagreement between these groups appears to be narrowing.

"We have to consider that the properties of Type IA supernovas may vary and that their brightness may not be uniform," Dr. Freedman said. "Our group is looking for possible systematic errors in everyone's estimates—the kind of errors that can skew results across the board. But I think so much data is accumulating so rapidly from instruments like the space telescope that we're going to be able to conclude this debate in the next three years."

—*March 5, 1996*

New Era Is Promised for Optical Telescopes

By MALCOLM W. BROWNE

Might it be possible for an observer on Earth to discern something no thicker than a pencil on the surface of the moon? Astronomers specializing in a burgeoning field of physics called optical interferometry believe so.

For more than a century, scientists have been trying to hone optical interferometry, intrinsically far more sensitive to detail than are ordinary optics, as a tool for getting detailed images of stars and other very distant objects. The obstacles were enormous, but some spectacular successes in recent months suggest that success is at hand and that a new era in observational astronomy has dawned.

Astronomers believe that it may soon be possible to build telescopes capable of imaging even small, Earth-sized planets orbiting distant stars, or the mysterious cores of violent "active" galaxies. Stellar images are expected to reveal star spots (similar to sunspots), flares and many other features.

In February 1996, Dr. John E. Baldwin and his colleagues at University of Cambridge's Mullard Radio Astronomy Observatory in England published the first detailed pictures ever obtained by any optical telescope of the double star Capella, which lies 40 light-years away. (A light-year is the distance light travels in one year, at the rate of 186,300 miles per second.)

The two stars in the Capella pair are only a little more than one million miles apart, far too close to be seen from Earth as separate objects by any conventional optical telescope, including the huge Keck telescope in Hawaii, the world's most powerful, and the Hubble Space Telescope.

The remarkable resolution, or sharpness, of the Cambridge group's images of Capella was achieved using three optically linked telescopes of only modest size at a site in England that would have been spurned by builders of large conventional telescopes. Cambridge is barely above sea level and is subject to England's notoriously unstable weather.

But the magic of interferometry made up for the site's shortcomings. The new Cambridge Optical Aperture Synthesis Telescope, known by its acronym, Coast, has made scientific history.

Dr. Baldwin's report of the feat in the journal *Astronomy and Astrophysics,* with accompanying pictures showing Capella's two stars rotating around each other, has caused a sensation.

At one of the other groups exploring the possibilities of optical interferometry, the Jet Propulsion Laboratory in Pasadena, California, Dr. Robert A. Laskin predicted that there would be "a quantum leap" this year in the technique's application to astronomy.

A dozen or more observatory groups in Europe, the United States, Chile and Australia are building optical interferometer telescopes.

The United States Naval Observatory recently completed its Navy Precision Optical Interferometer at Lowell Observatory near Flagstaff, Arizonia, an instrument with even greater potential sharpness than England's Coast. Dr. Nicholas M. Elias, an astronomer with the Naval Observatory, said he expected that by October 1996, the Naval Observatory interferometer would be producing even better images of stars than Coast's Capella pictures.

"The Coast group was the first to get optical interferometric images of stars," Dr. Elias said, "and they deserve great credit. But I must say that we're very proud of our own instrument, which has begun making important astronomical measurements. We've just measured the star Regulus. Soon we'll be making stellar images, and we're poised to do some great stuff."

Optical interferometric imaging is at about the same stage as that of radio astronomy 30 years ago, when it was emerging as one of the most powerful tools of astronomy, Dr. Elias said.

"It took mainstream astronomers a few years to realize the potential of radio astronomy," Dr. Elias said, "and I expect it will be the same with interferometry. But the potential is tremendous."

The Coast group credits its sensational success to long experience with the techniques of interferometry as applied to radio astronomy. The analysis of interference fringe patterns from radio telescopes and from optical telescopes is identical, except for one important difference: Radio waves are typically between one yard and a half mile in length, while light waves are only about one ten-thousandth of an inch long. The close matching of large-scale radio waves is much easier than the comparable matching of tiny light waves.

Air turbulence, heat or vibration can ruin the perfect matching of beams needed in optical interferometry to produce fringes and images. The Coast telescope system brings beams from three telescopes together for hours at a time in an underground bunker, which is designed to shield the air from abrupt changes in air temperature to keep the beams steady. The images are built up electronically during observing runs lasting entire nights.

But the design and manufacture of some necessary equipment has proved to be agonizingly slow; mirrors, beam splitters and lenses must move on miniature trolley tracks without vibrations to control the lengths of light paths. The Coast scientists developed special motors for moving the optical components along the tracks, using piezoelectric loudspeaker coils.

In the mid-1970s, astronomers at Kitt Peak National Observatory, near Tucson, Arizona, used a technique, called speckle interferometry, to combine a series of snapshots by one large telescope to produce an interference image of the giant star Betelgeuse. The blurry image revealed puzzling dark regions on the star's surface.

In 1989, Dr. Baldwin's group further demonstrated the potential of stellar interferometric imaging when it aimed the big Herschel telescope in the Canary Islands at Betelgeuse. The astronomers first drilled five holes in a mask placed over the telescope's mouth, which allowed them to get five similar but not identical beams of light derived from the huge star. By combining the wave fronts produced by the five separate beams and analyzing the resulting interference fringes, the Cambridge group obtained an impressive image of Betelgeuse's surface detail.

Then came the construction of Coast, which collects and merges light from three separate telescopes—soon to be four—spaced six meters, or nearly 20 feet, apart.

"We knew for several years that we were close," Dr. Baldwin, leader of the Coast group, said in a telephone interview, "but on the night of Sept. 13, 1995, everything quite suddenly clicked into place. We had resolved Capella's binary components into an image, and we had built an imaging optical interferometer for a price of only £850,000—about $1.3 million.

"Compare that with the billions spent on the Hubble Space Telescope," he added with a chuckle.

A dozen or so projects based on similar interferometric principles are under way in various parts of the world, and expectations of the new technology are running high. The projected resolving power of a Jet Propulsion Laboratory proposal called the Precision Optical Interferometer in Space would make it possible for an observer in New York, if he could look over the horizon, to discern something in Los Angeles no thicker than a human hair.

The Naval Observatory and a coalition of university groups are separately building interferometer telescopes in Arizona, and the Navy hopes to begin recording stellar images this year.

Dr. Laskin at the Jet Propulsion Laboratory predicted that an accelerated search for planetary systems far from Earth's solar system—including some that might support life—would soon begin to bear fruit.

The idea that interferometry might be applied to measure the size of stars was first proposed by a French astronomer, Armand Fizeau, in 1868. But the first real achievement of interferometry was a trailblazing series of experiments by Dr. Albert A. Michelson that began in April 1887. Dr. Michelson, a German-born American physicist who was awarded the Nobel Prize for his work, devised a system using mirrors and semitransparent mirrors (now known as beam splitters) for merging separated beams of light coming from the same object. The optical elements in the system were arranged in such a way that the beams "interfered" with each other; that is, the directions and distances of their light paths were so closely meshed that the beams could interact.

If the crests and troughs of the light waves in an interferometer coincide, the result is constructive interference, producing a brightening. But if the crests and troughs do not coincide, their combination produces dimming. The mixture of bright and dim regions in closely matched, combined beams is called "fringing."

Measurements of the fringes produced by two or more merged light beams can reveal differences in the lengths of their paths of as little as about one eight-millionth of a meter, or three hundred-thousandths of a foot.

Dr. Michelson and his colleague, Dr. Edward W. Morley, used interferometry in a long and futile attempt to detect the existence of "ether," a kind of all-pervasive fluid that many scientists believed extended throughout the universe, providing a medium through which light waves could propagate.

The Michelson and Morley experiment consisted of an L-shaped apparatus in which a beam of light was split in two, with the separated beams guided along perpendicular paths of identical length and then recombined. The scientists reasoned that if Earth was moving through a universal ocean of ether, the time needed by the two parts of the split light beam to traverse a sample of the ether current from perpendicular directions should differ slightly and that the difference should be detectable in the resulting pattern of interference fringes.

But Dr. Michelson never found any difference. This historic null result, as scientists call it, was proof that ether, as it was then imagined, does not exist and that Einstein's special theory of relativity, which offers an alternative explanation for the propagation of light, is correct.

For many years, interferometry has been used as the basis for forming images from the signals recorded by radio telescopes, including giants like the Very Large Array near Socorro, New Mexico, a Y-shaped array of 27 huge antenna dishes extending over a region 17 miles in diameter.

Interferometry using visible light or near-infrared radiation will not require such gigantic dimensions, astronomers say, because the wavelengths of light are far smaller than those of radio. But the disadvantage is that even tiny errors can spoil observations using optical interferometry.

The optical telescopes currently used in conjunction with interferometers must be quite small because when light is collected by large mirrors, the effects of atmospheric turbulence in disrupting the interfering merger of beams are too great. Eventually, however, this problem will be cured with the help of "adaptive optics" that automatically correct images within a telescope to compensate for atmospheric turbulence. That will allow large telescopes to function as interferometers.

Such an application is already planned for Keck I and Keck II—a matched pair of telescopes 10 meters, or about 30 feet, in diameter—on the summit of Mauna Kea, Hawaii. The newly completed pair, standing a few hundred feet apart, will function for several years as separate telescopes, but when adaptive optics for them are built and installed, the Keck twins, working in concert as a giant interferometer, should be capable of discoveries almost unimaginable by conventional standards.

"We have a way to go before we can start imaging things as small as planets," Dr. Elias said, "but we're getting there."

—April 30, 1996

At Other End of "Big Bang" May Simply Be a Big Sputter

By JOHN NOBLE WILFORD

The end is not near.

While others ponder the future in the approaching millennium, or rue some impending "end of history" or "end of science," a few astrophysicists specializing in cosmic vision are casting an eye so far ahead that they think they can see the fate of the universe. How, they ask and are prepared to answer, will the universe end?

Not with a bang, which presumably is how it all began: There will come a time, safely off the scale for workaday concern, when not only the sun will die, but the lights of all stars will also vanish. Left in the enveloping twilight will be stillborn stars like brown dwarfs, stellar ghosts like white dwarfs and neutron stars and those powerful gravitational sinks known as black holes.

In time even these will decay and disappear. All that will remain in this bleak, darkened future will be an increasingly diffuse sea of electrons, positrons, neutrinos and radiation.

This projected final dark era is expected to begin at an incomprehensibly distant time: 10,000 trillion trillion trillion trillion trillion trillion trillion trillion ($10^{10,100}$) years from now.

If this does not exactly spell the end of everything, it will effectively be the end of the universe as it is currently understood. At least this is how two astrophysicists at the University of Michigan at Ann Arbor described the fate of the universe in a report in Toronto at a meeting of the American Astronomical Society.

One of the scientists, Dr. Fred Adams, called the projections "a quantitative theory of the future of the universe," based on the assumption that the recognized laws of physics will continue to operate over the very long-term future and also the assumption, widely held, that the universe is destined to expand forever.

Dr. Adams and a colleague, Dr. Greg Laughlin, will publish a detailed analysis of their research, titled "A Dying Universe," in the April issue of the journal *Reviews of Modern Physics*.

Commenting on the research in an interview, Dr. P. James E. Peebles, a theoretical astrophysicist at Princeton University, said, "How can you resist speculating on the future of the universe?"

But like all projections, even of more readily conceivable futures, this one must be viewed with caution, Dr. Peebles said. Despite remarkable advances in recent years, cosmologists cannot be sure they understand the evolution of the universe well enough to extrapolate that knowledge far into the future.

"We have given up on the notion that the laws of physics are known absolutely," Dr. Peebles said. "So it's awfully hard to know if they are the real truth or be sure that they won't fail sometime in the distant future, leading to a completely different outcome."

Dr. Adams and Dr. Laughlin insisted that they had applied a conservative interpretation of current knowledge and conducted rigorous calculations in developing their far-reaching cosmic scenario. They also acknowledged their debt to separate studies, published at least 20 years ago, by Freeman J. Dyson of the Institute for Advanced Study in Princeton, New Jersey, and Sir Martin Rees of University of Cambridge in England, but said their own work benefited from more recently acquired insights.

In what Dr. Adams called "a more complete vision of the future," the two astrophysicists started with the present "stelliferous era," a time when stars dominate the universe. Star formation presumably began a few million years after the big bang, the hypothesized explosive origin of the universe, and has continued up to now, some 10 billion years later.

The following is how they see the future unfolding:

The sun is expected to die in about five billion years, reduced to an extinct remnant known as a white dwarf. Earth might survive the solar death throes, but its oceans would boil away from the heat, and life, if it endured that long, would no longer be possible. The sun will not be alone. Slowly, all the larger solar-mass stars will turn to white dwarfs, and the more slowly evolving small red-dwarf stars will assume more importance. The end of all star formation, the end of the stelliferous era, should come in 100 trillion years.

In the succeeding "degenerate era," most of the universe's mass will be in the form of brown dwarfs, starlike objects too small to shine from nuclear fusion; red dwarfs; white dwarfs; and black holes. The white dwarfs will capture most of the so-called dark matter, the hypothesized weakly interacting particles that constitute most of the cosmic mass. And the black holes will be gobbling up the white dwarfs.

"An occasional rare burst of energy will be generated when two brown dwarfs collide to create a new low-mass star," Dr. Laughlin said.

The decomposition of protons, basic particles of ordinary matter, will destroy what remains of the stellar relics, thus ringing down the curtain on the degenerate era in 10 trillion trillion trillion years. Although the concept of proton decay, somewhat analogous to radioactive decay, is widely accepted by physicists, experiments have failed to detect its supposedly slow rate of occurrence and so there is considerable uncertainty involving any predictions about its nature.

The third period, the black hole era, would be an even longer span of time, during which even these objects with powerful gravitational forces would slowly radiate away their mass and disappear. This apparently paradoxical concept of radiation from black holes, from which no matter can escape, has been advanced by Dr. Stephen W. Hawking of University of Cambridge, based on an interpretation of Einstein's general theory of relativity. A black hole's final moment might be an explosive blast of radiation.

After that, the dark era.

Even Dr. Adams and Dr. Laughlin do not presume to foresee what will happen after that. "This is not when the universe will end," Dr. Adams said, "but when any given physical processes in the universe will cease to be important."

An alternative to such a projected cosmic future, other astrophysicists say, is one in which the universe collapses of its own weight—not anytime soon, but long before it could enter some of the epochs envisaged by Dr. Adams and Dr. Laughlin. Such a theorized collapse is referred to as the Big Crunch. Most astrophysicists today think the density of the universe is sufficiently low that such a gravitational collapse is unlikely, but no one is really sure what the cosmic density is.

In any event, Dr. Adams offered the reassuring thought that the end was not coming any time soon.

January 16, 1997

Peek at Black Holes' Feast Reveals Awful Table Manners

By JOHN NOBLE WILFORD

Astronomers observing black holes, those gourmands with a boardinghouse reach at the cosmic table, are finding astonishing new evidence of their gustatory excesses, as well as hints of perplexing failures.

For one thing, it is apparently possible for some gas and dust spiraling within only 40 miles of the lips of a relatively small black hole to escape being its next feast, swallowed in a huge gravitational gulp. Every 30 minutes or so, like a colossal version of Old Faithful geyser in Yellowstone Park, the black hole seems to spit out the superhot gases as jets traveling at nearly the speed of light, more than 600 million miles per hour.

In another recent discovery, astronomers have traced how dust from outer parts of a galaxy is transported over distances of thousands of light-years to the galactic center and into the mouth of a black hole there.

And new evidence, the most compelling so far, reveals that apparently like many galaxies, the Milky Way has an enormous black hole at its core. Astronomers estimate that the mass of this compact object is 2.6 million times that of the sun. Under its powerful gravitational pull, some hot gases are drawn perilously close but at such great velocities that they escape being sucked in. Instead, the gases bend sharply around the black hole and are flung out the other side.

The Milky Way's black hole "is the gravitational anchor that all other objects in our galaxy, including our sun, are revolving around," said Dr. Farhad Yusef-Zaheh, an astronomer at Northwestern University in Evanston, Illinois.

The new observations of the black-hole phenomenon were reported recently at a meeting of the American Astronomical Society in Washington. They provide new clues to the nature of black holes, which are so dense that nothing, not even light, can escape once it crosses the gravitational boundary of no return known as the "event horizon."

Still, astronomers cannot observe the objects themselves, only their effects on nearby stars, gas and dust.

Not all black holes have masses on the scale of those at the galactic centers. Many smaller ones have a mass of only 10 to 30 suns compressed into a sphere about 25 miles wide, and some of these are known or suspected to be orbiting a close companion star. The black hole's gravity pulls gas from the companion into a swirling disk of material that spirals around the black hole like soapsuds swirling around a bathtub drain.

One such black hole, about 40,000 light-years away in the constellation Aquila, has been studied since 1992. In time, astronomers observed two strange phenomena. The inner part of the disk of gas and dust circling the black hole seemed to disappear periodically. Radio and X-ray observations also revealed sudden flares streaking out in opposite directions from the revolving disk. These jets were detected on many occasions, always traveling at velocities greater than 90 percent of the speed of light millions of miles out in space.

Now astronomers have simultaneously observed the two events and demonstrated that the jets of superhot gas are forming at the same time that the inner disk is vanishing, linking the two actions. The phenomenon occurs at intervals ranging from 20 to 40 minutes.

"We are very excited about these results," said Dr. Stephen Eikenberry, an astrophysicist at the California Institute of Technology in Pasadena. "The direct connection between the disappearance of the inner disk and the jet ejection has never been seen until now."

Though the findings show that all matter on the brink of a black hole is not doomed, scientists cannot explain what is going on. Somehow, soon after each ejection of hot gas from the inner disk, more matter fills the void, and the process is repeated. Scientists suspect this is a common occurrence in the vicinity of small black holes.

Dr. Ronald Remillard, a research scientist at the Massachusetts Institute of Technology, said, "The force of gravity pulling on the inner disk is incredibly strong, and it's amazing that the jets can rip this matter away from the black hole."

Dr. Jean Swank, an astronomer at the Goddard Space Flight Center in Greenbelt, Maryland, estimated that each jet throws off the mass of an asteroid. "This process clearly requires a lot of energy," she said. "Each cycle is equivalent to six trillion times the annual energy consumption of the entire United States."

In August 1997, Dr. Eikenberry and his colleagues, using the 200-inch telescope at Palomar Mountain in Southern California, observed the jets as infrared flares. At the same time, Dr. Remillard's team examined data from NASA's *Rossi X-Ray Timing Explorer* satellite and found significant dips in X-rays from the inner part of the disk around the black hole.

A month later, astronomers at the Goddard center and in France made similar observations and confirmed the findings.

In other research, astronomers at the Carnegie Institution of Washington have found spiral lanes that appear to bear dust for the feeding of voracious black holes at the centers of certain distant galaxies. It has not been clear how exactly mass from the outer parts of a galaxy is transported to the black hole at the core, although its powerful gravity obviously plays a dominant role.

The discovery, described by Dr. Michael W. Regan and Dr. John S. Mulchaey of the Carnegie Institution, was made with a combination of infrared and visible-light observations by the Hubble Space Telescope of the dense dust and gas in several Seyfert galaxies. These are galaxies with especially energetic cores, presumably occupied by black holes each as massive as 10 million suns. As material falls into a black hole, it is converted to energy.

The astronomers suspect that the orbits of stars around such massive black holes are sometimes aligned in such a way so as to direct dust and gas inward along certain thoroughfares. But from dust maps of two such galaxies, they could identify no such alignments.

"Our images suggest instead that material spirals in toward the center," Dr. Regan said. The mechanism for these spiraling dust lanes remains unexplained.

As sure as astronomers are of black holes occupying the centers of distant galaxies, they have struggled for years trying to peer into the dusty depths of the Milky Way for conclusive evidence of what is there.

Is its gravitational power coming from a thick cluster of stars? Or a black hole?

Improved observations of an unusual source of radio emissions called Sagittarius A at the center of the Milky Way have recently been made by the Very Long Baseline Array, a radio telescope at Socorro, New Mexico. If the radio source was a star cluster, it should be moving rapidly. If a black hole, it should be virtually stationary.

A team of astronomers led by Dr. Mark Reid of the Harvard-Smithsonian Center for Astrophysics in Cambridge, Massachusetts, studied the radio source for two years and could find no appreciable motion. It almost certainly is a black hole, they concluded.

Taking another research approach, scientists at the Max Planck Institute for Extraterrestrial Physics at Heidelberg, Germany, tracked velocities of stars orbiting near the Milky Way's core. The faster the stars moved, the more likely it was that they were orbiting a black hole. Infrared observations by European telescopes in Chile clocked velocities of such magnitude that scientists inferred that it was a black hole with a mass of almost three billion suns.

"This is one of the firmest cases we have" for the existence of a black hole in Earth's home galaxy, Dr. Andreas Eckart of the Max Planck Institute said.

What perplexes astrophysicists is why the energies radiating from the edge of this and some other suspected black holes are not more luminous. Considering the stellar matter they must be consuming, they are remarkably radiation-deficient.

In a report in a recent issue of *The Astrophysics Journal*, Dr. Ramesh Narayan of the Harvard-Smithsonian astrophysics center proposed an explanation in which most of the gravitational energy at a black hole's event horizon is converted to the thermal energy of hot electrified gases. Most of the energy is thus not radiated away by electrons. Instead, it flows across the boundary of no return.

Dr. Reinhard Genzel of the Max Planck Institute said the proposal "is the best answer yet to this strange paradox." Black holes may have huge appetites but they convert their food very inefficiently into radiation.

—January 13, 1998

In Chilean Desert, Observatory for 21st Century Takes Shape

By JOHN NOBLE WILFORD

In the oven heat of day on the high desert, cranes swing over the construction site, loaded trucks groan up the winding road to the summit and workers in blue hard hats bend to their tasks around a gleaming machine anchored inside one of the new buildings. It is the tableau of a mountaintop being transformed by builders in the tradition of the Inca at sacred Machu Picchu, but this time the lofty temples will be put to the service of a modern, scientific cosmology.

The worldwide building boom in powerful astronomical observatories has now reached this 8,000-foot mountain in the Atacama Desert of northern Chile. The Atacama is one of the driest places in the world, a vast expanse of barren hills and ridges baked to many shades of brown. The blue waters of the Pacific Ocean to the west and the snow-capped Andes to the east, both visible from the heights, only tease like mirages and bring no refreshment to the sere landscape.

But above, and this is what matters, the air is thin and virtually free of water vapor, offering ideally stable and transparent seeing conditions for astronomy. Nearly every night is full-starred. Beyond the ordinary range of the human eye, though manifest in technology's clever magnifications, distant galaxies hurtle outward in headlong flight, challenging the keepers of the new temples to decipher their messages about the beginning, evolution and future of the universe.

Here a consortium of European nations is constructing what is being called "a telescope for the 21st century." It is not one instrument but actually a planned complex of four equally large telescopes and three smaller ones. When all the pieces are in place and running in concert, in another five or six years perhaps, the prosaically named Very Large Telescope at Paranal Observatory will be the world's largest and technologically most advanced optical telescope array.

The first of the four big telescopes at Paranal is complete and has passed all of its early tests. The ceremonial moment of truth known as "first light" came in late May. For the first time, the telescope's huge 8.2-meter mirror, some 27 feet in diameter, was focused on familiar celestial objects to check its light-gathering qualities.

The successful test brought relief, elation and the tonic of rising self-confidence to the engineers and scientists of the European Southern Observatory, the eight-nation group building Paranal. They were meeting construction schedules and performance standards. Now, more than ever, their goal of creating a facility with observing powers in visible and infrared light that equal or surpass anything else in the world seemed attainable.

Their pride overflowed: For the first time in a century, Europe was stepping to the forefront of observational astronomy.

One day a few weeks later, Krister Wirenstrand, a Swedish engineer supervising the preparations, stood on a steel platform surrounding the completed telescope. He spoke of it as a big machine. The mirror was ringed and supported by heavy metal, 400 tons of weight resting on bearings in oil, and accompanied by a maze of hydraulic pipes and wires to microsensors and computer-operated motors.

"We are much farther along than we expected to be at this time," he told a visitor. "We have had first light. We have found no serious flaws or problems. Of course, there's still much to do, much tuning, but we feel good about everything."

At nightfall, as stiff winds chilled the summit, astronomers and computer specialists hovered over consoles in the dim control room. They put the telescope through its paces, sighting on galactic targets, tracking their motions and photographing them in sharp detail. These kinds of hardware and software tests are expected to continue until April 1999, when the telescope should be ready for regular research.

Dr. Roberto Gilmozzi, an Italian astronomer, remarked that he was "at the end of the food chain in the preparation of the telescope." His job was to oversee tests of the instrument's ability to make telling observations. To illustrate, he called up some recent pictures on a monitor.

There was Omega Centauri, a cluster of especially old stars. A long exposure demonstrated that the telescope was able to track objects continuously with high precision. It also produced sharp images of individual stars in the cluster, showing the near-perfect quality of the aluminized glass mirror.

A spectacular picture of Eta Carinae, a superluminous eruptive star more than one million times brighter than the sun, revealed fine details of structure that, European astronomers have said, were never before achieved with any ground-based telescope.

And the picture of the Butterfly nebula, taken through blue, green and red filters, was a sheer esthetic delight. This again demonstrated the telescope's ability to look into dense clusters and observe individual stars. "It was even better than we expected," Dr. Gilmozzi said.

Turning from the pictures, Dr. Gilmozzi gave voice to an emotion shared by many of his colleagues. Like any other science, astronomy can be fiercely competitive, and European astronomers have long chafed at their disadvantage in not having ready access to the most powerful telescopes.

"For a long time, Europe has not been in the forefront of astronomy," he said. "The VLT puts Europe back in the running, very close to the head of the pack in astronomy."

The Competition:
After Late Start, Heading South

European astronomy had been set back by two world wars and unfavorable geography. Low, damp and crowded, the continent had no remote, arid heights to compare with those in the American West and Hawaii. For much of the century, the 100-inch telescope on Mount Wilson near Los Angeles and then the 200-inch Hale Telescope on Palomar Mountain near San Diego set the standard, and American astronomers were favored in getting access to the observatories.

The Europeans were slow to look for their own sites far from home. The British established a joint Anglo-Australian observatory on Australia and set up telescopes in the Canary Islands. Eight other countries—Belgium, Denmark, France, Germany, Italy, the Netherlands, Sweden and Switzerland—formed the European Southern Observatory, with headquarters near Munich but nearly all of its telescopes here in Chile. Most are concentrated at La Silla, about 400 miles north of Santiago. Paranal is farther north, near the port city of Antofagasta. The United States also operates several large telescopes in Chile, making its high desert the primary platform for viewing the once-neglected southern skies.

Astronomers in the United States agree that, when in full operation, the Paranal instruments would in some respects be the world's most powerful telescope array. But they emphasize that this would not render others obsolete, especially not the Keck Observatory. The two Keck telescopes on Mauna Kea, in Hawaii, remain the largest individual instruments, each with 10-meter, or 33-foot, mirrors.

"Keck will remain more than completely competitive," said the observatory's director, Dr. Frederic Chaffee.

For one thing, he said, the greater light-gathering capability of the larger individual Keck mirrors is likely to continue affording the best views of especially faint phenomena like the most distant objects in the universe, the focus of much research in cosmology nowadays.

But Dr. Riccardo Giacconi, director-general of the European Southern Observatory, said that, once all the Paranal telescopes are operating, they should provide astronomers rare insights and discoveries into three of the great questions of modern astrophysics: the beginning and evolution of the universe; the formation in the remote past of stars and galaxies and other cosmic structure, and the formation and evolution of planets, especially planetary systems around other stars.

The Quarry:
Seeking New Worlds and Their Cousins

Most astronomers think that one advanced aspect of Paranal's telescope array will produce many discoveries of more of these extrasolar planets. One evening in Santiago, when asked what would be some of Paranal's first important discoveries, Dr. Maria Teresa Ruiz, an astronomer at the University of Chile, replied, "Planets."

The search for extrasolar planets and the enigmatic brown dwarfs, objects larger than planets but not big enough for stardom, will begin in earnest when at least two of the four large telescopes and two of the smaller ones are available for operation as an interferometric array, by 2002. The technique of interferometry effectively creates a poor man's supertelescope. Each telescope in the array must be kept extremely stable, better than a fraction of one wavelength of visible light over several hundred yards. Each then focuses simultaneously on a single object, with computers combining the incoming light into a composite image equivalent to what could have been produced by a telescope much larger than any of the individual ones used. The entire array of telescopes could be integrated for inter-ferometry by 2006.

Such a clever technique should enable astronomers to distinguish in the glare of nearby stars the presence of companion planets that might otherwise have gone undetected. They could be objects the size of a Jupiter or much smaller, perhaps planets with 10 times the mass of Earth. The European scientists are not ruling out the direct detection of infrared radiation originating from the planets themselves, which would provide data on their chemical composition, temperatures and surface structure. The few extrasolar planets found in recent years have not been observed directly, only inferred from their gravitational effects on their parent stars.

"There's a strong push in Europe to find planets," Dr. Gilmozzi said. "A good fraction of time will be dedicated to the search, especially after all four telescopes are ready with interferometry."

Other observatories, including Keck, have similar plans to put interferometry into practice. Dr. Stephen P. Maran, an astronomer and spokesman for the American Astronomical Society, said, "That's going to usher in a whole new era of astronomical observation."

Both Keck and Paranal epitomize astronomy's building boom. At the moment, new telescopes are being planned or constructed elsewhere in Chile and in Arizona, Hawaii and New Mexico. Spain recently announced plans for a 10-meter telescope in the Canary Islands, to be ready for first light in 2002; it is designed as a segmented mirror like the Kecks. South Africa is planning a 9.1-meter telescope, in a mosaic design comparable to the Hobby-Eberly Telescope recently completed at McDonald Observatory in Texas.

The New Mirror: Computers Control Huge Floppy Disk

Nearly all of the new observatories would have been inconceivable before new technology in telescope construction emerged a decade ago. Previously, the upper limit in effective mirror size seemed to be Palomar's 200-inch telescope; anything larger proved to be too heavy, unwieldy and expensive. Russian experiences with a larger mirror were not encouraging.

Then engineers began experimenting with lightweight mirrors made of honeycombed glass or segments of thin glass pieced together. Each of the 10-meter Keck mirrors is composed of 36 hexagonal segments fitted together like a jigsaw puzzle, a complicated design that would have been unworkable before the evolution of precise computer controls. Reacting to electronic sensors, the computers issue hundreds of commands a minute to make continuous adjustments of each segment to maintain the mirror's optimal shape and position at all times.

The Europeans have devised a way to apply similar control technologies, called active optics, to the largest telescope mirrors to be made from a single piece of glass. The only way this could be feasible was to keep the 8.2-meter-diameter glass unusually thin, only 10 centimeters. And the only way to keep such a thin mirror from flexing and warping like a giant floppy disk was to equip the back of the mirror with dozens of pistons for applying pressure here and there continuously to adjust the mirror shape.

In this case, the active-optics system depends on a high-speed analysis of the mirror's image quality. Any deviations, the result of slight changes in the telescope alignment, are registered, and signals are instantly sent to make the required adjustments. The Europeans first tested the system on its 3.5-meter New Technology Telescope at La Silla.

"This is the first time a mirror this big and this thin has been produced," Dr. Gilmozzi said. "Going from 3.5 meters to 8 meters is a big jump, and this requires a good deal of testing of the active optics. It works, but it still needs a lot of tuning."

Even more advanced controls, called adaptive optics, are being developed to compensate for atmospheric turbulence in telescope images. With corrective signals sent as often as 100 times a second, European astronomers said, adaptive optics applied to the Very Large Telescope should produce images at least 10 times sharper than the actual seeing conditions.

The Future: Throwing Light on the Invisible

Gathering more light, refining it to a sharpness unseen before, probing the dusty heart of distant galaxies and stellar nurseries, telescopes like those rising on the high desert of Chile are sending the same rush of excitement through astronomy as the opening of the first Keck telescope did eight years ago, but no one dares to say how the view from Paranal will change thinking about the universe.

"It's impossible to predict what's going to be discovered," Dr. Chaffee of the Keck Observatory said. "No one predicted that Mount Wilson would find the expanding universe or Palomar would see quasars."

In late June 1998, a month after first light, the Very Large Telescope gave astronomers a surprise, if not a discovery. A color photograph of a galaxy 165 million light-years away in the southern constellation of Centaurus revealed its strange crosslike configuration, as if two galaxies had intersected. In the strange motions of stars, gas and dust, astronomers said they could be seeing evidence of dark matter, the invisible stuff that presumably makes up most of the universe and is one mystery waiting for some telescope to solve, perhaps looking into the full-starred night in the Atacama.

—August 4, 1998

Pictures Give Hints of Universe at Its Dawn

By JOHN NOBLE WILFORD

Peering farther back in space and time than ever before, the Hubble Space Telescope has taken infrared pictures of what may be the most distant objects ever detected, early light from the galactic dawn of the universe.

Astronomers are not sure, but they think some of the faint objects could be galaxies formed in the first billion years of cosmic history, or more than 12 billion years ago. If so, they are crossing a new frontier of the cosmos, where they hope to find the very first galaxies and solve the mystery of when and how such congregations of stars emerged out of the formless dark.

"These images are the deepest images of distant galaxies that have ever been obtained," Dr. Rodger I. Thompson, a University of Arizona astronomer, said in announcing the new Hubble telescope findings yesterday.

Dr. Alan M. Dressler, an astronomer with the Carnegie Observatories in Pasadena, Calif., agreed that the infrared pictures were "our first tentative glimpse into the very remote universe."

Although these may represent "the first stages of galaxy formation," Dr. Dressler said, the objects are such faint blobs that "their true nature can only be explored with the advanced telescopes of the future." He was referring, in particular, to the more powerful replacement for the Earth-orbiting Hubble telescope that is being planned for launching in 2007.

The National Aeronautics and Space Administration made the infrared pictures public at a news conference in Washington. A catalogue of the observed objects and other research details will be published soon in *The Astronomical Journal.*

In January 1998, using its infrared spectrometer-camera, an instrument installed by visiting astronauts in early 1997, the telescope took a 36-hour exposure of a relatively narrow region of space that had previously been examined in visible and ultraviolet light. Three years ago, when the Hubble telescope took a 10-day exposure with its visible-light camera of the same field of view, astronomers found hundreds of galaxies that had never been seen before, inspiring them to raise their estimate of the total number of galaxies in the observable universe to more than 50 billion from 10 billion. The Hubble Deep Field, as the visible photography was known, had been until now the deepest view of galaxies.

Cameras sensitive to infrared wavelengths can often penetrate the dust that obscures many of the most distant cosmic phenomena. Some of these objects are so far away that their only detectable light is in the infrared wavelengths. Most galaxies radiate more infrared light than visible light.

The more than 300 galaxies in the main infrared picture included some that had been seen before. But the infrared camera revealed many new objects, the reddest and faintest of which may be the earliest galaxies ever detected. Astronomers said the distant galaxies seemed to be smaller than those from more recent epochs. But they said it was not yet possible to determine if the galaxies are truly small or whether these are merely the cores of larger structures that cannot be seen at these distances.

Dr. Thompson, the principal scientist working with the infrared instrument's observations, estimated that the most distant objects in the picture dated back to when the universe was 5 percent of its present age. That age is still a matter of debate, but some of the most recent data and theoretical models indicate that the universe began about 13 billion years ago.

"The fact that we have found new objects means we really have reached the edge of the universe," Dr. Thompson said. That is, the observations take astronomers close to the time when matter in the early universe first coalesced into stars and the stars clustered in galaxies. How this happened is one of the great unknowns of astronomy.

In analyzing the results, Dr. Thompson also found that faint red galaxies matched up with compact blue knots of light that had been seen in the earlier visible photography of the region.

"This means that some objects that appeared to be separate galaxies in the optical image are really hot star–forming regions in much larger, older galaxies," he said.

—*October 9, 1998*

Where Does the Time Go?
Forward, Physics Shows

By MALCOLM W. BROWNE

In Lewis Carroll's mirror world of *Through the Looking Glass,* it seems perfectly logical that the White Queen, who lives backward, first bandages her finger, then begins to bleed, then screams, and finally pricks her finger. On paper, if not in real life, the physics governing many natural phenomena permit time to run either forward, like a swimmer jumping from a diving board, or backward, like a reversed movie in which the swimmer leaps from the water and lands on the board.

But since a landmark experiment in 1964 by Dr. James W. Cronin and Dr. Val L. Fitch, both at Princeton University at the time, physicists have known that time reversal is not so neat in the microscopic world of particles. They found indirect but convincing evidence that sometimes a particle going backward in time fails to land on the metaphorical diving board; in other words, time, they found, could not be perfectly symmetrical.

Experimenters have now achieved direct confirmation of this unsettling inference.

To no one's surprise, physicists at two big particle accelerators, one in Switzerland and the other in Illinois, proved that when certain particles go backward in time, their behavior is somewhat different from what it is when they go forward.

If this sounds baffling to non-scientists, they are not alone; a member of the Nobel committee who sat on the panel that awarded Dr. Cronin and Dr. Fitch the 1980 prize in physics remarked, "It would take a new Einstein to say what it means."

But one important implication stands out: Time's slippery nature may explain why there was anything left after the big bang to build the universe as we know it. Theorists have concluded that there was a slight imbalance in the amounts of matter and antimatter created at the birth of the universe. The matter and anti-matter believed to have been created by the big bang presumably annihilated each other quickly, leaving only a slight excess of matter—just enough to create today's matter universe.

Time, which had a role in this, is probably the deepest of all enigmas in physics.

At the everyday level, physicists believe that the "arrow" of time points always in the direction of increasing disorder (or "entropy").

Natural processes run down, order yields to disorder, information disappears, and people grow old, die and decay. These processes mark the forward passage of time.

But a particle is not like a human being, and when physicists speak of a particle going backward in time, they do not mean that the particle is a tiny time machine capable of exploring the past.

A leading particle theorist, Dr. Chris Quigg of the Fermi National Accelerator Laboratory in Batavia, Illinois, explained: "It's not that antiparticles in my laboratory are actually moving backward in time. What's really meant by that is that if I think of a particle moving from one place to another forward in time, the physical process is the same as it would be if we imagine running the film backward and also changing the particle into an antiparticle."

Three fundamental transformations of particles are involved in all this: the reversal of electrical charge (C), which changes particles into antiparticles and vice versa; parity reversal (P), the mirror reversal of every dimension in a particle (turning it inside out, so to speak); and time reversal (T).

Physicists feel comfortable when things can be explained by balance sheets that show everything accounted for. They once believed that the symmetry of parity—the original form versus its inside-out version—was inviolate, meaning that physics in a mirror world would be identical to our own. But the 1957 Nobel Prize in physics honored Dr. Tsung-Dao Lee and Dr. Chen N. Ning Yang for discovering that the assumed symmetry of parity in particles did not exist; that when short-lived particles called K mesons (or kaons) decay, their transformations violate parity symmetry.

Many theorists expected that this asymmetry would be balanced out by another of the transformations, that of charge. Dr. Fitch and Dr. Cronin proved, however, that charge symmetry was also violated. That meant that to keep things in balance, the symmetry of time had to be violated to make up for the symmetry violations of charge and parity. In this way the total package of charge, parity and time, or CPT as physicists call this combination of interlocking components, would be "conserved," preserving the ideal of a universe that fits neatly together.

"If you believe that charge, parity and time taken together must balance out," Dr. Quigg said, "then if charge and parity are a little funny, and you divide them into the charge-parity-time package, then time must be a little funny, to compensate.

That was predicted by theory. But the two new experiments by Fermilab and CERN, the European accelerator group, show directly that time-reversed symmetry is violated in just the direction and amount predicted by theory. Now the CPT ledger book is in balance."

Both the CERN and Fermilab experiments measured decay processes of rare particles called neutral kaons and neutral antikaons, which consist of two quarks. (Protons and neutrons, the particles that make up the nuclei of ordinary atoms, contain three quarks.)

In the CERN experiment, detectors measured the oscillations of kaons into antikaons, and vice versa, as these fleeting particles sped away from their point of origin. If time were perfectly symmetrical, the rates at which kaons and antikaons are transformed into each other should be precisely equal. The experiment showed, however, that the rate at which antikaons (which are a form of antimatter) turn into kaons (which are normal matter) is higher than the time-reversed process in which kaons become antikaons.

Some similar asymmetry could help to explain the presumed excess of matter over antimatter when the universe was created.

But what, if anything, does a particle moving backward in time have to do with conventional time at larger scales?

Mathematical equations governing the laws of motion, electromagnetism and many other phenomena present no difficulty with time reversal. Nor is time reversal inconsistent with particle physics, which is governed by ordinary quantum mechanics, which for all its celebrated weirdness operates within a mathematical framework of classical three-dimensional space and time. Quantum mechanics requires no special direction of time, either forward or backward.

But cosmic relationships are governed by the laws of general relativity, Einstein's theory of gravity, and these have yet to be brought into consonance with quantum mechanics—the rules for the behavior of atoms and subatomic particles. In each of these domains, time has a somewhat different meaning.

Relativity decrees that time is not an absolute quantity. Among the surprising effects of relativity are that a moving clock runs slower than a clock at rest, and that time on a mountaintop runs faster than time at sea level, because gravity is stronger at sea level and gravity slows time down.

In theory, some scientists have suggested, it might be possible to travel in time using black holes, wormholes, cosmic strings or other distortions of space-time, but one of the problems with time machines is the "grandmother paradox":

if someone could go back to the past and kill his own grandmother, a paradoxical violation of the principle of causality would result. The possibility of this happening is one of the main objections to the idea of time travel, although some physicists have devised ingenious schemes for getting around the paradox.

But does a particle going backward in time pose the same kind of paradox? Physicists think not.

Noting that the physics of ordinary experience prevents time reversal and violations of causality, Dr. Fitch said in an interview that "Things in the everyday world are statistical in nature, and disorder always increases, fixing the direction of the arrow of time. But time asymmetry for particles applies to just a handful of individual particles, not to statistical aggregates."

Dr. J. Richard Gott 3d, a Princeton University cosmologist, envisions a possible universe that would be the opposite of ours in every sense.

"I can imagine living in a universe like ours, except that charge, parity and time are all reversed," Dr. Gott said in an interview. "Instead of expanding from the big bang, such a universe would be contracting toward a big crunch, with everything growing hotter."

In a recent paper in the journal *Physical Review D*, Dr. Gott and his colleague, Li-Xin Li, suggested that the laws of physics "may allow the universe to be its own mother."

In an antimatter, time-reversed universe, Dr. Gott said, people would remember what we call the future (but not what we think of as the past), and for them, the backward flow of time would seem as natural as does our sense of forward-flowing time.

But to come to terms with such things physicists need to deal with quanta—the discrete packets of energy that define the microscopic world: electrons, photons, quarks and so forth. Even empty space is believed to be quantized—subdivided into infinitesimal cells.

But so far, despite the best efforts of Albert Einstein and many other theorists, no one has been able to dissect gravity or time into their component quantum packets, if such exist.

"We're still children as far as quantum gravity is concerned," said Dr. Daniel E. Holz, a relativity theorist at the Max Planck Institute, Heidelberg, Germany.

"We don't know how to quantize time," Dr. Holz said. "You can't make heads or tails of it. When you try to quantize gravity, time is what sinks you. When we understand what to do with time in quantum gravity we'll have it done. Or turn it around: When we get quantum gravity, the big revelation will be, aha! So that's the way time works!"

Dr. John A. Wheeler of Princeton University, the cosmologist and astrophysicist who coined the term *black hole* to describe ultradense objects from which light cannot escape, believes that despite the puzzles and paradoxes posed by time, a fundamental simplicity underlies it.

"It's not so much that there's something strange about time," Dr. Wheeler said in an interview. "The thing that's strange is what's going on inside time.

"We will first understand how simple the universe is when we recognize how strange it is."

—December 22, 1998

Hubble Telescope Yields Data
for Recalculating Age of Universe

By JOHN NOBLE WILFORD

After eight years of measurements by the Hubble Space Telescope, astronomers think they have finally established a reliable value to the expansion rate of the universe, the indispensable number required for determining the age, size and fate of the universe.

It is a number that scientists have sought ever since 1929, when Edwin P. Hubble discovered that the galaxies are flying outward at velocities that increase with distance. Estimates of the expansion rate, also known as the Hubble constant, are fundamental to all theories of cosmic evolution. But they have fluctuated wildly between numbers that would put the age of the universe anywhere from 10 billion to 20 billion years.

By the new calculations, the universe has been expanding for at least 12 billion years since its theorized explosive creation in the big bang. Depending on the density of cosmic matter and the possible existence of a mysterious form of vacuum energy, the age of the universe could be closer to 13.5 billion or even 15 billion years.

Astronomers said they had 90 percent confidence in the reliability of the basic measurement underlying these age estimates. The number of the Hubble constant, they concluded, is 70, plus or minus 7. That means that a galaxy appears to be flying away at a rate of 70 kilometers per second per megaparsec, or 160,000 miles per hour faster for every 3.26 million light-years away from Earth.

A team of scientists led by Dr. Wendy L. Freedman of the Carnegie Observatories in Pasadena, California, determined the expansion rate by using the Earth-orbiting Hubble telescope in measuring distances to 800 stars of known brightness, called Cepheid stars, in 18 galaxies. Measuring the expansion rate was the telescope's top scientific objective.

"After all these years, we are finally entering an era of precision cosmology," Dr. Freedman said in announcing the findings at the National Aeronautics and Space Administration in Washington. "Now we can more reliably address the broader picture of the universe's origin, evolution and destiny."

Only a decade ago, astronomers argued over wide differences in expansion-rate estimates, which resulted in a range of cosmic ages running from 10 billion to 20 billion years. The first tentative results by Dr. Freedman's group, reported in 1994, seemed to defy logic. An expansion rate of 80 yielded age estimates ranging from 8 billion to 12 billion years, which made the universe younger than its oldest stars.

Dr. Freedman said in an interview that the lower value for the Hubble constant, as well as the team's confidence in the research, stemmed from an increase in observational data through different methods and a careful analysis of possible sources of uncertainty or error.

Now the estimates by rival teams are beginning to converge on numbers that lead to ages of the universe similar to or greater than the ages of the stars. But the controversy has not entirely ended.

"We used to disagree by a factor of two; now we are just as passionate about 10 percent," said Dr. Robert Kirshner of the Harvard-Smithsonian Center for Astrophysics in Cambridge, Massachusetts. "A factor of two is like being unsure if you have one foot or two. Ten percent is like arguing about one toe. It's a big step forward."

Dr. Freedman's adversary in the Hubble constant wars has been Dr. Allan R. Sandage, also at the Carnegie Observatories. His observations of distances to certain types of exploding stars have consistently led to lower values for the Hubble constant, usually less than 60, and thus greater ages for the universe, from 15 billion to 20 billion years.

At a recent science conference in Baltimore, Dr. Sandage, referring to his own results, reported: "At a 99 percent confidence level, the value is between 53 and 65. The curves are going to intersect in 2006."

Dr. Kirshner remarked, "We are no longer talking about a real conflict."

As Dr. Freedman and others observed, the margins for error in the two sets of Hubble constant estimates are such that they are not only drawing closer to each other, but are also beginning to overlap.

Dr. Michael S. Turner, a theoretical astrophysicist at the University of Chicago, said the refined value for the Hubble constant represented "the first act in a grand cosmological drama that's going to play out in the next two decades."

In the drama, cosmologists will be trying to answer some profound questions about the amount and nature of matter constituting the universe, the possibility that a strange form of energy is speeding up the cosmic expansion, what was the big bang itself and how will it all end. Uncertainties about the density of matter in the universe and the existence of a mysterious energy, called the cosmological constant, contribute to the wide range of age estimates drawn from the newly determined value of the Hubble constant.

Dr. Freedman, for example, explained that an estimated age of 12 billion years was valid even if the universe is relatively lightweight, as recent observations suggest. If there is indeed a cosmological constant, a kind of antigravity force acting to speed up cosmic expansion, that would increase the age estimate, perhaps up to 15 billion.

The expansion and age of the universe were also subjects of two articles in the current issue of the journal *Science*.

In an analysis of recent observations and other research, Dr. Charles H. Lineweaver, an astrophysicist at the University of New South Wales in Sydney, Australia, put the age for the universe at 13.4 billion years, plus or minus 1.6 billion. The estimate took into account low values for the mass density of the universe and the presence of a "missing energy" source like the cosmological constant.

In a review of "the state of the universe," Dr. Neta A. Bahcall of Princeton University and three colleagues noted growing evidence that is "forcing us to consider the possibility that some cosmic dark energy exists that opposes the self-attraction of matter and causes the expansion of the universe to accelerate."

Such findings suggest the future in store for the universe. It will probably keep on expanding, almost to the point of vanishing. As Dr. Turner of Chicago said, "The universe will become a bleaker and bleaker place," and as the galaxies speed away from one another and the stars grow dimmer, "in 400 to 500 billion years, we're only going to be able to see a few neighboring galaxies."

—*May 26, 1999*

Galaxies' Vastness Surprises Scientists

By JAMES GLANZ

Using a technique akin to overlaying thousands of faint X-ray images to create one sharp picture, astronomers have discovered that typical galaxies may be twice as large and contain twice as much mass as suggested by previous measurements. The new observations, which have emerged from a five-year census of the heavens called the Sloan Digital Sky Survey, indicate that an average galaxy extends invisibly for well over a million light-years into space and weighs the equivalent of at least five trillion suns.

By comparison, the glittering disk of stars in a typical galaxy like the Milky Way, where Earth is located, stretches only about 50,000 light-years from center to edge and contains about 100 billion stars. The finding raises the possibility that the Milky Way and its nearest galactic neighbor—Andromeda, more than two million light-years away—actually brush against each other in the remote darkness of space.

The observations probed not the bright disk but the mysterious cloud of non-luminous or "dark" matter, called a halo, that surrounds the disk. Although the dark matter is invisible and scientists have been unable to determine its precise nature, they know of its existence because of the gravitational pull it exerts on ordinary matter—the stars and gas in galaxy disks.

The Sloan measurements relied on another effect of a halo's gravity: Like a lens or a clear marble, it bends light rays that pass through it. One consequence is that extremely distant background galaxies, as seen through a foreground galaxy's halo, appear distorted. Because the effect is so minute, however, the Sloan astronomers observed it in about 30,000 foreground galaxies and added the data together.

Then, working backward to deduce what kind of halo would produce those distortions, the astronomers found that the dark matter trails deep into space, even farther than earlier studies by other methods had suggested.

"The result from the Sloan analysis is that the halos of galaxies are enormous," said Dr. Tereasa Brainerd, an astronomer at Boston University, whose earlier work helped develop the technique, "perhaps extending 10 to 20 times the size of the visible regions of the galaxies."

Dr. Vera C. Rubin, the astronomer at the Carnegie Institution of Washington who discovered that galaxies must contain dark matter, added that the Sloan observations had taken a "conceptually brilliant" idea, whose application had been disappointing in the past, and turned it into a practical technique.

"For years, when I would talk about dark matter, I had to explain why it hadn't been seen that way," Dr. Rubin said, adding that the technique was "observationally very difficult."

The group's report on the finding, which has been submitted to the *Astronomical Journal,* is also likely to be seen as an early sign that the unusual $80 million project will help change the way astronomy is done.

The Sloan survey is an ambitious effort to collect and categorize hundreds of millions of objects over a vast swath of the northern sky. To complete the project—which involves nine institutions in the United States, Germany and Japan—scientists and engineers have constructed a special telescope, the largest electronic camera ever built, and other apparatus at Apache Point, New Mexico.

Rather than picking out interesting objects and studying them one by one, the Sloan telescope sweeps across the sky, scooping up high-quality data on everything in its view. Later, any astronomer who wishes can obtain data on particular objects and then analyze them.

"We've gotten into a mode in our country, and pretty much worldwide, of using a telescope for a night or three nights, looking at an object and going home," said Dr. Tony Tyson, the astronomer at Bell Laboratories, Lucent Technologies who first proposed the idea of using the lensing of background galaxies to determine the size of halos.

With its entirely different philosophy, Dr. Tyson said, the Sloan survey "is going to come up with some fantastic answers by the time the experiment is complete."

Indeed, the galaxy results have emerged from just 3 percent of the data that the survey is expected to compile over five years. And that is not the only discovery to pop out of the data, which is eventually expected to yield a tremendous haul of celestial oddities, a detailed "field guide" to all the varieties of ordinary objects in the heavens, and a three-dimensional map of a million galaxies in the Milky Way's corner of the cosmos.

Dr. Michael Turner, a cosmologist at the University of Chicago and the Fermi National Accelerator Laboratory, who is the Sloan survey's spokesman, said: "It's kind of like a wall of water coming at you. When you feel the spray, you won't have much time to catch your breath before the wall of water comes." He added that with the first results, "we're feeling the spray."

The results on galaxy halos give some insight into the multifarious ways astronomers will use the coming tidal wave of Sloan data.

In the more than two decades since Dr. Rubin discovered the galactic dark matter from studying the rotation patterns of spiral galaxies, astrophysicists and cosmologists have learned indirectly, through studies of pristine traces of the big bang in distant space and other methods, that at least 90 percent of the matter in the universe is dark.

Some astronomers have suggested on general grounds that all the dark matter might be found in galaxy halos, while others have theorized that it might be spread out in the space between galaxies. But turning up direct evidence has proved elusive. Since galaxy disks are comparatively small, it is difficult to use their motions as probes of anything but the inmost portions of the surrounding halos.

Dr. Penny Sackett of the Kapteyn Astronomical Institute in the Netherlands, who recently published a review on the topic, said that a different kind of observation, involving small "satellite" galaxies that felt the tug of a larger galaxy's halo, could probe its existence farther into space.

That study, led by Dr. Dennis F. Zaritsky of the University of Arizona, suggested that halos extend to at least 600,000 light-years, but could not rule out the possibility that they go farther.

"We had a situation where maybe they're large, and maybe they're not so large," Dr. Sackett said.

The work that formed the basis for the new Sloan report, which was led by Dr. Philippe Fischer and Dr. Timothy McKay, both of the University of Michigan, sought to probe the dark matter halos more directly and deeper into the space around galaxies.

The researchers turned to the effect called gravitational lensing. According to Einstein's theory of relativity, gravity bends light just as an ordinary glass lens does. The effect was first seen in 1919, when astronomers saw a slight deflection in the apparent position of a star over the limb of the sun during an eclipse, an early step in verifying the now-accepted theory.

"Now that it's well understood, instead of using the lensing to test the theory, we use the lensing to measure the mass of objects," Dr. McKay said.

When rays from a distant source of light pass through a distributed mass, like a halo, rather than past its edge, the source gets smeared into little arcs rather than simply deflected. So the Sloan team searched their data for relatively nearby galaxies with more distant galaxies behind them, lying along almost the same line of sight.

The astronomers immediately faced a colossal technical challenge: The expected smearing would be so slight that it would stretch out any individual background galaxy by only about half a percent. Because galaxies are irregularly shaped anyway, the only option was to add up the tiny smearing in a statistical sample of many background galaxies shining through many halos.

And it is in exactly such a study that the Sloan survey excels, since it collects great numbers of both kinds of galaxies: those in the foreground whose halos do the lensing, and those in the background whose shapes are slightly smeared. In the early data, the team found an average of 50 background galaxies whose light pierced the halo of each of 30,000 foreground galaxies.

"That's where the Sloan wins out—it's simply the vast number of objects in their data set," said Dr. Brainerd of Boston University.

The results showed that typical halos extended to at least 1.3 million light-years into space from the luminous galactic disks, more than double the previous limit.

In one sense, the results only deepen the mystery of the dark matter. They shed no new light on exactly what it is, although Dr. Turner said that the new findings were consistent with the dark matter's being a strange sort of subatomic particle left over from the explosive birth of the universe.

But aside from the light the results should shed on the formation of galaxies and galaxy clusters, they also let scientists complete a bit of cosmic bookkeeping, Dr. Turner said.

Unrelated studies of elements that were produced through nuclear reactions in the big bang, the explosion in which the universe is thought to have been born, combined with observations of large galaxy clusters, have allowed cosmologists to estimate the total amount of matter in the universe. Astronomers have had a hard time finding that much matter in the heavens, but the larger halos may mean that all the matter has now been accounted for. The vast majority of that matter appears to be of the dark variety, inhabiting the halos.

"We can now balance the books and say that we know that the dark matter in the universe is associated with galaxies," Dr. Turner said. "It's a bit mind-blowing. The halos of galaxies are really big."

"I'm quite excited about this result," said Dr. Joshua Frieman, a Sloan collaborator at Fermilab. "Considering it's based on a few percent of the total amount of data we'll have, it's a real indicator of what we'll be able to accomplish in this area of the survey."

That same statistical power is what astronomers in other fields of study will be looking forward to mining as the Sloan survey scans the heavens, Dr. Rubin said.

In addition to the University of Chicago and Fermilab, institutions taking part in the Sloan survey include Princeton, Johns Hopkins, the Institute for Advanced Study, the United States Naval Observatory, a collaboration called the Japan Participation Group, the University of Washington, the Max Planck Institute for Astronomy in Heidelberg, Germany, and several institutions involved in only selected parts.

Already, said Dr. Donald G. York, a collaborator at the University of Chicago, who is a founder of the project, discoveries in the Sloan data have doubled the number of extremely distant quasars, or beacons near the edge of the visible universe that burn with the intensity of hundreds of billions of suns.

"I like to say it took 35 years to find the first dozen and 6 months to find the second dozen," Dr. York said of the discoveries, which have emerged as the Sloan telescope has operated only sporadically as part of a commissioning period.

"It's really a look at the way astronomers of the future will work," Dr. Rubin said.

—December 14, 1999

In the Dark Matter Wars, Wimps Beat Machos

By JAMES GLANZ

The last hopes for a universe filled with familiar stuff behaving in comprehensible ways died in Marina del Rey in Los Angeles County. It showed in the head-turning attire of at least two of the scientists who carried the news to a major conference on cosmology in a hotel across the street from a restaurant called Killer Shrimp.

But never mind the shrimp. Dr. Joel Primack spoke about the overall contents of the universe while wearing a midnight-blue jacket and a tie that bore the likeness of Van Gogh's "Starry Night," in which some ominous presence between the stars overwhelms the visible bodies themselves. In her own talk, Dr. Katherine Freese heralded "The Death of Baryonic Dark Matter" in an all-black pantsuit.

Baryons are the ordinary particles, like protons and neutrons, of which stars, asteroids, comets, planets, people and textiles are made. By measuring the gravitational pull of some unknown "dark matter" on the visible stars and galaxies, astronomers have determined that this mysterious material which seems to permeate the universe has a weight that is 60 times that of the stars and 7 times that of all baryons, including gas and solid material in space.

By combining calculations with observational data, Dr. Freese reported that she had eliminated the last shred of possibility that the dark matter is ordinary, baryonic material. Dr. Primack, his tie providing mute acknowledgment, added that an unknown form of dark energy permeating space is apparently pushing against gravity on large scales. There is so much of this dark energy, sometimes called quintessence or the cosmological constant, that it in effect has a weight almost twice that of all matter, dark or visible.

Cosmologists have seen evidence for the dark matter and dark energy for some time. What has changed is that, for now, there seems to be little chance of escaping the conclusion that the stuff of which we are made amounts to no more than cuff links on the cosmic tuxedo, as it were.

"The bulk of the energy density in the universe seems to be dark energy," said Dr. Primack, who is at the University of California at Santa Cruz. "And the big question is, What is the dark energy?"

A similar state of affairs holds for the dark matter, he said. But despite the lingering mysteries, he said, most cosmologists are far from put off by a universe filled with dark, strange and unknown stuff. "Most professionals in this field have had a long time to get used to these ideas," Dr. Primack said.

He added that the professionals were pleased that reigning theories of how the universe was born, generated chemical elements and formed structures like galaxies mesh remarkably well with the observed amounts of ordinary matter, dark matter and dark energy in space. Many of those theories deal only with the overall balance sheets of matter and energy and not their details.

While it is true that those dark details remain to be explained, he said, "I tend to think of this more as a problem for fundamental physicists, and less for me."

As recently as a few years ago, a group of astronomers called the Macho collaboration thought they had found the dark matter in chunks of presumably baryonic material orbiting in a huge cloud, or halo, around the Milky Way. (*Macho* is an acronym for the "massive compact halo objects" in question.)

The main evidence came from observing brief brightenings of stars outside the Milky Way. The Macho collaboration concluded that these brightenings had occurred when the objects, assumed to be in the halo, passed directly between Earth and distant stars and magnified the light of the stars behind them. The gravity of the objects, each roughly the mass of our sun, acted as a kind of lens, just as Einstein's equations of relativity say that they should.

The discoveries seemed to fit nicely with the long-standing observation that galaxies like the Milky Way spin too fast to be held together only by the gravity of visible matter—a main argument for the existence of dark matter in the universe. Was it possible that the dim objects, suggested by the lensing observations, could be the galactic dark matter?

The answer turns out to be no, Dr. Freese said in her talk. The so-called "white dwarf" stars, which those objects would almost certainly have to be, would have to be the remnants of a population of stars that over their lifetimes would have spewed out substantial amounts of gas, heavy elements and infrared radiation—none of which are seen.

In addition, she said, such a large quantity of baryons in galaxies like the Milky Way could not be reconciled with accepted theories of how elements formed in the big bang explosion in which the universe is thought to have been born.

"It's looking very likely that 50 to 90 percent of our galaxy is nonbaryonic," said Dr. Freese, a University of Michigan physicist.

In the same session, which was held at the Fourth International Symposium on Sources and Detection of Dark Matter in the Universe, members of the Macho collaboration and a related French collaboration called EROS conceded that the high rate at which they initially observed lensing events was probably a statistical fluke, at best.

"There are not enough machos in the galactic halo," said Dr. Eric Aubourg of EROS. "It looks like everybody agrees on this."

So what is the dark matter? As tempting as the supposition may be, it could not be killer shrimp, which would presumably be baryonic. Theorists believe it could exist in clouds of nonbaryonic particles called WIMPs, for weakly interacting massive particles. At the same conference, two experimental groups—one based in Rome and one at 10 American institutions—presented conflicting evidence on whether WIMPs have been detected. But that argument remains for the future.

While some astronomers presented fresh evidence that dark energy is filling the space between galaxies, counteracting their gravity and accelerating the expansion of the universe, Dr. Edwin L. Turner of Princeton University pointed out that a few observations can still be found that do not fit the concordance that cosmologists have found with their theories.

And not everyone is impressed with the elegance of the dark new universe. "It's a bit ugly, isn't it?" said Dr. Turner. "A little of this, a little of that—I think this would have been considered appalling a decade ago."

—*February 29, 2000*

Before the Big Bang, There Was . . . What?

By DENNIS OVERBYE

What was God doing before he created the world? The philosopher and writer (and later saint) Augustine posed the question in his *Confessions* in the fourth century, and then came up with a strikingly modern answer: Before God created the world, there was no time and thus no "before." To paraphrase Gertrude Stein, there was no "then" then.

Until recently no one could attend a lecture on astronomy and ask the modern version of Augustine's question—What happened before the big bang?—without receiving the same frustrating answer, courtesy of Albert Einstein's general theory of relativity, which describes how matter and energy bend space and time.

If we imagine the universe shrinking backward, like a film in reverse, the density of matter and energy rises toward infinity as we approach the moment of origin. Smoke pours from the computer, and space and time themselves dissolve into a quantum "foam." "Our rulers and our clocks break," explained Dr. Andrei Linde, a cosmologist at Stanford University. "To ask what is before this moment is a self-contradiction."

But lately, emboldened by progress in new theories that seek to unite Einstein's lordly realm with the unruly quantum rules that govern subatomic physics—so-called quantum gravity—Dr. Linde and his colleagues have begun to edge their speculations closer and closer to the ultimate moment and, in some cases, beyond it.

Some theorists suggest that the big bang was not so much a birth as a transition, a "quantum leap" from some formless era of imaginary time, or from nothing at all. Still others are exploring models in which cosmic history begins with a collision with a universe from another dimension.

All this theorizing has received a further boost of sorts from recent reports of ripples in a diffuse radio glow in the sky, thought to be the remains of the big bang fireball itself. These ripples are consistent with a popular theory, known as inflation, that the universe briefly speeded its expansion under the influence of a violent antigravitational force, when it was only a fraction of a fraction of a nanosecond old. Those ripples thus provide a useful check on theorists' imaginations. Any theory of cosmic origins that does not explain this phenomenon, cosmologists agree, stands little chance of being right.

Fortunately or unfortunately, that still leaves room for a lot of possibilities.

"If inflation is the dynamite behind the big bang, we're still looking for the match," said Dr. Michael Turner, a cosmologist at the University of Chicago. The only thing that all the experts agree on is that no idea works—yet. Dr. Turner likened cosmologists to jazz musicians collecting themes that sound good for a work in progress: "You hear something and you say, oh yeah, we want that in the final piece."

One answer to the question of what happened before the big bang is that it does not matter because it does not affect the state of our universe today. According to a theory known as eternal inflation, put forward by Dr. Linde in 1986, what we know as the big bang was only one out of many in a chain reaction of big bangs by which the universe endlessly reproduces and reinvents itself. "Any particular part of the universe may die, and probably will die," Dr. Linde said, "but the universe as a whole is immortal."

Dr. Linde's theory is a modification of the inflation theory that was proposed in 1980 by Dr. Alan Guth, a physicist. He considered what would happen if, as the universe was cooling during its first violently hot moments, an energy field known as the Higgs field, which interacts with particles to give them their masses, was somehow, briefly, unable to release its energy.

Space, he concluded, would be suffused with a sort of latent energy that would violently push the universe apart. In an eyeblink the universe would double some 60 times over, until the Higgs field released its energy and filled the out-rushing universe with hot particles. Cosmic history would then ensue.

Cosmologists like inflation because such a huge outrush would have smoothed any gross irregularities from the primordial cosmos, leaving it homogeneous and geometrically flat. Moreover, it allows the whole cosmos to grow from next to nothing, which caused Dr. Guth to dub the universe "the ultimate free lunch."

Subsequent calculations ruled out the Higgs field as the inflating agent, but there are other inflation candidates that would have the same effect. More important, from the pre–big bang perspective, Dr. Linde concluded, one inflationary bubble would sprout another, which in turn would sprout even more. In effect each bubble would be a new big bang, a new universe with different characteristics and perhaps even different dimensions. Our universe would merely be one of them.

"If it starts, this process can keep happening forever," Dr. Linde explained. "It can happen now, in some part of the universe."

The greater universe envisioned by eternal inflation is so unimaginably large, chaotic and diverse that the question of a beginning to the whole shebang becomes almost irrelevant. For cosmologists like Dr. Guth and Dr. Linde, that is in fact the theory's lure.

"Chaotic inflation allows us to explain our world without making such assumptions as the simultaneous creation of the whole universe from nothing," Dr. Linde said in an e-mail message.

Questions for Eternity:
Trying to Imagine the Nothingness

Nevertheless, most cosmologists, including Dr. Guth and Dr. Linde, agree that the universe ultimately must come from somewhere, and that nothing is the leading candidate.

As a result, another tune that cosmologists like to hum is quantum theory. According to Heisenberg's uncertainty principle, one of the pillars of this paradoxical world, empty space can never be considered really empty; subatomic particles can flit in and out of existence on energy borrowed from energy fields. Crazy as it sounds, the effects of these quantum fluctuations have been observed in atoms, and similar fluctuations during the inflation are thought to have produced the seeds around which today's galaxies were formed.

Could the whole universe likewise be the result of a quantum fluctuation in some sort of primordial or eternal nothingness? Perhaps, as Dr. Turner put it, "Nothing is unstable."

The philosophical problems that plague ordinary quantum mechanics are amplified in so-called quantum cosmology. For example, as Dr. Linde points out, there is a chicken-and-egg problem. Which came first—the universe or the law governing it? Or, as he asks, "If there was no law, how did the universe appear?"

One of the earliest attempts to imagine the nothingness that is the source of everything came in 1965 when Dr. John Wheeler and Dr. Bryce DeWitt, now at the University of Texas, wrote down an equation that combined general relativity and quantum theory. Physicists have been arguing about it ever since.

The Wheeler-DeWitt equation seems to live in what physicists have dubbed "superspace," a sort of mathematical ensemble of all possible universes, ones that live only five minutes before collapsing into black holes and ones full of red stars that live forever, ones full of life and ones that are empty deserts, ones in which the constants of nature and perhaps even the number of dimensions are different from our own.

In ordinary quantum mechanics, an electron can be thought of as spread out over all of space until it is measured and observed to be at some specific location. Likewise, our own universe is similarly spread out over all of superspace until it is somehow observed to have a particular set of qualities and laws. That raises another of the big questions: Since nobody can step outside the universe, who is doing the observing?

Dr. Wheeler has suggested that one answer to that question may be simply us, acting through quantum-mechanical acts of observation, a process he calls "genesis by observership."

"The past is theory," he once wrote. "It has no existence except in the records of the present. We are participators, at the microscopic level, in making that past, as well as the present and the future." In effect, Dr. Wheeler's answer to Augustine is that we are collectively God and that we are always creating the universe.

Another option, favored by many cosmologists, is the so-called many worlds interpretation, which says that all of these possible universes actually do exist. We just happen to inhabit one whose attributes are friendly to our existence.

The End of Time:
Just Another Card in the Big Deck

Yet another puzzle about the Wheeler-DeWitt equation is that it makes no mention of time. In superspace everything happens at once and forever, leading some physicists to question the role of time in the fundamental laws of nature. In his book *The End of Time,* published to coincide with the millennium, Dr. Julian Barbour, an independent physicist and Einstein scholar in England, argues that the universe consists of a stack of moments, like the cards in a deck, that can be shuffled and reshuffled arbitrarily to give the illusion of time and history.

The big bang is just another card in this deck, along with every other moment, forever part of the universe. "Immortality is here," he writes in his book. "Our task is to recognize it."

Dr. Carlo Rovelli, a quantum gravity theorist at the University of Pittsburgh, pointed out that the Wheeler-DeWitt equation doesn't mention space either, suggesting that both space and time might turn out to be artifacts of something deeper. "If we take general relativity seriously," he said, "we have to learn to do physics without time, without space, in the fundamental theory."

While admitting that they cannot answer these philosophical questions, some theorists have committed pen to paper in attempts to imagine quantum creation mathematical rigor.

Dr. Alexander Vilenkin, a physicist at Tufts University in Somerville, Massachusetts, has likened the universe to a bubble in a pot of boiling water. As in water, only bubbles of a certain size will survive and expand; smaller ones collapse. So, in being created, the universe must leap from no size at all—zero radius, "no space and no time"—to a radius large enough for inflation to take over without passing through the in-between sizes, a quantum-mechanical process called "tunneling."

Dr. Stephen Hawking, the University of Cambridge cosmologist and best-selling author, would eliminate this quantum leap altogether. For the last 20 years he and a series of collaborators have been working on what he calls a "no boundary proposal." The boundary of the universe is that it has no boundary, Dr. Hawking likes to say.

One of the keys to Dr. Hawking's approach is to replace time in his equations with a mathematical conceit called imaginary time; this technique is commonly used in calculations regarding black holes and in certain fields of particle physics, but its application to cosmology is controversial.

The universe, up to and including its origin, is then represented by a single conical-shaped mathematical object, known as an instanton, that has four spatial dimensions (shaped roughly like a squashed sphere) at the big bang end and then shifts into real time and proceeds to inflate. "Actually it sort of bursts and makes an infinite universe," said Dr. Neil Turok, also from University of Cambridge. "Everything for all future time is determined, everything is implicit in the instanton."

Unfortunately the physical meaning of imaginary time is not clear. Beyond that, the approach produces a universe that is far less dense than the real one.

The Faith of Strings:
Theorists Bring on the "Brane" Worlds

But any real progress in discerning the details of the leap from eternity into time, cosmologists say, must wait for the formulation of a unified theory of quantum gravity that succeeds in marrying Einstein's general relativity to quantum mechanics—two views of the world, one describing a continuous curved space-time, the other a discontinuous random one—that have been philosophically and mathematically at war for almost a century. Such a theory would be able to deal with the universe during the cauldron of the big bang itself, when even space and time, theorists say, have to pay their dues to the uncertainty principle and become fuzzy and discontinuous.

In the last few years, many physicists have pinned their hopes for quantum gravity on string theory, an ongoing mathematically labyrinthine effort to portray nature as comprising tiny wiggly strings or membranes vibrating in 10 or 11 dimensions.

In principle, string theory can explain all the known (and unknown) forces of nature. In practice, string theorists admit that even their equations are still only approximations, and physicists outside the fold complain that the effects of "stringy physics" happen at such high energies that there is no hope of testing them in today's particle accelerators. So theorists have been venturing into cosmology, partly in the hopes of discovering some effect that can be observed.

The big bang is an obvious target. A world made of little loops has a minimum size. It cannot shrink beyond the size of the string loops themselves, Dr. Robert Brandenberger, now at Brown, and Dr. Cumrun Vafa, now at Harvard, deduced in 1989. When they used their string equations to imagine space shrinking smaller than a certain size, Dr. Brandenberger said, the universe acted instead as if it were getting larger. "It looks like it is bouncing from a collapsing phase."

In this view, the big bang is more like a transformation, like the melting of ice to become water, than a birth, explained Dr. Linde, calling it "an interesting idea that should be pursued." Perhaps, he mused, there could be a different form of space and time before the big bang. "Maybe the universe is immortal," he said. "Maybe it just changes phase. Is it nothing? Is it a phase transition? These are very close to religious questions."

Work by Dr. Brandenberger and Dr. Vafa also explains how it is that we only see 3 of the 9 or 10 spatial dimensions the theory calls for. Early in time the strings, they showed, could wrap around space and strangle most of the spatial dimensions, keeping them from growing.

In the last few years, however, string theorists have been galvanized by the discovery that their theory allows for membranes of various dimensions ("branes" in string jargon) as well as strings. Moreover, they have begun to explore the possibility that at least one of the extra dimensions could be as large as a millimeter, which is gigantic in string physics. In this new cosmology, our world is a three-dimensional island, or brane floating in a five-dimensional space, like a leaf in a fish tank. Other branes might be floating nearby. Particles like quarks and electrons and forces like electromagnetism are stuck to the brane, but gravity is not, and thus the brane worlds can exert gravitational pulls on each other.

"A fraction of a millimeter from you is another universe," said Dr. Linde. "It might be there. It might be the determining factor of the universe in which you live."

Worlds in Collision:
A New Possibility Is Introduced

That other universe could bring about creation itself, according to several recent theories. One of them, called branefall, was developed in 1998 by Dr. Georgi Dvali of New York University and Dr. Henry Tye, of Cornell. In it the universe emerges from its state of quantum formlessness as a tangle of strings and cold empty membranes stuck together. If, however, there is a gap between the branes at some point, the physicists said, they will begin to fall together.

Each brane, Dr. Dvali said, will experience the looming gravitational field of the other as an energy field in its own three-dimensional space and will begin to inflate rapidly, doubling its size more than a thousand times in the period it takes for the branes to fall together. "If there is at least one region where the branes are parallel, those regions will start an enormous expansion while other regions will collapse and shrink," Dr. Dvali said.

When the branes finally collide, their energy is released and the universe heats up, filling with matter and heat, as in the standard big bang.

In spring 2001, four physicists proposed a different kind of brane clash that they say could do away with inflation, the polestar of big bang theorizing for 20 years, altogether. Dr. Paul Steinhardt, one of the fathers of inflation, and his student Justin Khoury, both of Princeton, Dr. Burt Ovrut of the University of Pennsylvania and Dr. Turok call it the ekpyrotic universe, after the Greek word *ekpyrosis,* which denotes the fiery death and rebirth of the world in Stoic philosophy.

The ekpyrotic process begins far in the indefinite past with a pair of flat empty branes sitting parallel to each other in a warped five-dimensional space—a situation, they say, that represents the simplest solution of Einstein's equations in an advanced version of string theory. The authors count it as a point in their favor that they have not assumed any extra effects that do not already exist in that theory. "Hence we are proposing a potentially realistic model of cosmology," they wrote in their paper.

The two branes, which form the walls of the fifth dimension, could have popped out of nothingness as a quantum fluctuation in the even more distant past and then drifted apart.

At some point, perhaps when the branes had reached a critical distance apart, the story goes, a third brane could have peeled off the other brane and begun falling toward ours. During its long journey, quantum fluctuations would ripple the drifting brane's surface, and those would imprint the seeds of future galaxies all across our own brane at the moment of collision. Dr. Steinhardt offered the theory at an astronomical conference in Baltimore in April 2001.

In the subsequent weeks the ekpyrotic universe has been much discussed. Some cosmologists, particularly Dr. Linde, have argued that in requiring perfectly flat and parallel branes the ekpyrotic universe required too much fine-tuning.

In a critique, Dr. Linde and his coauthors suggested a modification they called the "pyrotechnic universe."

Dr. Steinhardt admitted that the ekpyrotic model started from a very specific condition, but that it was a logical one. The point, he said, was to see if the universe could begin in a long-lived quasi-stable state "starkly different from inflation." The answer was yes. His coauthor, Dr. Turok, pointed out, moreover, that inflation also requires fine-tuning to produce the modern universe, and physicists still don't know what field actually produces it.

"Until we have solved quantum gravity and connected string theory to particle physics, none of us can claim victory," Dr. Turok said.

In the meantime, Augustine sleeps peacefully.

—*May 22, 2001*

A New View of Our Universe: Only One of Many

By DENNIS OVERBYE

Astronomers have gazed out at the universe for centuries, asking why it is the way it is. But lately a growing number of them are dreaming of universes that never were and asking, why not?

Why, they ask, do we live in 3 dimensions of space and not 2, 10 or 25? Why is a light ray so fast and a whisper so slow? Why are atoms so tiny and stars so big? Why is the universe so old? Does it have to be that way, or are there places, other universes, where things are different?

Once upon a time (only a century ago), a few billion stars and gas clouds smeared along the Milky Way were thought to encompass all of existence, and the notion of understanding it was daunting—and hubristic—enough. Now astronomers know that galaxies are scattered like dust across the cosmos. And understanding them might require recourse to an even broader canvas, what they sometimes call a "multiverse."

For some cosmologists, that means universes sprouting from one another in an endless geometric progression, like mushrooms upon mushrooms upon mushrooms, or baby universes hatched inside black holes.

For example, Dr. Max Tegmark, a University of Pennsylvania cosmologist, has posited at least four different levels of universes, ranging from the familiar (impossibly distant zones of our own universe) to the strange (space-times in which the fundamental laws of physics are different).

Dr. Martin Rees, a University of Cambridge cosmologist and the Astronomer Royal, said contemplating these alternate universes could help scientists distinguish which features of our own universe are fundamental and necessary and which are accidents of cosmic history. "It's all science, but science for the 21st century, to seek the answers to these questions," Dr. Rees said, adding that he is often accused of believing in other universes.

"I don't believe," he said, "but I think it's part of science to find out."

Some cosmologists now say the realm we call the observable universe—roughly 14 billion light-years deep of galaxies and stars—could be only a small patch of a vast bubble or "pocket" in a much vaster ensemble bred endlessly in a chain of big bangs.

The idea, they say, is a natural extension of the theory of inflation, introduced by Dr. Alan Guth, now at the Massachusetts Institute of Technology, in 1980. That theory asserts that when the universe was less than a trillionth of a trillionth of a second old it underwent a brief hyperexplosive growth spurt fueled by an antigravitational force embedded in space itself, a possibility suggested by theories of modern particle physics.

Because inflation can grow a whole universe from about an ounce of primordial stuff, Dr. Guth likes to refer to the universe as "the ultimate free lunch." But Dr. Guth and various other theorists—including Dr. Andrei Linde of Stanford, Dr. Alexander Vilenkin of Tufts and Dr. Paul Steinhardt of Princeton—have suggested that it may be an endless one as well. Once inflation starts anywhere, it will keep happening over and over again, they say, spawning a chain of universes, bubbles within bubbles, in a scheme that Dr. Linde called "eternal inflation."

"Once you've discovered it's easy to make a universe out of an ounce of vacuum, why not make a bunch of them?" asked Dr. Craig Hogan, a cosmologist at the University of Washington.

In fact, Dr. Guth said, "Inflation pretty much forces the idea of multiple universes upon us."

Moreover, there is no reason to expect that these universes will be identical. Even within our own bubble, tiny random nonuniformities in the primordial raw material would cause the cosmos to look different from place to place. If the universe is big enough, Dr. Tegmark and others say, everything that can happen will happen, so that if we could look out far enough we would eventually discover an exact replica of ourselves.

Moreover, cosmologists say, the laws of physics themselves, as experienced by creatures like ourselves, confined to four dimensions and the energy scales of ordinary life, could evolve differently in different bubble universes.

"Geography is now a much more interesting subject than you thought," Dr. John Barrow, a physicist at the University of Cambridge, observed.

Inflation has gained much credit with cosmologists, despite its strangeness, Dr. Guth noted, because it plays a vital role in calculations of the big bang that have been vindicated by the detection of the radio waves it produced. The prediction of other universes must therefore be taken seriously, he said.

Lucky Numbers:
Adjusting the Dials of Nature's Console

The prospect of this plethora of universes has brought new attention to a philosophical debate that has lurked on the edges of science for the last few decades, a debate over the role of life in the universe and whether its physical laws are unique—or, as Einstein once put it, "whether God had any choice."

Sprinkled through the Standard Model, the suite of equations that describe all natural phenomena, are various mysterious constants, like the speed of light or the masses of the elementary particles, whose value is not specified by any theory now known.

In effect, the knobs on nature's console have been set to these numbers. Scientists can imagine twiddling them, but it turns out that nature is surprisingly finicky, they say, and only a narrow range of settings is suitable for the evolution of complexity or Life as We Know It.

For example, much of the carbon and oxygen needed for life is produced by the fusion of helium atoms in stars called red giants.

But a change of only half a percent in the strength of the so-called strong force that governs nuclear structure would be enough to prevent those reactions from occurring, according to recent work by Dr. Heinz Oberhummer of Vienna University of Technology. The result would be a dearth of the raw materials of biology, he said.

Similarly, a number known as the fine structure constant characterizes the strength of electromagnetic forces. If it were a little larger, astronomers say, stars could not burn, and if it were only a little smaller, molecules would never form.

So is this a lucky universe, or what?

In 1974, Dr. Brandon Carter, a theoretical physicist then at Cambridge, now at the Paris Observatory in Meudon, pointed out that these coincidences were not just luck, but were rather necessary preconditions for us to be looking at the universe.

After all, we are hardly likely to discover laws that are incompatible with our own existence.

That insight is the basis of what Dr. Carter called the anthropic principle, an idea that means many things to many scientists. Expressed most emphatically, it declares that the universe is somehow designed for life. Or as the physicist Freeman Dyson once put it, "The universe in some sense must have known that we were coming."

This notion horrifies some physicists, who feel it is their mission to find a mathematical explanation of nature that leaves nothing to chance or "the whim of the Creator," in Einstein's phrase.

"It touches on philosophical issues that scientists oftentimes skirt," said Dr. John Schwarz, a physicist and string theorist at the California Institute of Technology. "There should be mathematical ways of understanding how nature works."

Dr. Steven Weinberg, the University of Texas physicist and Nobel laureate, referred to this so-called "strong" version of the anthropic principle as "little more than mystical mumbo jumbo" in a recent article in *The New York Review of Books*.

Sorting Universes: Finding a Home for the "A-Word"

Nevertheless, the "A-word" is popping up more and more lately, at conferences and in the scientific literature, often to the groans of particle physicists. The reason is the multiverse.

"It is possible that, as theoretical physics develops, it will present us with multiple universes," Dr. Weinberg said.

If different laws or physical constants prevail in other bubble universes, the conditions may not allow the existence of life or intelligence, he explained.

In that case the anthropic principle loses its mysticism and simply becomes a prescription for deciding which bubbles are capable of supporting life.

But many string theorists still resent the principle as an abridgment on their ambitions. The result has been a spirited debate about what physicists can expect from their theories.

"They have the pious hope that string theory will uniquely determine all the constants of nature," said Dr. Barrow, who wrote the 1984 book *The Cosmological Anthropic Principle* with Dr. Frank Tipler, a Tulane University physicist. The book argued that once life emerges in the universe it will never die.

In a recent paper titled "The Beginning of the End of the Anthropic Principle," three physicists—Dr. Gordon Kane of the University of Michigan, and Dr. Malcolm Perry and Dr. Anna Zytkow, both of Cambridge—argued that a unified theory of physics, as string theory purports to be, when finally formulated, would specify most of the constants of nature or specify relationships between them, leaving little room for anthropic arguments.

"The anthropic principle isn't as anthropic as people wanted," Dr. Kane said in an interview.

But in a rejoinder titled "Why the Universe Is Just So," Dr. Hogan of Washington argued that physics was replete with messy processes like quantum effects, which leave some aspects of reality and the laws of physics to chance.

According to string theory, he pointed out, the laws of physics that we mortals experience are low-energy, 4-dimensional shadows, of sorts, of a 10- or 11-dimensional universe. As a result, the so-called "fundamental constants" could look different in different bubbles.

Dr. Hogan admitted that this undermined some of the traditional aspirations of physics, writing, "at least some properties of the world might not have an elegant mathematical explanation, and we can try to guess which ones these are."

Even string theorists like Dr. Kane admit that, in the absence of a final form of the theory, they have no idea how many solutions there may be—one, many or even an infinite number—to the "final" string equations. Each one would correspond to a different condition of space-time, with a different set of physical constants.

"Any set that allows life to happen will have life," he said.

Dark Energies:
When the Numbers Just Don't Add Up

But even some of the most hard-core physicists, including Dr. Weinberg, suggest they may have to resort to the anthropic principle to explain one of the deepest mysteries looming like a headache over science: the discovery that the expansion of the universe seems to be speeding up, perhaps in a kind of low-energy reprise of an inflation episode 14 billion years ago.

Cosmologists suspect a repulsion or antigravity associated with space itself is propelling this motion. This force, known as the cosmological constant, was first proposed by Einstein back in 1917, and has been a problem ever since— "a veritable crisis," Dr. Weinberg has called it.

According to astronomical observations, otherwise undetectable energy— "dark energy"—accounts for about two-thirds of the mass-energy of the universe today, outweighing matter by two to one. But according to modern quantum physics, empty space should be seething with energy that would outweigh matter in the universe by far, far more, by a factor of at least 10^{60}. This mismatch has been called the worst discrepancy in the history of physics.

But that mismatch is crucial for life, as Dr. Weinberg first pointed out in 1987.

At the time there was no evidence for a cosmological constant and many physicists presumed that its magnitude was, in fact, zero.

In his paper, Dr. Weinberg used so-called anthropic reasoning to pin the value of any cosmological constant to between about one-tenth and a few times the density of matter in the universe. If it were any larger, he said, the universe would blow apart before galaxies had a chance to form, leaving no cradle for the stellar evolution of elements necessary for life or other complicated structures.

The measured value of the constant is about what would be expected from anthropic arguments, Dr. Weinberg said, adding that nobody knows enough about physics yet to tell whether there are other universes with other constants. He called the anthropic principle "a sensible approach" to the cosmological constant problem.

"We may wind up using the anthropic principle to satisfy our sense of wonder about why things are the way they are," Dr. Weinberg said.

Beyond the Dark:
Searching for Proof from Better Theories

For Dr. Rees, the Astronomer Royal, it is not necessary to observe other universes to gain some confidence that they may exist. One thing that will help, he explained, is a more precise theory of how the cosmological constant may vary and how it will affect life in the universe. We should live in a statistically typical example of the range of universes compatible with life, he explained. For example, if the cosmological constant was, say, 10 percent of the maximum value consistent with life, that would be acceptable, he said.

"If it was a millionth, that would raise eyebrows."

Another confidence builder would be more support for the theory of inflation, either in the form of evidence from particle physics theory or measurements of the cosmic big bang radiation that gave a more detailed model of what theoretically happened during that first trillionth of a trillionth of a second.

"If we had a theory then we would know whether there were many big bangs or one, and then we would know if the features we see are fixed laws of the universe or bylaws for which we can never have an explanation," Dr. Rees said.

In a talk in September 2002, at a cosmology conference in Chicago, Dr. Joseph Polchinski, from the Institute for Theoretical Physics at the University of California at Santa Barbara, speculated that there could be 10^{60} different solutions to the basic string equations, thus making it more likely that at least one universe would have a friendly cosmological constant.

Reminded that he had once joked about retiring if a cosmological constant was discovered, on the ground that the dreaded anthropic principle would be the only explanation, he was at first at a loss for words.

Later he said he hoped the range of solutions and possible universes permitted by string theory could be narrowed by astronomical observations and new theoretical techniques to the point where the anthropic principle could be counted out as an explanation.

"Life is still good," he said.

But Dr. Hogan said that multiple universes would have to be taken seriously if they came out of equations that science had faith in.

"You have to be open-minded," he said. "You can't impose conditions.

"It's the most scientific attitude," he added.

—*October 29, 2002*

Radio Telescope Proves
a Big Bang Prediction

By DENNIS OVERBYE

After 271 20-hour nights of staring at the Antarctic sky, a radio telescope at the South Pole has confirmed a critical prediction of the big bang theory of the origin of the universe, astronomers from the University of Chicago and the University of California announced here today.

The result reassured cosmologists that their theories of the universe were on track and pioneered a new technique that greatly increases cosmologists' ability to know what was going on in the early universe.

Using their telescope in effect as a pair of Polaroid sunglasses, the team, headed by Dr. John Carlstrom of Chicago, discovered that a faint radio haze thought to be the fading remnant of the big bang itself is slightly polarized. That is to say, its flickering electromagnetic fields that constitute light waves were not completely jumbled, vibrating in all different planes as they sped to Earth, like feathers sticking out at all angles at the end of an arrow. Rather, they showed a slight preference for one plane of vibration, as if all the feathers lined up.

That, theorists say, is the predicted signature of the last bounce of light from hot, electrified cosmic gases just as the universe was cooling to the point where atoms could form, 400,000 years after the universe was born.

"The prediction is bang-on," Dr. Carlstrom told a rapt audience of more than 200 cosmologists at a conference at the Adler Planetarium in Chicago. "We think we know the universe, but if the polarization is not there at the predicted level we're back to the drawing board."

The findings lend more credence to many of the more preposterous-sounding conclusions that have made headlines over the last few years, which have united famously fractious cosmologists in a view of a universe born 14 billion years ago—a view in which ordinary matter is swamped with mysterious dark matter and dark energy and is apparently accelerating into the cosmic night.

"We're stuck with a preposterous universe," Dr. Carlstrom said. Papers describing the work will be posted on the team's Web site, astro.uchicago.edu/dasi, he said.

Other cosmologists responded with a mixture of glee and relief, saying it would have been bigger news if the polarization had not been found. "If you had any doubts that this radiation is from the big bang, this should quash them," said Dr. Michael Turner, a cosmologist at the University of Chicago.

Dr. Martin Rees, a cosmologist at the University of Cambridge and the Astronomer Royal, who suggested looking for polarization in the cosmic radiation 34 years ago, said it was "gratifying" but emphasized that it was only the first step in a campaign to understand the early universe.

Dr. Max Tegmark, a cosmologist at the University of Pennsylvania, said, "In my opinion, this is huge and marks the beginning of a new era in cosmic microwave background research."

The observations also represented another notch in the belt for one of the most controversial ingredients in the standard model of the big bang. Known as *inflation,* it posits that the universe underwent a hyperexplosive growth spurt early in its existence. Although it is successful as a theory, astronomers admit that they do not know what caused inflation or precisely when it happened.

The new results are consistent with inflation but do not establish it as unquestionable truth. Future polarization measurements, with more sensitive radio telescopes, Dr. Carlstrom said, might detect the roiling of space left behind by its wrenching outburst and confirm that it had happened and when. "We can go from checking inflation to actually testing it," Dr. Carlstrom said.

The cosmic radiation has transfixed astronomers since it was discovered in 1965 by two Bell Laboratories radio astronomers, Dr. Arno Penzias and Dr. Robert Wilson; they later received the Nobel Prize. According to theoretical calculations, this radiation is in effect a snapshot of the universe at an age of 400,000 years. But water vapor in the atmosphere and the faintness of the microwaves makes them hard to study except from satellites or balloons or very high altitude observatories. The South Pole instrument, known as DASI, for Degree Angular Scale Interferometer, is about two miles high, Dr. Carlstrom said, and since it is so cold in Antarctica, the air holds very little water vapor.

Since 1992 NASA's COBE (COsmic Background Explorer) satellite and a host of smaller instruments have examined the cosmic background and confirmed that faint ripples or blotches—the seeds of future cosmic structures like galaxies—follow the pattern predicted by inflationary models. But while mapping these blotches reveals the distribution of matter at that time in the universe, Dr. Carlstrom said, mapping the polarization of the cloud tells how the material is actually moving and has the potential of containing much more information. "It's like going from a black-and-white television to color," he said.

For the last year, polarization measurements have been the Next Big Thing in cosmology, with telescopes like DASI being converted to that purpose. Dr. Carlstrom's group won the race today, but the bigger race is only beginning, cosmologists say. At least a tenfold increase in sensitivity is needed to detect the signature of inflation in the cosmic background, so-called gravitational waves that would ripple space itself, according to Einstein's general theory of relativity. Dr. Turner has called these the "smoking gun" signature of inflation. Most experts agree that a new generation of satellites and telescopes will be needed to pull it off, but they seemed invigorated by the challenge.

"It's so exciting to see if we could test inflation," Dr. Carlstrom said, "to be able to look at the universe at 10^{-30} seconds."

—September 20, 2002

Cosmos Sits for Early Portrait, Gives Up Secrets

By DENNIS OVERBYE

The most detailed and precise map yet produced of the universe just after its birth confirms the big bang theory in triumphant detail and opens new chapters in the early history of the cosmos, astronomers said yesterday.

It reveals the emergence of the first stars in the cosmos, only 200 million years after the big bang, some half a billion years earlier than theorists had thought, and gives a first tantalizing hint at the physics of the "dynamite" behind the big bang.

Astronomers said the map results lent impressive support to the strange picture that has emerged recently: the universe is expanding at an ever-faster rate, pushed apart by a mysterious "dark energy."

By comparing their data with other astronomical observations, the astronomers have also made far more precise calculations of the basic parameters that characterize the universe, including its age, geometry, composition and weight.

In a nutshell, the universe is 13.7 billion years old, plus or minus 1 percent; a recent previous estimate had a margin of error three times as much. By weight it is 4 percent atoms, 23 percent dark matter—presumably undiscovered elementary particles left over from the big bang—and 73 percent dark energy. And it is geometrically "flat," meaning that parallel lines will not meet over cosmic scales.

The result, the astronomers said, is a seamless and consistent history of the universe, from its first few seconds, when it was a sizzling soup of particles and energy, to the modern day and a sky beribboned with chains of pearly galaxies inhabited by at least one race of puzzled and ambitious bipeds.

The map, compiled by a satellite called the Wilkinson Microwave Anisotropy Probe, shows the slight temperature variations in a haze of radio microwaves believed to be the remains of the fires of the big bang. Cosmologists said the map would serve as the basis for studying the universe for the rest of the decade.

"We have laid the cornerstone of a unified coherent theory of the cosmos," said Dr. Charles L. Bennett, an astronomer at the Goddard Space Flight Center in Greenbelt, Maryland, who led an international team that built the satellite and analyzed the results.

The satellite was launched on June 30, 2001, and has been orbiting Earth and recording cosmic emanations from a point on the other side of the moon.

The satellite is the successor to NASA's Cosmic Background Explorer, or COBE, which first mapped the cosmic radiation in broad brushstrokes in 1992. The new satellite can resolve features one-fortieth the size of those in the COBE map.

The results were announced at a news conference at NASA headquarters in Washington and posted online at map.gsfc.nasa.gov/.

Cosmologists hailed the new map and said it had exceeded their expectations. Dr. Max Tegmark, a cosmologist at the University of Pennsylvania, called the results "wild," and said they had put the ball in the court of regular astronomy to match its precision. "WMAP will be the foundation of all cosmology in the next five years," he said.

Dr. Michael Turner, a cosmologist at the University of Chicago, hailed the results as having something for everyone, confirmation of the New Cosmology that his generation had put together and "hints of surprises" for the next generation to figure out. "This is a great time to be a cosmologist," he said.

Dr. John Bahcall, an astrophysicist at the Institute for Advanced Study in Princeton, New Jersey, said the results were a "rite of passage" for cosmology from philosophical uncertainty to precision. "The motley mixture of strange elements that astronomers have put together over the last two or three decades is confirmed to remarkable accuracy," he said, referring to the entry of dark energy and dark matter into the astronomers' world.

Dr. David N. Spergel, a Princeton astrophysicist and member of the WMAP team, said: "We've answered the set of questions that have driven the field of cosmology for the last two decades. How many atoms in the universe? How old is the universe?"

The task now, he and others agreed, is to understand those motley elements, the dark stuff that apparently makes up 96 percent of everything, and what happened in the big bang that gave birth to it all.

Cosmologists do not know what dark energy is. One leading candidate is a repulsive force called the cosmological constant, which Einstein created as a fudge factor to keep the universe from collapsing in his equations, and later disavowed. But some theories of modern physics postulate mysterious force fields called quintessence as the dark energy. While the new analysis has not solved the problem, Dr. Spergel said its data seemed to favor Einstein's fudge factor.

The cosmic microwaves have mesmerized astronomers ever since they were discovered in 1965 as a faint radio hiss filling the sky by Dr. Arno Penzias and Dr. Robert Wilson, radio astronomers at Bell Laboratories who won Nobels for their work. The microwaves represent a snapshot of the universe as it was cooling to the point where atoms could form, at an age of about 380,000 years. But water vapor in the atmosphere obscures the microwaves, and so astronomers have had to be satisfied with glimpses from mountaintops or high-flying balloons.

In 1992, COBE confirmed that this cosmic gravy has lumps, the seeds from which galaxies and other cosmic structures would grow. Since then a series of smaller experiments have studied these lumps, which can be used to diagnose properties like the geometry and matter density of the cosmos, on finer and finer scales. These experiments suggested that the universe is flat and dominated by dark energy, but they only glimpsed small portions of the sky for limited times.

The new satellite scans the whole sky every six months. It is designed to operate for four years. The new map was based on the first year's worth of data.

Originally known as MAP, the satellite was renamed yesterday in honor of Dr. David Wilkinson, a Princeton cosmologist and leader of the MAP project who died last September.

In addition to measuring the brightness or temperature of the microwaves, the satellite's instruments, like a pair of Polaroid sunglasses, can also measure the polarization of the microwaves. That ability was crucial to the discovery of the era of the first stars. Like light skipping off a lake, the electric and magnetic fields that constitute light bouncing off an electrified gas are not jumbled but show a preference to vibrate in a particular plane. Last year astronomers showed that a polarization had been imparted to the cosmic microwaves at the moment that the first atoms formed, and the cosmic fireball thus lost its free electrons.

But astronomers thought there should be another polarization episode. When the lights went on in the universe, blazing ultraviolet from the first stars would have stripped the electrons from hydrogen atoms in space. Those electrons, which scatter the cosmic microwaves, would also polarize them again.

Most astronomers suspected that this had happened at about the time of the most distant and early quasars, around 800 million years of age. It was a surprise, astronomers said, to find the stars had formed so early.

The first stars, Dr. Bennett explained in an interview, were probably monsters 100 times as massive as the sun and burned out rapidly and violently, transmuting primordial hydrogen and helium into heavy elements like carbon and oxygen and spewing them out into space to form the basis for future generations of stars and eventually life.

The scientists also said that their data was beginning to shed light on a theory of what might have been going on during the big bang.

That theory, known as *inflation,* hypothesizes that the universe underwent an enormous growth spurt during the first trillionth of a trillionth of a second of time under the influence of a brief but powerful antigravitational field that permeated space. Such behavior is allowed by the laws of physics, and it has formed the core of big bang theorizing, but the details depend on the unknown physics that prevails at the energies of the early universe—far beyond the capacity of modern particle accelerators. And so inflation, as Dr. Bennett noted, is often called a paradigm instead of a theory.

By analyzing the bumps in the cosmic microwaves, which according to inflation are the result of microscopic fluctuations in the mysterious force field that drove inflation, along with other data, Dr. Spergel said, the scientists have ruled out one simple version of inflation that is often seen in textbooks. Other versions, he added, fit the data quite well.

"The data are good enough to rule out whole classes of inflationary theories," Dr. Spergel said. That is a boon, he said, for particle physicists, who want to know what laws governed the universe at the beginning of time.

"It really is a big hint for them," he said.

Dr. Andrei Linde, a cosmologist at Stanford and one of the fathers of inflation theory, and the inventor of the model that was ruled out, said that it was "great" that theories were getting culled.

He said that it was "painful" for him that one of his theories got killed, but that it was good news that several of his other versions were doing well.

Dr. Turner said: "This is the door to precision cosmology being opened. It's the first step in a long march."

—*February 12, 2003*

Astronomers Report Evidence of "Dark Energy" Splitting the Universe

By DENNIS OVERBYE

By comparing maps of heat emanating from the fading remnants of the big bang to maps of the modern universe, astronomers say they have uncovered evidence that some "dark energy" is wrenching the universe apart.

The new work, they said, provides independent confirmation of one of the strangest astronomical findings in recent years, that based on studies of distant exploding stars the expansion of the universe is speeding up.

The simplest explanation, astrophysicists say, is that space is imbued with a repulsive, or antigravitational, force first hypothesized in 1917 by Einstein and known as the cosmological constant. But nobody understands this so-called dark energy, although speculations have blossomed in the physics literature in the last few years.

Using the maps, a multinational team of 33 astrophysicists, led by Dr. Ryan Scranton of the University of Pittsburgh, found what the members called "the shadow of dark energy" in the form of a slight boost in the energy of the radiation from the big bang as it passed through huge clouds of galaxies.

The astronomers said their results represented an important validation of dark energy and the emerging consensus of a universe dominated by mysterious dark matter and even more mysterious dark energy, which is geometrically "flat." That means that parallel lines drawn across the cosmos will not meet.

"This result is the piece of physical evidence that really closes the door," said Dr. Robert C. Nichol, an astrophysicist at Carnegie Mellon. Many physicists had taken a wait-and-see attitude about the dark energy acceleration, Dr. Nichol said.

An astronomer at the Space Telescope Science Institute in Baltimore, Dr. Adam Riess, said dark energy was becoming "the great cosmological detective story of today."

"If we can just keep collecting a few more clues about it," Dr. Riess added, "we might actually be able to figure out what the heck it is."

The results were obtained by combining information from the Sloan Digital Sky Survey, which is mapping the distances and positions of more than a million galaxies, with the Wilkinson Microwave Anisotropy Probe of NASA. The probe, a satellite, is busy mapping the intensity of a faint cosmic microwave radiation that fills the sky and is presumed to represent heat emanating from the remains of the big bang when the universe was only 380,000 years old.

The cosmic radiation is rippled with hot and cool spots. Some are a result of lumps in the primordial cosmic gravy and are the seeds of galaxies and other conglomerations of matter. But other hot spots, theorists point out, may be generated by the passage of microwaves through the modern universe.

As a microwave passes through a large cloud of galaxies, its energy will first increase, as a rolling marble speeds up when it hits a dip in the road. Later, as the microwave leaves the cloud, gravity will take away some energy, as the marble climbs out of the dip.

In a universe that is geometrically "flat" and with no dark energy, those effects will cancel each other out. No net change in the energy of the microwaves will occur.

But in an accelerating universe, the effects will not always cancel each other out. In the largest agglomerations of matter, so-called superclusters that are forming, the microwaves will gain energy and thus appear hotter.

In such systems, tens of millions of light-years across, the force of dark energy that is trying to push apart the cloud is winning the battle over the gravity trying to pull together the galaxies. As a result, the cloud becomes less dense rather than more dense as the microwaves pass through it, explained Dr. Andrew J. Connolly, a team member from the University of Pittsburgh.

It takes less energy for them to climb back out than they acquired falling in.

Dr. Max Tegmark, a cosmologist at the University of Pennsylvania, compared the effect to racking up credit card debts in an inflationary era.

"The payback is less than what is borrowed," Dr. Tegmark said.

So the microwaves should be slightly hotter, by a minuscule fraction of a degree.

The effect is known as the Integrated Sachs-Wolfe effect, after Dr. Arthur M. Wolfe, who is now at the University of California at San Diego, and Dr. Rainer K. Sachs, who is now at the University of California at Berkeley, who first investigated the effects of lumps in the universe on the cosmic microwaves in 1967.

In recent months, several groups, including those led by Dr. Stephen Boughn of Haverford College in Pennsylvania, Dr. Michael R. Nolta of Princeton and Dr. Pablo Fosalba of the Institute of Astrophysics in Paris, have reported promising correlations between cosmic hot spots and sky catalogues of radio sources and X-rays, as well as galaxy maps.

In an e-mail message referring to the findings of the multinational team, Dr. Fosalba, who used part of the Sloan information, said, "Despite the fact that we are using different galaxy samples, results from both analyses are in very good agreement and provide strong evidence for dark energy in the universe."

Dr. Scranton, also in an e-mail message, said his team's work was important in validating the dark energy because it relied on sky survey data not available to other teams.

The Sloan survey aims to map more than a million galaxies, out to a distance of 1.5 billion light-years, over a quarter of the sky.

—July 22, 2003

Remembrance of Things Future:
The Mystery of Time

By DENNIS OVERBYE

There was a conference for time travelers at MIT in the spring of 2005.

I'm still hoping to attend, and although the odds are slim, they are apparently not zero, despite the efforts and hopes of deterministically minded physicists who would like to eliminate the possibility of your creating a paradox by going back in time and killing your grandfather.

"No law of physics that we know of prohibits time travel," said Dr. J. Richard Gott, a Princeton astrophysicist.

Dr. Gott, author of the 2001 book *Time Travel in Einstein's Universe: The Physical Possibilities of Travel through Time,* is one of a small breed of physicists who spend part of their time (and their research grants) thinking about wormholes in space, warp drives and other cosmic constructions, that "absurdly advanced civilizations" might use to travel through time.

It's not that physicists expect to be able to go back and attend Woodstock, drop by the Bern patent office to take Einstein to lunch, see the dinosaurs or investigate John F. Kennedy's assassination.

In fact, they're pretty sure those are absurd dreams and are all bemused by the fact that they can't say why. They hope such extreme theorizing could reveal new features, gaps or perhaps paradoxes or contradictions in the foundations of Physics As We Know It and point the way to new ideas.

"Traversable wormholes are primarily useful as a *'gedanken* experiment' to explore the limitations of general relativity," said Dr. Francisco Lobo of the University of Lisbon.

If general relativity, Einstein's theory of gravity and space-time, allows for the ability to go back in time and kill your grandfather, asks Dr. David Z. Albert, a physicist and philosopher at Columbia University, "how can it be a logically consistent theory?"

In his recent book *The Universe in a Nutshell,* Dr. Stephen W. Hawking wrote, "Even if it turns out that time travel is impossible, it is important that we understand why it is impossible."

When it comes to the nature of time, physicists are pretty much at as much of a loss as the rest of us who seem hopelessly swept along in its current. The mystery of time is connected with some of the thorniest questions in physics, as well as in philosophy, like why we remember the past but not the future, how causality works, why you can't stir cream out of your coffee or put perfume back in a bottle.

But some theorists think that has to change.

Just as Einstein needed to come up with a new concept of time in order to invent relativity 100 years ago, so physicists say that a new insight into time—or beyond it—may be required to crack profound problems like how the universe began, what happens at the center of a black hole or how to marry relativity and quantum theory into a unified theory of nature.

Space and time, some quantum gravity theorists say, are most likely a sort of illusion —or less sensationally, an "approximation"—doomed to be replaced by some more fundamental idea. If only they could think of what that idea is.

"By convention there is space, by convention time," Dr. David J. Gross, director of the Kavli Institute for Theoretical Physics and a winner of the 2004 Nobel Prize, said recently, paraphrasing the Greek philosopher Democritus, "in reality there is . . . ?" his voice trailing off.

The issues raised by time travel are connected to these questions, Dr. Lawrence Krauss, a physicist at Case Western Reserve University in Cleveland and author of the book *The Physics of Star Trek,* said. "The minute you have time travel you have paradoxes," Dr. Krauss said, explaining that if you can go backward in time you confront fundamental issues like cause and effect or the meaning of your own identity if there can be two of you at once. A refined theory of time would have to explain "how a sensible world could result from something so nonsensical."

"That's why time travel is philosophically important and has captivated the public, who care about these paradoxes," he said.

At stake, said Dr. Albert, the philosopher and author of his own time book, *Time and Chance,* is "what kind of view science presents us of the world."

"Physics gets time wrong, and time is the most familiar thing there is," Dr. Albert said.

We all feel time passing in our bones, but ever since Galileo and Newton in the 17th century began using time as a coordinate to help chart the motion of cannonballs, time—for physicists—has simply been an "addendum in the address of an event," Dr. Albert said.

"There is a feeling in philosophy," he said, "that this picture leaves no room for locutions about flow and the passage of time we experience."

Then there is what physicists call "the arrow of time" problem. The fundamental laws of physics don't care what direction time goes in, he pointed out. Run a movie of billiard balls colliding or planets swirling around in their orbits in reverse and nothing will look weird, but if you run a movie of a baseball game in reverse, people will laugh.

Einstein once termed the distinction between past, present and future "a stubborn illusion," but as Dr. Albert said, "It's hard to imagine something more basic than the distinction between the future and the past."

The Birth of an Illusion

Space and time, the philosopher Augustine famously argued 1,700 years ago, are creatures of existence and the universe, born with it, not separately standing features of eternity. That is the same answer that Einstein came up with in 1915 when he finished his general theory of relativity.

That theory explains how matter and energy warp the geometry of space and time to produce the effect we call gravity. It also predicted, somewhat to Einstein's dismay, the expansion of the universe, which forms the basis of modern cosmology.

But Einstein's theory is incompatible, mathematically and philosophically, with the quirky rules known as quantum mechanics that describe the microscopic randomness that fills this elegantly curved expanding space-time. According to relativity, nature is continuous, smooth and orderly; in quantum theory the world is jumpy and discontinuous. The sacred laws of physics are correct only on average.

Until the pair are married in a theory of so-called quantum gravity, physics has no way to investigate what happens in the big bang, when the entire universe is so small that quantum rules apply.

Looked at closely enough, with an imaginary microscope that could see lengths down to 10^{-33} centimeters, quantum gravity theorists say, even ordinary space and time dissolve into a boiling mess that Dr. John Wheeler, the Princeton physicist and phrasemaker, called "space-time foam." At that level of reality, which exists underneath all our fingernails, clocks and rulers as we know them cease to exist.

"Everything we know about stops at the big bang, the Big Crunch," said Dr. Raphael Bousso, a physicist at the University of California at Berkeley.

What happens to time at this level of reality is anybody's guess. Dr. Lee Smolin, of the Perimeter Institute for Theoretical Physics in Waterloo, Ontario, said, "There are several different, very different, ideas about time in quantum gravity."

One view, he explained, is that space and time "emerge" from this foamy substrate when it is viewed at larger scales. Another is that space emerges but that time or some deeper relations of cause and effect are fundamental.

Dr. Fotini Markopoulou Kalamara of the Perimeter Institute described time as, if not an illusion, an approximation, "a bit like the way you can see the river flow in a smooth way even though the individual water molecules follow much more complicated patterns."

She added in an e-mail message: "I have always thought that there has to be some basic fundamental notion of causality, even if it doesn't look at all like the one of the space-time we live in. I can't see how to get causality from something that has none; neither have I ever seen anyone succeed in doing so."

Physicists say they have a sense of how space can emerge, because of recent advances in string theory, the putative theory of everything, which posits that nature is composed of wriggling little strings.

Calculations by Dr. Juan Maldacena of the Institute for Advanced Study in Princeton and by others have shown how an extra dimension of space can pop mathematically into being almost like magic, the way the illusion of three dimensions can appear in the holograms on bank cards. But string theorists admit they don't know how to do the same thing for time yet.

"Time is really difficult," said Dr. Cumrun Vafa, a Harvard string theorist. "We have not made much progress on the emergence of time. Once we make progress we will make progress on the early universe, on high-energy physics and black holes.

"We are out on a limb trying to understand what's going on here."

Dr. Bousso, an expert on holographic theories of space-time, said that in general relativity time gets no special treatment.

He said he expected both time and space to break down, adding, "We really just don't know what's going to go.

"There is a lot of mysticism about time," Dr. Bousso said. "Time is what a clock measures. What a clock measures is more interesting than you thought."

A Brief History of Time Travel

"If we could go faster than light, we could telegraph into the past," Einstein once said. According to the theory of special relativity—which he proposed in 1905 and which ushered $E = mc^2$ into the world and set the speed of light as the cosmic speed limit—such telegraphy is not possible, and there is no way of getting back to the past.

But, somewhat to Einstein's surprise, in general relativity it is possible to beat a light beam across space. That theory, which Einstein finished in 1916, said that gravity resulted from the warping of space-time geometry by matter and energy, the way a bowling ball sags a trampoline. And all this warping and sagging can create shortcuts through space-time.

In 1949, Kurt Gödel, the Austrian logician and mathematician then at the Institute for Advanced Study, showed that in a rotating universe, according to general relativity, there were paths, technically called "closed timelike curves," you could follow to get back to the past. But it has turned out that the universe does not rotate very much, if at all.

Most scientists, including Einstein, resisted the idea of time travel until 1988 when Dr. Kip Thorne, a gravitational theorist at the California Institute of Technology, and two of his graduate students, Dr. Mike Morris and Dr. Ulvi Yurtsever, published a pair of papers concluding that the laws of physics may allow you to use wormholes, which are like tunnels through space connecting distant points, to travel in time.

These holes, technically called Einstein-Rosen bridges, have long been predicted as a solution of Einstein's equations. But physicists dismissed them because calculations predicted that gravity would slam them shut.

Dr. Thorne was inspired by his friend, the late Cornell scientist and author Carl Sagan, who was writing the science fiction novel *Contact*, later made into a Jodie Foster movie, and was looking for a way to send his heroine, Eleanor Arroway, across the galaxy. Dr. Thorne and his colleagues imagined that such holes could be kept from collapsing and thus maintained to be used as a galactic subway, at least in principle, by threading them with something called Casimir energy (after the Dutch physicist Hendrik Casimir), which is a sort of quantum suction produced when two parallel metal plates are placed very close together. According to Einstein's equations, this suction, or negative pressure, would have an antigravitational effect, keeping the walls of the wormhole apart.

If one mouth of a wormhole was then grabbed by a spaceship and taken on a high-speed trip, according to relativity, its clock would run slow compared with the other end of the wormhole. So the wormhole would become a portal between two different times as well as places.

Dr. Thorne later said he had been afraid that the words "time travel" in the second paper's title would create a sensation and tarnish his students' careers, and he had forbidden Caltech to publicize it.

In fact, their paper made time travel safe for serious scientists, and other theorists, including Dr. Frank Tipler of Tulane University and Dr. Hawking, jumped in. In 1991, for example, Dr. Gott of Princeton showed how another shortcut through space-time could be manufactured using pairs of cosmic strings—dense tubes of primordial energy not to be confused with the strings of string theory, left over by the big bang in some theories of cosmic evolution—rushing past each other and warping space around them.

Harnessing the Dark Side

These speculations have been bolstered (not that time machine architects lack imagination) with the unsettling discovery that the universe may be full of exactly the kind of antigravity stuff needed to grow and prop open a wormhole. Some mysterious "dark energy," astronomers say, is pushing space apart and accelerating the expansion of the universe. The race is on to measure this energy precisely and find out what it is.

Among the weirder and more disturbing explanations for this cosmic riddle is something called phantom energy, which is so virulently antigravitational that it would eventually rip planets, people and even atoms apart, ending everything. As it happens, this bizarre stuff would also be perfect for propping open a wormhole, Dr. Lobo of Lisbon recently pointed out. "This certainly is an interesting prospect for an absurdly advanced civilization, as phantom energy probably comprises 70 percent of the universe," Dr. Lobo wrote in an e-mail message. Dr. Sergey Sushkov of Kazan State Pedagogical University in Russia has made the same suggestion.

In a paper posted on the physics Web site arxiv.org/abs/gr-qc/0502099, Dr. Lobo suggested that as the universe was stretched and stretched under phantom energy, microscopic holes in the quantum "space-time foam" might grow to macroscopic usable size. "One could also imagine an advanced civilization mining the cosmic fluid for phantom energy necessary to construct and sustain a traversable wormhole," he wrote.

Such a wormhole he even speculated, could be used to escape the "big rip" in which a phantom energy universe will eventually end.

But nobody knows if phantom, or exotic, energy is really allowed in nature and most physicists would be happy if it were not. Its existence would lead to paradoxes, like negative kinetic energy, where something could lose energy by speeding up, violating what is left of common sense in modern physics.

Dr. Krauss said, "From the point of view of realistic theories, phantom energy just doesn't exist."

But such exotic stuff is not required for all time machines, Dr. Gott's cosmic strings for example. In another recent paper, Dr. Amos Ori of the Technion-Israel Institute of Technology in Haifa describes a time machine that he claims can be built by moving around colossal masses to warp the space inside a doughnut of regular empty space into a particular configuration, something an advanced civilization may be able to do in 100 or 200 years.

The space inside the doughnut, he said, will then naturally evolve according to Einstein's laws into a time machine.

Dr. Ori admits that he doesn't know if his machine would be stable. Time machines could blow up as soon as you turned them on, say some physicists, including Dr. Hawking, who has proposed what he calls the "chronology protection" conjecture to keep the past safe for historians. Random microscopic fluctuations in matter and energy and space itself, they argue, would be amplified by going around and around boundaries of the machine or the wormhole, and finally blow it up.

Dr. Gott and his colleague Dr. Li-Xin Li have shown that there are at least some cases where the time machine does not blow up. But until gravity marries quantum theory, they admit, nobody knows how to predict exactly what the fluctuations would be.

"That's why we really need to know about quantum gravity," Dr. Gott said. "That's one reason people are interested in time travel."

Saving Grandpa

But what about killing your grandfather? In a well-ordered universe, that would be a paradox and shouldn't be able to happen, everybody agrees.

That was the challenge that Dr. Joe Polchinski, now at the Kavli Institute for Theoretical Physics in Santa Barbara, Calif., issued to Dr. Thorne and his colleagues after their paper was published.

Being a good physicist, Dr. Polchinski phrased the problem in terms of billiard balls. A billiard ball, he suggested, could roll into one end of a time machine, come back out the other end a little earlier and collide with its earlier self, thereby preventing itself from entering the time machine to begin with.

Dr. Thorne and two students, Fernando Echeverria and Gunnar Klinkhammer, concluded after months of mathematical struggle that there was a logically consistent solution to the billiard matricide that Dr. Polchinski had set up. The ball would come back out of the time machine and deliver only a glancing blow to itself, altering its path just enough so that it would still hit the time machine. When it came back out, it would be aimed just so as to deflect itself rather than hitting full-on. And so it would go like a movie with a circular plot.

In other words, it's not a paradox if you go back in time and save your grandfather. And, added Dr. Polchinski, "It's not a paradox if you try to shoot your grandfather and miss."

"The conclusion is somewhat satisfying," Dr. Thorne wrote in his book *Black Holes and Time Warps: Einstein's Outrageous Legacy.* "It suggests that the laws of physics might accommodate themselves to time machines fairly nicely."

Dr. Polchinski agreed. "I was making the point that the grandfather paradox had nothing to do with free will, and they found a nifty resolution," he said in an e-mail message, adding, nevertheless, that his intuition still tells him time machines would lead to paradoxes.

Dr. Bousso said, "Most of us would consider it quite satisfactory if the laws of quantum gravity forbid time travel."

—June 28, 2005

A Trip Forward in Time.
Your Travel Agent: Einstein

By DENNIS OVERBYE

When H. G. Wells published his epochal novel *The Time Machine* in 1895, time travel was outlawed by the laws of physics. But that was Newtonian physics, and everything changed 10 years later with Einstein's theory of relativity.

That theory—which ushered in the age of $E = mc^2$ and set the speed of light, 186,300 miles per second, as the cosmic speed limit—allows for time travel to the future, physicists say. Here's how:

One consequence of Einstein's theory is that a clock in motion will always appear to run slowly compared with one at rest (and since all motion is relative, the clock at rest will appear to go slowly from the vantage pointof the one moving).

This leads to the famous "twin paradox" in which one twin is rocketed at high speed across the galaxy and back home. Even at a velocity close to the speed of light, the journey would take tens of thousands of years from the vantage point of Earth, but because of his high relative motion the astronaut would age more slowly than he or she would on Earth, and would return home only a few years older. His twin would be long dead.

In effect, the astronaut would have traveled into the future, said Dr. J. Richard Gott, a Princeton astrophysicist.

The slowing clock prediction has been confirmed by flying atomic clocks around Earth on jets.

"If you take a plane east around the world, you will come back 59 nanoseconds younger than if you had stayed home," Dr. Gott said.

The record holder for this type of travel, he said, is the Russian astronaut Sergei Krikalev, who came back from 748 days orbiting in the *Mir* space station a full one-fiftieth of a second younger than he would have if he had stayed on the ground.

In his 1905 paper Einstein predicted that because of the rotation speed of Earth, clocks would also run slower at the Equator than the poles, but that turned out to be wrong.

In a recent article in *Physics Today*, Dr. Alex Harvey of Queens College in New York and Dr. Engelbert Schucking of New York University pointed out that Einstein had not taken account of an effect of general relativity, then 10 years in the future, which says that clocks run slower the more deeply in a gravitational field they sit.

The rotation of Earth causes it to bulge at the Equator, lifting clocks there and making them run slightly faster relative to those at the poles by just enough to compensate for the extra speed.

So the two effects just cancel each other out, and clocks at the Equator and poles run at the same speed. "It's a deep coincidence," Dr. Gott said.

The two effects could be combined for an even deeper trip into the future by going to Mercury, which is both deep in the gravitational field of the sun and also zooming around it at high speed. A 30-year stay there, Dr. Gott said, would save 22 seconds of an astronaut's life.

A few seconds might not sound like much of a trip in time, but Dr. Gott points out that astronauts haven't been that far into space, either. The moon, humanity's most distant destination so far, is only about 1.3 light-seconds away, about like hopping over the Atlantic, he said.

"The astronauts are the Lindberghs of time travel," Dr. Gott said.

—June 28, 2005

Dark, Perhaps Forever

By DENNIS OVERBYE

Mario Livio tossed his car keys in the air.

They rose ever more slowly, paused, shining, at the top of their arc, and then in accordance with everything our Galilean ape brains have ever learned to expect, crashed back down into his hand.

That was the whole problem, explained Dr. Livio, a theorist at the Space Telescope Science Institute in Baltimore on the Johns Hopkins campus.

A decade ago, astronomers discovered that what is true for your car keys is not true for the galaxies. Having been impelled apart by the force of the big bang, the galaxies, in defiance of cosmic gravity, are picking up speed on a dash toward eternity. If they were keys, they would be shooting for the ceiling.

"That is how shocking this was," Dr. Livio said.

It is still shocking. Although cosmologists have adopted a cute name—dark energy—for whatever is driving this apparently antigravitational behavior on the part of the universe, nobody claims to understand why it is happening, or its implications for the future of the universe and of the life within it, despite thousands of learned papers, scores of conferences and millions of dollars' worth of telescope time. It has led some cosmologists to the verge of abandoning their fondest dream: a theory that can account for the universe and everything about it in a single breath.

"The discovery of dark energy has greatly changed how we think about the laws of nature," said Edward Witten, a theorist at the Institute for Advanced Study in Princeton, New Jersey.

In the fall of 2008, NASA and the Department of Energy plan to invite proposals for a $600 million satellite mission devoted to dark energy. But some scientists fear that might not be enough. When astronomers and physicists gathered at the Space Telescope Science Institute recently to take stock of the revolution, their despair at getting to the bottom of the dark energy mystery anytime soon, if ever, was palpable, even as they anticipate a flood of new data from the sky in coming years. When it came time for one physicist to discuss new ideas about dark energy, he showed a blank screen.

The institute's director, Matt Mountain, said that dark energy had given this generation of astronomers a rare opportunity, and he admonished them to use it wisely.

"We are placing a large bet," Dr. Mountain said, "using our credibility as collateral, that we as a community know what we are doing."

But many stressed that it was going to be a long march with no clear end in sight. Lawrence Krauss of Case Western Reserve University told them, "In spite of the fact that you are liable to spend the rest of your lives measuring stuff that won't tell us what we want to know, you should keep doing it."

Scuffling in the Dark

Through myriad techniques and observations, cosmologists have recently arrived, after decades of strife, at a robust but dark consensus regarding a cosmos in which stars and galaxies, as well as the humans who gawk at them, amount to barely more than a disputatious froth. It was born 13.7 billion years ago in the big bang. By weight it is 4 percent atoms and 22 percent so-called dark matter of unknown identity—perhaps elementary particles that will be discovered at the Large Hadron Collider starting up outside Geneva this year. That leaves 74 percent for the weight of whatever began causing the cosmos to accelerate about five billion years ago.

As far as astronomers can tell, there is no relation between dark matter, the particles, and dark energy other than the name, but you never know. Some physicists are even willing to burn down their old sainted Einstein and revise his theory of gravity, general relativity, to make the cosmic discrepancies go away. There is in fact a simple explanation for the dark energy, Dr. Witten pointed out, one whose tangled history goes all the way back to Einstein, but it is also the most troubling.

"Dark energy has the somewhat unusual property that it was embarrassing before it was discovered," he said.

In 1917, Einstein invented a fudge factor known as the cosmological constant, a sort of cosmic repulsion to balance gravity and keep the universe in balance. He abandoned his constant when the universe was discovered to be expanding, but quantum physics resurrected it by showing that empty space should be foaming with energy that had the properties of Einstein's constant.

Alas, all attempts to calculate the amount of this energy come up with an unrealistically huge number, enough energy to blow away the contents of the cosmos like leaves in a storm before stars or galaxies could form. Nothing could live there.

Dr. Witten and other physicists used to think this conundrum "would somehow go away." Something was missing in physicists' understanding of physics, the logic went. The constant was really zero for deep reasons that, when revealed, would lead physicists closer to an understanding of what they call "the vacuum," that is to say, the structure of reality.

"It seems now that the answer is not really zero," Dr. Witten said.

Requiem for a Dream

Einstein's constant is the most economical explanation for dark energy, Dr. Witten said. The others, involving new force fields or tinkering with Einstein's gravity, are hard to make work and raise more questions than they answer. But if dark energy is the cosmological constant, it is smaller than predicted by a shocking factor of 10^{60}. No fundamental principles can explain why Einstein's constant, or any physical parameter, could be so small without being zero, Dr. Witten said. Zero can be a fundamental number, he said, but not a 1 with 59 zeroes between it and the decimal point.

As a result, he said, maybe physicists should give up trying to explain that number and look instead for a theory that generates all kinds of universes, a so-called multiverse.

That idea has been given mathematical form by string theory, which portrays the constituents of nature as tiny wriggling strings, an elegant idea that in principle explains all the forces of nature but in practice leads to at least 10^{500} potential universes.

This maze was an embarrassment for string theory. As Dr. Witten, one of the leaders of the field, said, "I am tempted to say this was an embarrassment of my youth."

"Who needs that mess?" he recalled thinking. "There is just one world we live in."

Now, Dr. Witten allowed, dark energy might have transformed this fecundity from a vice into a virtue, a way to generate universes where you can find any cosmological constant you want. We just live in one where life is possible, just as fish only live in water.

"This interpretation of string theory might be close to the truth," Dr. Witten said. But that truth comes at a cost.

"Before the discovery of the dark energy, quantum physicists tended to assume that the 'vacuum' we live in has some deep meaning that reflects nature's deepest secrets," Dr. Witten said. But if ours is only one of a zillion in a haystack, there is nothing special about it, no secret to be found.

It could still turn out that dark energy is some as-yet-undiscovered "fifth force," say, or the result of not understanding gravity. In that case, Dr. Witten said, "All the old viewpoints would be correct," and physicists could go back to dreaming of a final theory.

"I'd be happy if that happened," he said. "Our reward would be to go back to where we were, not understanding the cosmological constant."

The notion that there are a zillion universes, whose individual properties are just a cosmic dice throw, is a story that has been told before and "raises the blood pressure of many physicists seriously," as Dr. Livio put it. But the idea has rarely been mentioned by Dr. Witten, who is seen in the community as a symbol of the old Einsteinian ideal.

Dr. Witten said he was just doing his duty to explain what dark energy meant to physics.

"As for how I feel personally, I am not sure what to say," he said in an e-mail message. "I wasn't terribly enthusiastic the first, or even second, time I heard the proposal of a multiverse. But none of us were consulted when the universe was created."

Astronomy of the Invisible

The trouble started in 1998 when two competing teams of astronomers, one led by Saul Perlmutter of the Lawrence Berkeley National Laboratory in California and the other by Brian Schmidt of the Australian National University, discovered that the expansion of the universe was inexplicably accelerating.

Both teams were using a kind of exploding star known as a Type IA supernova as standard candles—objects whose distance can be inferred from their apparent brightness and a few other tricks of the trade—to investigate the history and fate of the universe. They found, on the basis of a few dozen of these stars, that the more distant ones were dimmer than expected, meaning that they had been carried farther away by the cosmic expansion than expected, meaning that the universe was speeding up. The car keys were streaking for the ceiling.

The groups quibble about who saw and said what first, but they have shared in a cavalcade of awards and prizes, among them the $1 million Shaw Prize in 2006 and the $500,000 Gruber Cosmology Prize, awarded in 2007 at the University of Cambridge in England, where Dr. Perlmutter and Dr. Schmidt lectured jointly, trading sentences.

Since then, myriad collaborations have joined in the hunt for these exploding stars. In Baltimore, Dr. Perlmutter reported on a new analysis of "the world's data set," more than 300 supernovas observed by various groups, which he said would provide the tightest constraints on the nature of dark energy "for at least the next 15 minutes."

Dr. Perlmutter's results, along with all the others that were presented over the next four days, were consistent with Einstein's cosmological constant, plus or minus 10 percent, but with just about everything else the theorists can throw into the pot, as well.

Nor is there any solid evidence yet that dark energy is or is not varying with time—if it is not constant, it cannot be Einstein's constant. Adam Riess of the Johns Hopkins space telescope institute, a key member of Dr. Schmidt's team, said, "The biggest thing we could learn is by ruling that out."

He added, "We have a suspect, but we're not ready to convict anyone yet."

Dr. Perlmutter said, "The challenge is to make dramatic improvements in the quality of the data," adding, "The next decade should be a very fertile time."

Astronomers have developed a smorgasbord of other ways of tracking the effect of dark energy. They have learned how to map the growth of clusters of galaxies, by analyzing how their gravity distorts the light from galaxies far behind them. Gravity makes the clusters grow; dark energy holds them back.

"We can see dark matter, and in principle even invisible clusters," said Henk Hoekstra of the University of Victoria in Canada.

Another technique is to simply count the clusters at different times in the cosmic past, the way one might count trees to gauge the growth of a forest. Yet another method is to use sound waves from the hot, early days of the universe, which have left an imprint on the distribution of galaxies today—a 500-million-light-year "bump"—as a cosmic yardstick for measuring the universe as it grew.

Each of these methods has its own strengths and weaknesses, and experts agree that it will be necessary to marry the results from many methods to zoom in on the properties of dark energy. They also agree that the best place to do that is in space.

The Big Bake-Off

Last year a committee from the National Academy of Sciences recommended that a dark energy observatory be the next mission in an astrophysics program called Beyond Einstein.

There are now three competitors angling for the job: Dr. Perlmutter's SNAP, for Supernova Acceleration Probe; Adept, or Advanced Dark Energy Physics Telescope, led by Charles Bennett of Johns Hopkins; and Destiny, for Dark Energy Space Telescope, led by Tod Lauer of the National Optical Astronomy Observatory in Tucson.

Also in the works, just to add spice, is a European mission known as Euclid, which could fly in 2017, if it is approved by the European Space Agency. NASA and the Department of Energy, working together, expect to make a final selection for the dark energy mission—known colloquially as J-dem for Joint Dark Energy Mission—in the spring of 2009 and launch it in the middle of the next decade.

That sounds like progress, but some astronomers, including the former members of the academy committee itself, have complained that $600 million is less than half of the $1.2 billion to $1.5 billion the academy committee estimated was necessary to do the job. In a recent letter to Michael Salamon, NASA scientist in charge of the project, 11 of the committee members, including both of its chairmen, urged NASA to raise the cost cap on the mission, writing, "Cutting the budget in half would probably make the attainment of these goals impossible."

NASA's $600 million does not include the cost of launching the satellite, so the discrepancy is not as big as it looks. But in Baltimore, Jon Morse, director of astrophysics at NASA headquarters, warned that if the astronomers wanted to spend a billion dollars, some other astronomy mission would have to come off the table.

NASA has to live within its means, Dr. Morse said in an interview.

"Otherwise," he said, "Beyond Einstein becomes beyond reality."

A Hole in the Future?

Whatever proposal is eventually selected, the dark energy satellite will return a tidal wave of data about the universe and its weird denizens, both visible and invisible. This data is likely to transform astronomy in unpredictable ways, but there is no guarantee that it will nail the mystery of dark energy.

Both alternatives to the constant—some weird energy field in space, or a modification to Einstein's theory of gravity—could vary wildly over the course of history. But Paul Steinhardt, a theorist from Princeton University, argued that they would tend to mimic the cosmological constant so closely that the different models cannot be distinguished within the projected error limits, of a few percent. He called this blur of ignorance "the J-dem hole." The specter of the J-dem hole dominated a panel discussion devoted to the question, "How well do we have to do?"

The answer, said Dr. Krauss of Case Western Reserve, was "better than you will be able to do."

The only real job, he said, is to distinguish dark energy from the cosmological constant. "If we don't answer that question, we won't have learned a thing," Dr. Krauss said.

He compared the present situation with the development of quantum mechanics, the paradoxical-sounding rules that govern inside the atom, which overturned science in the 1920s.

That revolution, he pointed out, stemmed from theorists' inability to explain the so-called black body radiation emitted from a hot glowing object. The solution did not come from more and more precise measurements of the black body spectrum, but rather from the heads of people like Niels Bohr and Werner Heisenberg, who envisioned new ways that atoms could work and weird new laws of nature.

"We really need new theory, and we have none," Dr. Krauss said.

In the meantime, astronomers could get lucky. Despite Dr. Steinhardt's analysis, measurements of dark energy's strength could converge on a value not quite the same as Einstein's constant. Or it could turn out that it has changed over cosmic time and is not constant. Einstein and Dr. Witten would be off the hook.

Michael Turner, a University of Chicago cosmologist who coined the term *dark energy*, said you could measure the health of a field by the big questions it takes on, and addressing Dr. Morse of NASA, who was moderating the discussion, as well as his colleagues, he said, "You have a job, to go knock on everyone's door and say this is the opportunity of a lifetime."

Dr. Krauss said, "It would be crazy to talk ourselves out of this."

He added: "You have to do what you can. You would be crazy not to look."

—*June 3, 2008*

The Struggle to Measure Cosmic Expansion

By DENNIS OVERBYE

Hoping to understand why the universe seems to be coming apart at its seams, a young astronomer and his colleagues have embarked on one of the oldest quests in cosmology, to measure how fast the universe is growing, how big it is and how old it is.

That information is encoded in the value of an elusive number known as the Hubble constant that has led astronomers on a merry chase for three-quarters of a century. "It is the most fundamental number in cosmology," said Adam Riess, 38, an astronomer at the Space Telescope Science Institute and Johns Hopkins University, and one of the discoverers 10 years ago that some kind of "dark energy" is speeding up the expansion of the universe.

In the spring of 2008, in what he called "a triumph of metrology," Dr. Riess announced that he and his comrade, Lucas Macri of Texas A&M University, had used the Hubble Space Telescope to make the newest and most precise measurement yet of this parameter.

Expressed in the quaint terms astronomers favor, the Hubble constant, Dr. Riess reported, is 74 kilometers per second per megaparsec. It means that for every additional million parsecs (about 3.26 million light-years) a galaxy is from us, it is going 74 kilometers per second faster.

The news was not in Dr. Riess's value, which, reassuringly, agreed roughly with the result from an earlier space telescope team led by Wendy Freedman, the director of Carnegie Observatories, and with calculations based on measurements of relic radiation surmised to be left from the big bang, but in the precision with which his group claimed to have measured it: an uncertainty of only 4.3 percent.

Only 30 years ago, distinguished astronomers could not agree within a factor of two on the value of Hubble's constant, leaving every other parameter in cosmology uncertain by at least the same factor and provoking snickers from other fields of science.

But this is the age of so-called precision cosmology.

"I'm not saying we're going to get to 1 percent," Dr. Riess said, "but we might."

Dr. Riess's announcement was regarded as a hopeful beginning by other astronomers and cosmologists concerned with the fate of the universe and of physics. Knowing the precise value of the rate of expansion of the universe, they explain, has emerged as a key to understanding dark energy. The more precisely they can pin down the value of the Hubble constant, the more precisely they can pin down the properties of that enigmatic, cinematic-sounding force.

"I think Adam's work is nice," said Dr. Freedman, who has led a large space telescope effort to measure the constant. But she and others added that some parts of Dr. Riess's scheme could be vulnerable to the sorts of so-called systematic errors that have embroiled previous generations of astronomers in controversy—the effects of dust and galactic chemistry, for example, on the brightness of distant stars.

In an e-mail message, John Huchra of the Harvard-Smithsonian Center for Astrophysics wrote, "we know of several big bugaboos."

The stakes are bigger than just dark energy. Cosmologists would like to know whether their so-called Standard Model of the universe makes sense. Is the universe in fact 13.7 billion years old, full of dark matter and dark energy, and speckled with galaxies that grew by gravity from random microscopic fluctuations in the big bang?

That universe is described mathematically by half a dozen fundamental parameters, from which the Hubble constant can be calculated. But to test the model "at a physically interesting level," in the words of Dr. Huchra, the Hubble constant, as well as other parameters, needs to be actually measured to high accuracy.

Both the telescope and the "constant" are named after Edwin Hubble, the Mount Wilson astronomer who discovered in 1929 that the universe was expanding. It is not really constant. Over cosmic time, gravity tries to slow the expansion while dark energy, as astronomers discovered to their surprise 10 years ago, tries to speed it up. The history of the Hubble constant has seen many hopeful beginnings that have subsequently floundered on the difficulty of divining accurate distances to dim blurry lights in the sky, that is to say, galaxies. Both the 200-inch Hale telescope on Palomar Mountain in California, inaugurated in 1948, and the Hubble Space Telescope, launched 42 years later, were supposed to solve the problem.

Allan Sandage, also of Carnegie Observatories, who inherited the universe when Mr. Hubble died in 1953 and has been measuring and remeasuring the Hubble constant ever since, likes to say that astronomy is an impossible science. "It's marvelous to get a distance," he said once, "because it's almost impossible to believe that you can do it."

Astronomers can triangulate to determine the distances to the nearest stars, looking to see how they shift against background stars as the Earth goes from one side of its orbit around the sun to the other side, but to gauge deeper distances they depend on finding so-called *standard candles*. These are stars or other objects whose intrinsic luminosities are known and thus their distances can be inferred from their apparent brightness.

Among the most reliable of these candles are Cepheid variables, pulsating stars that dim and fade in a sawtooth pattern. The more luminous they are, the longer their cycle. So such a star in a distant galaxy, by its winking, is broadcasting its luminosity and distance.

Unfortunately, the more luminous and thus more useful such candles are, the rarer they are and the harder it is to find enough examples to calibrate them. So the blue-water sailors of the cosmos have to step outward by a "distance ladder," calibrating stars nearby and then using them to calibrate brighter but rarer standard candles in more distant galaxies, stepping ever outward. The standard candles of choice for many astronomers are exploding stars known as Type IA supernovae, brilliant enough to be seen across the universe.

But as astronomers step outward, small errors multiply and their candles get more uncertain.

According to a recent compilation by Dr. Huchra, more than 500 values of the Hubble constant have been published over the years. Astronomers are now within shouting distance of agreement. In recent years the two main teams using the Hubble telescope to measure the constant, one led by Dr. Freedman and the other led by Dr. Sandage, have arrived at answers 15 percent apart, 70 and 62, respectively, with 10 percent error bars that slightly overlap.

And there things might have stayed, Dr. Riess said. Few Americans are lying awake at night waiting to know the expansion rate of the universe to 1 percent accuracy. For most people, that the universe is expanding is baffling enough.

But dark energy has upped the ante. Many physicists and astronomers are wondering if the dark energy driving this behavior is a fudge factor that Einstein invented in 1917 to keep the universe static, and then abandoned.

Dr. Riess likens the different kinds of cosmological observations that go into making the standard cosmological model to the spokes of a bicycle wheel. To home in on dark energy, "We have to go around the wheel, tightening the spokes," he said.

One of the easiest spokes to tighten, Dr. Riess said he realized a couple of years ago, is the Hubble constant. In one typical calculation, for example, an uncertainty in the Hubble constant translates into twice as much uncertainty in a crucial measure of dark energy's oomph. So cutting the Hubble uncertainty goes a long way toward sharpening the estimates of dark energy.

All the measurements of the Hubble constant have suffered from the fact that there are too many rungs on the distance ladder, and thus chances for error, Dr. Riess said.

Measuring the Hubble constant this way, he said, is like measuring a room by laying a ruler end over end. Every time you pick it up and lay it down again you can make a mistake. "What you need is a tape measure," he said.

Dr. Riess's tape measure is the Hubble Space Telescope, and its workhorse instrument, the Advanced Camera for Surveys. The pair can find and measure the gold standard Cepheid stars much farther out in space than other telescopes, he said, and thus skip several intermediate calibration steps and the attendant chances for error.

"Better data and techniques come along in time whether anyone likes it or not," Dr. Riess said via e-mail. "I want to make clear that the Hubble constant can be measured to better precision than in the past and should be no more controversial than any other physical parameter we measure."

Dr. Riess's distance ladder has only three rungs and one telescope, leaping from the Milky Way's neighborhood to supernova explosions as distant as a billion light-years.

It starts with a galaxy known as NGC 4258 (aka Messier 106 in Ursa Major), where astronomers have found clouds emitting radio waves at a frequency characteristic of water vapor circling the center of the galaxy, as well as the all-important Cepheid stars. By tracking the speeds and motion across the sky of these clouds with high-resolution radio observations, a team led by James Herrnstein of the National Radio Astronomy Observatory in Socorro, New Mexico, in 1999 determined its distance as 23.5 million light-years.

Knowing the distance to that galaxy allowed Dr. Riess and his team to calibrate the Cepheids, which they then used to calibrate supernovas.

Several astronomers said it was worrisome that Dr. Riess's calibration of the Cepheids and thus the whole distance ladder rests on that one galaxy. It would be desirable, they say, to have accurate distances to more such galaxies, a project being pursued by Jim Braatz of the National Radio Astronomy Observatory in Virginia. In the meantime, as a backup technique for calibrating the Cepheids, Dr. Riess and colleagues have used Hubble's fine guidance sensors, which help the telescope find and track stars, to triangulate the distances to Cepheid variable stars in our galaxy. The results for the Hubble constant are the same, he said.

So, Dr. Riess's trek along the Hubble trail is just beginning. The results will probably get "a smidgen better" over the coming months, he said.

Astronauts are going to try to repair the advanced camera, which broke down during the final Hubble service call in October 2007. If they are successful, Dr. Riess and his team will be using the camera to extend their search for more supernovas and Cepheid stars. On the other end of the ladder, new telescope surveys, including Pan-STARRS, for Panoramic Survey Telescope and Rapid Response System, whose prototype is in operation on top of Haleakala on Maui, Hawaii, are expected to find thousands of supernovas far out in space, greatly increasing the accuracy of measurements both of dark energy and of Hubble's troublesome constant.

It's never going to be "Yup, now we've nailed it," Dr. Riess said. "This is humankind's quest: to be always doing this. We're looking to always make a cleaner handoff."

Showing no effect of the weight of history, he said, "This is still early days."

—*August 19, 2008*

Particle Hunt Nets Almost Nothing; the Hunters Are Almost Thrilled

By DENNIS OVERBYE

This could have been the day they discovered dark matter.

On the morning of April 4, 2011, a dozen or so graduate students and postdoctoral fellows gathered in the offices of Elena Aprile, a physics professor at Columbia University, to get their first look at the data from an experiment on the other side of the world. In a tunnel deep under Gran Sasso, Italy, Dr. Aprile and an international team of scientists had wired a vat containing 134 pounds of liquid xenon to record the pit-pat of invisible particles, the so-called dark matter that astronomers say constitutes a quarter of the universe.

Photographers were on hand to record the action—after all, you never know—although theoretical calculations suggested that with only 100 days of observation, the xenon experiment was probably still shy of the time necessary to see dark matter. "We will not discover dark matter today," Dr. Aprile said. "We will be doing this again and again."

Dark matter has teased and tantalized physicists since the 1970s when it was demonstrated that some invisible material must be providing the gravitational glue to hold galaxies together. Knowing what it is would provide a road map to new particles and forces, a new view of what happened in the big bang, and more Nobel Prizes than you can count. Failure to find it would mean that Einstein did not get the laws of gravity quite right.

The best guess is that this dark matter consists of clouds of exotic subatomic particles left over from the big bang and known generically as WIMPs, for weakly interacting massive particles, which can pass through the Earth like smoke through a screen door.

Some particle physicists hope to produce them in the Large Hadron Collider outside Geneva or to read their signature in cosmic rays from outer space. An experiment to do just that, the Alpha Magnetic Spectrometer, is scheduled to be launched into space and installed on the International Space Station at the end of April 2011. Other physicists, including Dr. Aprile's team, have been trying to catch the putative particles in detectors set far underground to guard against contamination from cosmic rays.

For the last year the eyes of the physics world have been on Dr. Aprile's experiment in the Gran Sasso National Laboratory, part of Italy's National Institute of Nuclear Physics, which is widely acknowledged as the biggest and most sensitive detector out there. She hopes to record the characteristic signal—a bump and a flash—of the rare collision of a WIMP with a xenon nucleus. The experiment began in 2010 and ran for 100 days.

At the push of a button the data, unseen until now to guard against unconscious bias, would begin flowing through an analysis pipeline and show up as red dots on a big computer screen.

On a table in the corner was a stack of folded yellow notepapers, on which collaboration members had written their bets on how many events—putative dark matter detections—would be recorded. They ranged from 20, by an optimistic graduate student, to 2 from a skeptical astrophysicist. The tension and giddiness in the room rose as the 10:30 deadline came and went, due to computer glitches.

Finally, the promised graph appeared on the screen, showing the first of 91 batches of data. A red dot appeared, the first event signal. It was rapidly joined by another, and then another, each accompanied by a sharp intake of breath in the room.

"Oh, God," Dr. Aprile said as the count rose to four. "I can't sit anymore." She got up from her chair.

There were more oohs and ahs as the count climbed to six, more than would be expected from background radioactivity in the detector, and finally stopped.

Everybody clapped, and Dr. Aprile went around the room offering hugs and kissing cheeks. But the results, she admitted, were ambiguous.

"Six points mean nothing until they have been analyzed," she said. "I feel optimistic about the future. We have a lot more to do."

Indeed, the collaborators soon threw out three of those points, concluding that they had been caused by noise in the electronics.

"We knew within 10 minutes," said Rafael Lang of Columbia. "It was totally obvious."

That left them with three events, compared with two expected from background, not a large enough disparity to claim evidence of a WIMP. On Wednesday Dr. Aprile's group posted a paper on the physics Web site www.arXiv.com saying it had not detected any WIMPs yet; the paper has also been submitted to the journal *Physical Review Letters*.

But the group refused to be disappointed. The results, members said, had set new and stringent limits on the nature of the putative dark matter particles, eliminating some theoretical models, as well as showing that their detector was performing up to snuff. Dr. Aprile called it "a spectacular result."

Neal Weiner, a particle theorist at New York University, agreed, noting that these were only the first results from an experiment that will go on for years and get more sensitive. If there is any dark matter in their data set, they will not have to wait years to find out, he said. "We just to have to wait for later this year."

Dr. Lang said, "It's the feeling of the community that something new and big is just around the corner. We are not there just yet but maybe we are not far from it, and this is very exciting."

Dr. Aprile said they would definitely be doing this again.

In an e-mail from Italy, she wrote, "I know there is nothing more exciting than a signal, but when we are searching for the unknown, the more we probe the closer we get to truth."

—*April 13, 2011*

There's More to Nothing Than We Knew

By DENNIS OVERBYE

Why is there something, rather than nothing at all?

This is, perhaps, the mystery of last resort. Scientists may be at least theoretically able to trace every last galaxy back to a bump in the big bang, to complete the entire quantum roll call of particles and forces. But the question of why there was a big bang or any quantum particles at all was presumed to lie safely out of scientific bounds, in the realms of philosophy or religion.

Now even that assumption is no longer safe, as exemplified by a new book by the cosmologist Lawrence M. Krauss. In it he joins a chorus of physicists and cosmologists who have been pushing into sacred ground, proclaiming more and more loudly in the last few years that science can explain how something—namely our star-spangled cosmos—could be born from, if not nothing, something very close to it. God, they argue, is not part of the equation. The book, *A Universe from Nothing,* is a best seller and follows recent popular tomes like *God Is Not Great,* by the late Christopher Hitchens; *The God Delusion,* by Richard Dawkins; and *The Grand Design,* by the British cosmologist Stephen Hawking (with Leonard Mlodinow), which generated headlines two years ago with its assertion that physicists do not need God to account for the universe.

Dr. Krauss is a pint-size spark plug of erudition and ambition, who often seems to be jetting off in several directions at once on more missions than can be listed on a business card. Among other things he is foundation professor and director of the Origins Project at Arizona State University.

And he knows his universe. In 1995, he and Michael S. Turner of the University of Chicago made waves by arguing that many of the paradoxes regarding cosmology could be resolved if a large portion of the cosmos resided in the form of a hitherto-undiscovered energy, known then as the cosmological constant. Three years later astronomers discovered that the expansion of the universe was being accelerated by some "dark energy" that behaves exactly like the cosmological constant.

Dr. Krauss is also a prolific author of popular science books, including *The Physics of Star Trek*. And he has been an outspoken critic of attempts to introduce creationist ideas and to censor the teaching of evolution in schools and textbooks.

The new book grew out of a talk he gave in 2009 that got more than a million hits on YouTube.

The point of the book, Dr. Krauss, a self-described nonbeliever, writes at the outset, is not to try to make people lose their faith, but to illuminate how modern science has changed the meaning of nothingness from a vague philosophical concept to something we can almost put under a lab microscope.

How well you think he succeeds might depend on how far you yourself want to go down the rabbit hole of nonbeing. Why, for example, should we assume that nothingness is more natural than somethingness? Indeed, you might ask why it is that we think there is something here at all. The total energy of the universe might actually be zero, according to the strange bookkeeping of Einstein's general theory of relativity, as Dr. Krauss points out. "The universe," Alan H. Guth, a physicist at MIT, likes to say, "might be the ultimate free lunch." Even space and time themselves might be a kind of holographic illusion, string theorists say.

You might think to dispute this by kicking a rock, but remember that both the rock and your foot are mostly empty space, prevented from intermingling by electric fields.

Dr. Krauss delineates three different kinds of nothingness. First is what may have passed muster as nothing with the ancient Greeks: empty space. But we now know that even empty space is filled with energy, vibrating with electromagnetic fields and so-called virtual particles dancing in and out of existence on borrowed energy courtesy of the randomness that characterizes reality on the smallest scales, according to the rules of quantum theory.

Second is nothing, without even space and time. Following a similar quantum logic, theorists have proposed that whole universes, little bubbles of space-time, could pop into existence, like bubbles in boiling water, out of this nothing.

There is a deeper nothing in which even the laws of physics are absent. Where do the laws come from? Are they born with the universe, or is the universe born in accordance with them? Here Dr. Krauss, unhappily in my view, resorts to the newest and most controversial toy in the cosmologist's toolbox: the multiverse, a nearly infinite assemblage of universes, each with its own randomly determined rules, particles and forces, that represent solutions to the basic equations of string theory—the alleged theory of everything, or perhaps, as wags say, anything.

Within this landscape of possibilities, almost anything goes.

But even the multiverse is not totally lawless, as Dr. Krauss acknowledged. We are not quite there yet. At the very least, there would still be the string equations and those quantum principles that undergird them. Is quantum randomness the secret of existence?

"Maybe in the true eternal multiverse there are truly no laws," Dr. Krauss said in an e-mail. "Maybe indeed randomness is all there is and everything that can happen happens somewhere."

It would be silly to think that we won't have better answers and better questions 50 or 100 years from now, but for the moment this is the story science can tell. If you find it bleak, that is your problem. "The universe is the way it is, whether we like it or not," Dr. Krauss writes.

It gets worse.

If nothing is our past, it could also be our future. As the universe, driven by dark energy—that is to say, the negative pressure of nothing—expands faster and faster, the galaxies will become invisible, and all the energy and information will be sucked out of the cosmos. The universe will revert to nothingness.

Nothing to nothing.

One day it's all going to seem like a dream.

But who is or was the dreamer?

—February 20, 2012

At the End of the Earth, Seeking Clues to the Universe

By SIMON ROMERO

Trucks stall on the road to the plateau called Llano de Chajnantor in Chile, 16,597 feet up in the Atacama Desert, where scientists are installing one of the world's largest ground-based astronomical projects. Heads ache. Noses bleed. Dizziness overcomes the researchers toiling near the in the shadow of the Licancabur volcano.

"Then there's what we call 'jelly legs,'"said Diego García-Appadoo, a Spanish astronomer studying galaxy formation. "You feel shattered, as if you ran a marathon."

Still, the same conditions that make the Atacama, Earth's driest desert, so inhospitable make it beguiling for astronomy. In northern Chile, it is far from big cities, with little light pollution. Its arid climate prevents radio signals from being absorbed by water droplets. The altitude, as high as the Himalaya base camps for climbers preparing to scale Mount Everest, places astronomers closer to the heavens.

Opened in October 2011, the Atacama Large Millimeter/submillimeter Array, known as ALMA, will have spread 66 radio antennas near the spine of the Andes by the time it is completed next year. Drawing more than $1 billion in funding mainly from the United States, European countries and Japan, ALMA will help the oxygen-deprived scientists flocking to this region to study the origins of the universe.

The project also strengthens Chile's position in the vanguard of astronomy. Observatories are already scattered throughout the Atacama, including the Cerro Paranal Observatory, where scientists discovered in 2010 the largest star observed to date, and the Cerro Tololo Inter-American Observatory, which was founded in 1961 and endured Chile's tumult of revolution and counterrevolution in the 1970s.

But ALMA opens a new stage for astronomy in Chile, which is now favored by international research organizations for the stability of its economy and legal system. Like other radio telescopes, ALMA does not detect optical light but radio waves, allowing researchers to study parts of the universe that are dark, like the clouds of cold gas from which stars are formed.

With ALMA, astronomers hope to see where the first galaxies were formed, and perhaps even detect solar systems with the conditions to support life, like water-bearing planets. But the scientists here express caution about their chances of finding life elsewhere in the universe, explaining that such definitive proof is likely to remain elusive.

"We won't be able to see life, but perhaps signatures of life," said Thijs de Graauw, a Dutch astronomer who is ALMA's director.

Still, scientists believe ALMA will make transformational leaps possible in the understanding of the universe, enabling a hunt for so-called cold gas tracers, which are ashes of exploded stars from a time about a few hundred million years after the big bang that astronomers call "cosmic dawn."

ALMA's construction, said Jesús Mosterín, a prominent Spanish philosopher who writes about the frontier between science and philosophy, and who visited the observatory last year, is taking place at "the only time in history that windows into the universe are being thrown wide open."

Chile is not the only country luring big investments in astronomical projects. South Africa and Australia are competing to host an even bigger radio telescope, the Square Kilometre Array, which would be fully operational by 2024. China has begun building its own large radio telescope in a craterlike setting in the southern province of Guizhou.

At the same time, the financial crisis in rich industrialized countries has raised concerns that funding for some ambitious astronomy projects could face constraints. In the United States, a congressional panel proposed killing NASA's James Webb Space Telescope in 2011 before a compromise spending plan saved the project.

"It would be very sad for humankind if we were so spiritually decadent to forgo the pleasures of consciousness and of knowledge," said Mr. Mosterín, reflecting on the funding choices political leaders need to make. "These things make human beings a very interesting animal indeed."

ALMA's Operations Support Facility, an outpost built for the scientists here in the Atacama, offers a glimpse into the lengths to which people go for astronomical discoveries. From Chajnantor, where dust devils dance across the plain, and unusually extreme weather in recent months has included rains and sand-storms, a dirt road runs to the facility past towering cactuses and herds of wild donkeys and vicuñas.

The facility, at an altitude of about 9,500 feet, houses about 500 researchers and other staff in shipping containers turned living quarters. In a system similar to that on offshore oil platforms, scientists have daily shifts lasting up to 12 hours for 8 days straight. Many toil through the night.

"Quiet Zone," reads one sign in an area of containers for ALMA's so-called day sleepers.

Supervisors enforce other rules, ensuring a work environment almost as austere as the surrounding Mars-esque landscape. Alcohol is prohibited, and those found drinking after trips into San Pedro de Atacama, a town about 30 minutes away by car, must dry out at a security checkpoint before entering the futuristic complex.

In the control room, where astronomers spend hours peering into screens displaying the array of antennas, some gallows humor prevails. "We are well in the control room, the 17," reads one message scribbled on a piece of roofing and posted on the wall after fire alarms went off by accident at the facility in 2011, sealing those inside the control room until they broke a door to escape.

The note riffs on the 2010 mine accident and subsequent rescue of 33 miners in the Chilean desert, during which the trapped men sent a note to the surface saying, "We are well in the shelter, the 33."

Developments elsewhere in Chile occasionally raise eyebrows in Chajnantor, like antigovernment protests that have rocked remote regions of the country this year and spread in March to the nearby mining city of Calama. "The protests are not directly a concern," said Mr. de Graauw, ALMA's director. "They are part of a democratic process, not a revolution."

Still, it seems at times that the astronomers stationed here are as far removed from the world around them as the miners working beneath other parts of the Atacama. English predominates as the observatory's language, tying together scientists from dozens of countries.

A sense of awe still accompanies the installation of each new antenna. Two giant German-manufactured transporters, each with 28 tires and engines equivalent to two Formula 1 racing vehicles, are used to transport the antennas. Called Otto and Lore, they look like massive mechanized centipedes making their way across the arid landscape.

"There's a quietness that comes to you at Chajnantor," said Lutz Stenvers, a German engineer who came here in 2008 to lead a team from General Dynamics building the antennas. "I can see why this place was chosen."

—*April 7, 2012*

American Physics Dreams Deferred

By DENNIS OVERBYE

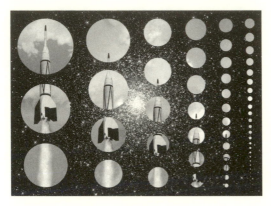

When three American astronomers won the Nobel Prize in physics in 2011, for discovering that the expansion of the universe was speeding up in defiance of cosmic gravity—as if change fell out of your pockets onto the ceiling—it reaffirmed dark energy, the glibly named culprit behind this behavior, as the great cosmic surprise and mystery of our time.

And it underscored the case, long urged by American astronomers, for a NASA mission to measure dark energy—to determine, for example, whether the cosmos would expand forever or whether, perhaps, there might be something wrong with our understanding of gravity.

In 2019, a spacecraft known as *Euclid* will begin such a mission to study dark energy. But it is being launched by the European Space Agency, not NASA, with American astronomers serving only as very junior partners, contributing $20 million and some infrared sensors.

For some scientists, this represents an ingenious solution, allowing American astronomers access to the kind of data they will not be able to obtain on their own until NASA can mount its own, more ambitious mission in 2024.

But for others, it is a setback. It means that for at least the next decade, Americans will be relegated to a minor role in following up on their own discovery.

"While it's great to support other missions," said Adam Riess of Johns Hopkins and the Space Telescope Science Institute, who shared that Nobel last year, "it would be disappointing to see the U.S. lose or outsource its own leading role in one of the hottest areas of research."

For Dr. Riess and his colleagues, this turn of events is another example of a worrying trend in which American scientists, facing budget deficits and political gridlock, have had to pull back from or delay promising projects while teams based in Europe hunt down the long-sought Higgs boson or rocket scientists in China plan a moon landing in 2025.

Michael Turner, a cosmologist at the University of Chicago, called dark energy "an example of how the U.S. seems to misplay its science hand these days."

"We predicted and discovered dark energy," he said. "We have the biggest dark-energy community and the best ground game; we have been designing a space mission since 1998; and now the Europeans will fly it with our minor participation. Something is wrong with this picture."

Saul Perlmutter, of the University of California at Berkeley, another of the dark energy Nobel winners, said, "The danger, of course, is that we will watch the science [and scientists—and good students] move on to other countries and continents, where projects are being begun and completed."

With them, scientists say, could go the cultural excitement and innovative spark that invigorates the economy. The World Wide Web, for example, was invented at CERN, the European Organization for Nuclear Research (home to the world's most powerful particle accelerator, the Large Hadron Collider), to help particle physicists communicate.

By contrast, the United States' flagship lab for high-energy physics, the Fermi National Accelerator Laboratory, known as Fermilab, had to close down its accelerator, the Tevatron, last fall, and learned from the Energy Department in March 2012 that the agency could not afford to follow through for now on a $1.3 billion underground experiment to study the spooky shape-shifting properties of particles known as neutrinos in an effort to investigate why the universe is made of matter and not antimatter.

At the same time, the department also canceled money for studies for the world's next big physics machine, the International Linear Collider, which would be the successor to CERN's giant collider. American scientists are resigned to the likelihood that it will not be built in the United States.

American physicists are now rethinking how to carry out the neutrino experiment, which was to have been the centerpiece of a plan to convert the old 8,000-foot-deep Homestake gold mine in Lead, S.D., into a national laboratory for underground science.

Similar facilities in Italy, Canada and Japan have become centers in the search for dark matter, neutrino experiments and other delicate work that requires shelter from cosmic rays. But in 2010, the National Science Foundation walked away from the $875 million project, citing unease about safety and "stewardship" of the old mine, which is half full of water. Meanwhile, with support from a philanthropist, T. Denny Sanford, the state of South Dakota has reopened parts of the mine for a pair of physics experiments.

Of course, there is no achievement of modern American science—from the Manhattan Project to the Hubble Space Telescope to the decoding of the human genome—that does not owe a debt to hard and even heroic bargaining in the formerly smoke-filled rooms of Congress and the White House. Complaints and grim prognostications about the federal research budget are part of the background music of science. The situation is always fluid.

Given all that, other scientists say that basic research is doing as well as can be expected, given severe budgetary restraints. Among other things, NASA's Webb telescope, the successor to the Hubble, is on target for launching in 2018, at a cost of $8 billion.

The science office in the Energy Department actually got an increase in the budget released by President Obama in February 2012, but the money went to more applied research, into areas like energy and scientific computing.

Only last month, April 2012, Congress added money to the neutrino experiment, although not enough to get the whole project back on track, said Katie Yurkewicz, Fermilab's spokeswoman, and the Association of American Universities issued a statement applauding appropriators "for their bipartisan actions thus far to sustain the nation's investment in scientific research."

Debra Elmegreen, a professor of astronomy at Vassar and president of the American Astronomical Society, has spent a lot of time in Washington lately. "Congress seems supportive of science," she said. "I'm encouraged that people recognize the need for science and technology to continue."

She added: "U.S. leadership is at risk in practically every area. We'll get by for now, but we can't be complacent."

Indeed, all bets could be off if the automatic budget cuts, called "sequestration," decreed by the failure of the deficit reduction negotiations in the summer of 2011, go into effect in January 2013. A certain amount of gloom has reached the most prestigious levels of American science. Writing in a recent issue of *The New York Review of Books,* Steven Weinberg, a Nobel laureate at the University of Texas in Austin, decried the fact that science was increasingly having to compete with other worthy causes, like health care and education, for money. The solution, he said, was to raise taxes.

Frank Wilczek, a Nobel laureate at the Massachusetts Institute of Technology, described himself as "less gloomy" than Dr. Weinberg, but nonetheless concerned that, with the end of the cold war and bad economic times, the traditional political bases of government support for basic research had dwindled, leaving only "curiosity and desire to participate, even indirectly, in doing something great." He added, "We should be doing that anyway, of course."

"This is all a great pity, because tremendous scientific opportunities are available," Dr. Wilczek said.

The United States gave up its leadership in high-energy particle physics 20 years ago, when Congress canceled the Superconducting Supercollider, a particle accelerator that was under construction in Texas. That move cleared the way for CERN's eventual supremacy.

Could the same thing happen in space? Those fears were aroused in 2010 when NASA announced that the James Webb Space Telescope project needed $1.6 billion more and several years to complete. It was canceled by the House Appropriations Committee in the summer of 2011, but later restored. The price of that rescue, however, was to delay NASA's dark energy mission, and to withdraw from a couple of upcoming joint Mars missions.

The travails of the dark energy researchers provide a window into the labyrinthine process by which big projects live or die. Whether you find the story hopeful or depressing depends on who you are.

Dark energy is, according to Dr. Wilczek, "the most mysterious fact in all of physical science, the fact with the greatest potential to rock the foundations."

The discovery that the expansion of the cosmos was speeding up came from observing exploding stars known as Type IA supernovas, luminous and uniform enough to serve as distance markers. Realizing that only a telescope in space could find and measure supernovas distant enough to shed light on the genesis of this behavior, Dr. Perlmutter early on urged the construction of a special space probe.

In 2010, after a decade of wrangling among astronomers and NASA and the Energy Department, a blue-ribbon panel from the National Academy of Sciences charged with determining astronomical priorities endorsed a version of this idea as the highest-priority space science mission for the coming decade. The billion-dollar mission, called Wfirst, for Wide-Field Infrared Survey Telescope, would search for exoplanets as well as measure the effects of dark energy on the history and evolution of the universe. The academy's deliberations were ambushed, however, by the subsequent announcement of the Webb telescope's problems, which have pushed the Webb launch all the way to 2018.

Lia LaPiana, a program executive at NASA, said that, from 2013 to 2018, there is "zero money" in the president's budget for the mission.

As a down payment on an eventual mission, NASA suggested spending about $200 million for a 20 percent stake in Euclid, the European Space Agency's mission. The academy rejected that idea, saying that Euclid, which would not measure supernovas at all, did not meet the academy's requirements and could undermine the possibility of an eventual Wfirst project.

Recently, however, a specially convened committee of the academy has given the nod to a $20 million investment in Euclid in the form of equipment, saying it would give American astronomers a seat on the Euclid science team and access to its data.

What changed in the last year? David Spergel, an astrophysicist at Princeton, who was chairman of the committee, said, "For 20 percent, we were offered a modest participation. For 2 percent, we were offered a modest participation."

Paul Schechter of MIT, cochairman of a team planning Wfirst, said he had endorsed the deal on the grounds that it would not impinge on the Wfirst project. When the president's budget was released, however, there was $9 million for *Euclid* and no money for beginning the Wfirst project. The budget is only the first step in this dance, however.

In March 2012, Dr. Schechter told a House Appropriations subcommittee that the budget did not represent the "strong U.S. commitment" to Wfirst that the academy had recommended. He asked for $8 million to get things going.

The House panel agreed and responded in its report by directing NASA to explain how its plans were consistent with the academy's recommendations. In the meantime, the corresponding Senate Appropriations subcommittee has put $10 million for the NASA dark energy mission into its proposed budget report, citing last year's physics Nobelists by name.

There are more dance moves to make in this year's budget and the rest of the decade before the Wfirst mission can become a reality, but Dr. Schechter said he felt encouraged. "I suspect Wfirst wasn't very high on anyone's list of priorities," he said, referring to turmoil at NASA last year.

He added, "I think that is changing."

—May 21, 2012

Appendices

NOBEL LAUREATES IN PHYSICS

The Nobel Prize, established under the will of Alfred Nobel, the Swedish entrepreneur who made his fortune in dynamite, has been awarded in physics and other scientific fields since 1901. Here is a list of the physics laureates with brief descriptions of the work for which they were honored.

1901   William Conrad Röntgen of Germany, for the discovery of X-rays.

1902   Hendrik Antoon Lorentz and Pieter Zeeman of the Netherlands, for research on magnetism and radiation.

1903   Antoine Henri Becquerel of France, for the discovery of radioactivity and Pierre Curie of France and Marie Sklodowska Curie of Poland and France, for their work on radiation.

1904   John William Strutt (Lord Rayleigh) of the United Kingdom, for research on gases.

1905   Philipp Eduard Anton von Lenard of Germany, for work on cathode rays.

1906   Joseph John Thompson of the United Kingdom, for discovering the electron and for research on the way gases conduct electricity.

1907   Albert A. Michelson of the United States, for development of precision optical instruments and for "meteorological investigations" done with them.

1908   Gabriel Lippmann of France, for work on reproducing colors photographically.

1909   Gugliemo Marconi of Italy and Karl Ferdinand Braun of Germany, for work on wireless telegraphy.

1910   Johannes Diderik van der Waals of the Netherlands, for work on gases and liquids.

1911   Wilhelm Wien of Germany, for research on heat.

1912   Nils Gustaf Dalén of Sweden, for the invention of valves for use with gases.

1913   Heike Kamerlingh Onnes of the Netherlands, for studies of matter at low temperatures, including the invention of liquid helium.

1914   Max von Laue of Germany, for the discovery of X-ray diffraction by crystals.

1915   William Henry Bragg and William Lawrence Bragg of the United Kingdom, for the use of X-rays on the analysis of crystal structures.

1916   No prize given.

1917   Charles Glover Barkla of the United Kingdom, for work on X-ray spectroscopy.

1918   Max Karl Ernst Ludwig Planck of Germany, for his discovery of energy quanta.

1919   Johannes Stark of Germany, for his research on spectra in electric fields.

1920   Charles Édouard Guillaume of Switzerland, for his studies of anomalies in nickel alloys.

1921   Albert Einstein of Germany, "for his services to theoretical physics."

1922   Niels Hendrik David Bohr of Denmark, for his elucidation of the structure of atoms.

1923   Robert Andrews Millikan of the United States, for research on electricity and the photoelectric effect.

1924   Manne Georg Siegbahn of Sweden, for work on X-ray spectroscopy.

1925   James Franck and Gustav Hertz of Germany, for work on the impact of electrons on atoms.

1926   Jean Baptiste Perrin of France, for work on the structure of matter.

1927   Arthur Holly Compton of the United States for research on the scattering of electrons in matter and Charles Thomson Rees Wilson of the United Kingdom, for developing a method of detecting the movements of electrically charged particles through vapor.

1928   Owen Willans Richardson of the United Kingdom, for research on heat-induced flow of electrons.

1929   Prince Louis-Victor Pierre Raymond de Broglie of France, for discovering the wave nature of electrons.

1930   Sir Chandrasekhara Venkata Raman of India, for research on light.

1931   No prize given.

1932   Werner Heisenberg of Germany, "for the creation of quantum mechanics."

1933   Erwin Schrödinger of Austria and Paul Dirac of the United Kingdom, for work on the atomic theory.

1934   James Chadwick of the United Kingdom, for discovering the neutron.

1935 Victor Francis Hess of Austria, for discovering cosmic radiation.

1936 Carl David Anderson of the United States, for the discovery of the positron.

1937 Clinton Joseph Davisson of the United States and George Paget Thomson of the United Kingdom, for research on the diffraction of electrons by crystals.

1938 Enrico Fermi of Italy, for his discovery of new elements produced by neutron irradiation and the related discovery of nuclear reactions brought about by the action of neutrons.

1939 Ernest Lawrence of the United States, for the invention of the cyclotron particle accelerator and related research.

1940 No prize given.

1941 No prize given.

1942 No prize given.

1943 Otto Stern of the United States, for studies of molecules, atoms and atomic nuclei, particularly the action of magnetic fields on them.

1944 Isidor Isaac Rabi of the United States, for developing a way of recording the magnetic properties of atomic nuclei.

1945 Wolfgang Pauli of Austria, for discovering that no two electrons in an atom can have the same quantum number.

1946 Percy Williams Bridgman of the United States, for inventing ways to produce high pressure, and for research in high-pressure physics.

1947 Edward Victor Appleton of the United Kingdom, for research on the physics of the upper atmosphere.

1948 Patrick Maynard Stuart Blackett of the United Kingdom, for discoveries in nuclear physics and cosmic radiation.

1949 Hideki Yukawa of Japan, for predicting the existence of mesons.

1950 Cecil Frank Powell of the United Kingdom, for developing photographic methods of studying subatomic particles.

1951 John Douglas Cockcroft of the United Kingdom and Ernest Thomas Sinton Walton of Ireland, for work on the effects of bombarding atomic nuclei with accelerated particles.

1952 Felix Bloch and Edward Mills Purcell of the United States, for developments in nuclear magnetic precision measurements.

1953 Frits Zernike of the Netherlands, for inventions in microscopy.

1954 Max Born of the United Kingdom, for research on quantum mechanics, and Walther Bothe of West Germany, for work confirming the existence of radiation quanta.

1955 Willis Eugene Lamb of the United States, for discoveries concerning the structure of the hydrogen spectrum, and Polykarp Kusch of the United States, for research on the electron.

1956 John Bardeen, Walter Houser Brattain and William Bradford Shockley, for inventing the transistor.

1957 Tsung-Dao Lee and Chen Ning Yang of China, for research on elementary particles.

1958 Pavel Alekseyevich Cherenkov, Il'ja Mikhailovich Frank and Igor Yevgenyevich Tamm of the Soviet Union, for research on electromagnetic radiation.

1959 Owen Chamberlain of the United States and Emilio Gino Segrè of Italy, for discovering the antiproton.

1960 Donald Arthur Glaser of the United States, for inventing the bubble chamber.

1961 Robert Hofstadter of the United States, for studies of subatomic particles, and Rudolf Ludwig Mössbauer of West Germany, for research on the absorption of gamma radiation.

1962 Lev Davidovich Landau of the Soviet Union, for his theories on condensed matter, especially liquid helium.

1963 Eugene Paul Wigner of Hungary and the United States, for discoveries relating to the symmetry found in elementary particles, and Maria Goeppert Mayer of the United States and J. Hans D. Jensen of West Germany, for discoveries relating to the structure of the atomic nucleus.

1964 Nicolay Gennadiyevich Basov and Alexander Prokhorov of the Soviet Union and Charles Hard Townes of the United States, for work on quantum electronics, including the construction of oscillators and amplifiers.

1965 Richard Phillips Feynman and Julian Schwinger of the United States and Shinichiro Tomonaga of Japan, for fundamental work in quantum electrodynamics.

1966 Alfred Kastler of France, for developing ways to study resonance in atoms.

1967 Hans Albrecht Bethe of the United States, for research on nuclear reactions, especially energy production in stars.

1968 Luis Walter Alvarez of the United States, for research in elementary particle physics.

1969 Murray Gell-Mann of the United States, for work on the classification of elementary particles and their interactions.

1970 Hannes Olof Gösta Alfvén of Sweden, for research in plasma physics and Louis Néel of France, for work in solid-state physics.

1971 Dennis Gabor of the United Kingdom, for inventing the holographic process.

1972 John Bardeen, Leon Neil Cooper and John Robert Schrieffer of the United States, for developing the theory of superconductivity.

1973 Leo Esaki of Japan and Ivar Giaever of the United States and Norway, for their work on semiconductors and superconductors, and Brian David Josephson of the United Kingdom, for work on tunneling phenomena in matter.

1974 Sir Martin Ryle and Antony Hewish of the United Kingdom, for research in radio astrophysics.

1975 Aage Niels Bohr and Ben Roy Mottelson of Denmark and Leo James Rainwater of the United States, for research on the structure of the atomic nucleus.

1976 Burton Richter and Samuel Chao Chung Ting of the United States, for work on heavy elementary particles.

1977 Philip Warren Anderson and John Hasbrouck van Vleck of the United States and Sir Nevill Francis Mott of the United Kingdom, for work on the electronic structure of magnetic systems.

1978 Arno Allan Penzias and Robert Woodrow Wilson of the United States, for discovering cosmic microwave background radiation, and Pyotr Leonidovich Kapitsa of the Soviet Union, for his work on low-temperature physics.

1979 Sheldon Lee Glashow and Steven Weinberg of the United States and Abdus Salam of Pakistan, for work on the weak electromagnetic interactions between elementary particles.

1980 James Watson Cronin and Val Logsdon Fitch of the United States, for work on elementary particles.

1981 Nicolaas Bloembergen and Arthur Leonard Schawlow of the United States, for contributions to laser spectroscopy, and Kai M. Siegbahn of Sweden, for work on electron spectroscopy.

1982 Kenneth G. Wilson of the United States, for work on phase transitions in matter.

1983 Subrahmanyan Chandrasekhar of India and the United States, for research on the structure and formation of stars, and William Alfred Fowler of the United States, for studies of the formation of elements in the universe.

1984 Carlo Rubbia of Italy and Simon van der Meer of the Netherlands, for research on the weak interaction of subatomic particles.

1985 Klaus von Klitzing of West Germany, for findings on the movement of electrons in semiconductors.

1986 Ernst Ruska of West Germany, for inventing the electron microscope, and Gerd Binnig of West Germany and Heinrich Rohrer of Switzerland, for developing the scanning tunneling microscope.

1987 Johannes Georg Bednorz of West Germany and K. Alexander Müller of Switzerland, for discovering superconductivity in ceramic materials.

1988 Leon M. Lederman, Melvin Schwartz and Jack Steinberger of the United States, for research on subatomic particles.

1989 Norman F. Ramsey of the United States, for work that contributed to, among other things, the development of the atomic clock, and Hans Georg Dehmelt of the United States and Wolfgang Paul of West Germany, for developing a method of studying a single electron or ion.

1990 Jerome I. Friedman and Henry W. Kendall of the United States and Richard E. Taylor of Canada, for work developing the quark model in particle physics.

1991 Pierre-Gilles de Gennes of France, for devising ways of studying complex forms of matter.

1992 Georges Charpak of France, for work on particle detectors.

1993 Russell Alan Hulse and Joseph H. Taylor Jr. of the United States, for discovering a new type of pulsar.

1994 Bertram Brockhouse of Canada and Clifford G. Shull of the United States, for development of neutron-scattering techniques for studying condensed matter.

1995 Martin Lewis Perl and Frederick Reines of the United States, for research on the physics of leptons.

1996 David Morris Lee, Douglas D. Osheroff and Robert Coleman Richardson of the United States, for the discovery of superfluidity in helium-3.

1997   Steven Chu and William Daniel Phillips of the United States and Claude
       Cohen-Tannoudji of France, for developing ways to cool and trap atoms
       with lasers.
1998   Robert B. Laughlin and Daniel Chee Tsui of the United States and Horst
       Ludwig Störmer of Germany, for work on quantum fluids.
1999   Gerard 't Hooft and Martinus J. G. Veltman of the Netherlands, for work
       on the quantum structure of electroweak interactions in physics.
2000   Zhores I. Alferov of Russia and Herbert Kroemer of Germany, for
       developing semiconductors for opto-electronics, and Jack S. Kilby of the
       United States, for work on the integrated circuit.
2001   Eric A. Cornell and Carl E. Wieman of the United States and Wolfgang
       Ketterle of Germany, for fundamental studies of Bose-Einstein
       condensates.
2002   Raymond Davis Jr., of the United States and Masatoshi Koshiba of
       Japan, for detecting cosmic neutrinos, and Riccardo Giacconi of the
       United States, for work leading to the discovery of cosmic X-ray sources.
2003   Alexei A. Abrikosov and Vitaly L. Ginzburg of Russia and Anthony
       J. Leggett of the United Kingdom and the United States, for work on
       superconductors and superfluids.
2004   David J. Gross, H. David Politzer and Frank Wilczek of the United
       States, for the discovery of asymptotic freedom in the theory of the strong
       interaction.
2005   Roy J. Glauber of the United States, for work on optical coherence, and
       John L. Hall of the United States and Theodor W. Hänsch of Germany,
       for work in laser-based spectroscopy.
2006   John C. Mather and George F. Smoot of the United States, for work on
       cosmic microwave background radiation.
2007   Albert Fert of France and Peter Grünberg of Germany, for discovering a
       technique for reading data on hard disks.
2008   Makoto Kobayashi, Toshihide Maskawa and Yoichiro Nambu, all of
       Japan, for discoveries relating to symmetry in subatomic particles.
2009   Charles Kuen Kao of the United Kingdom and the United States, for work on
       fiberoptic communication, and Willard S. Boyle of Canada and the United
       States and George E. Smith of the United States, for work on semiconductors.
2010   Andre Geim of Russia and the Netherlands and Konstantin Novoselov of
       Russia and the United Kingdom, for work on two-dimensional graphene.

2011   Saul Perlmutter and Adam G. Riess of the United States and Brian
       P. Schmidt of Australia and the United States, for the discovery of
       the accelerating expansion of the universe through observations of
       supernovae.

2012   Serge Haroche and David J. Wineland, for ground-breaking
       experimental methods that enable measuring and manipulation of
       individual quantum systems.

PHYSICS TIMELINE

Selected major developments in physics and astrophysics since *The New York Times* began publishing in 1851.

1851	William Thomson (Lord Kelvin) calculates the temperature of absolute zero—the total absence of heat. It is –459.67 degrees Fahrenheit or –273.15 degrees Celsius.
1868	James Clark proposes the electromagnetic nature of light, assuming waves move in an ether.
1887	Heinrich Hertz generates electromagnetic waves.
1888	Albert A. Michelson and E. W. Morley demonstrate that the ether, a substance believed to fill all of space, does not exist.
1895	Wilhelm Röntgen discovers X-rays.
1896	Henri Becquerel discovers radioactivity.
1897	J. J. Thompson discovers the electron.
1898	Pierre and Marie Curie discover radium and polonium, radioactive elements.
1900	Max Planck describes "quanta" of light—energy has discrete levels or quanta.
1905	Albert Einstein's "miracle year" in which, among other things, he publishes his special theory of relativity and postulates the equivalence of mass and energy ($E = mc^2$).
1909	Robert Millikan measures the charge of the electron.
1911	Ernest Rutherford discovers the nucleus of the atom.
1912	Victor Hess, in a hot air balloon, discovers penetrating, ionizing (cosmic) radiation from space.
1913	Niels Bohr proposes the quantum theory of electron orbits in the atom.
1915	Einstein publishes his general theory of relativity, a description of gravity in terms of curvature of space-time.
1919	Researchers observe light from a star bent by the sun's gravitational field.
1923	Working with the 100-inch telescope at Mount Wilson, in California, Edwin Hubble shows that galaxies exist outside the Milky Way.

1923   Louis de Broglie predicts the wave nature of particles.

1924   Satyendranath Bose and Einstein theorize about subatomic particles, called bosons, which act on atoms.

1925–26 The basic theory of subatomic particles, called quantum mechanics, is developed in two forms—one in 1925 by Werner Heisenberg, another in 1926 by Erwin Schrödinger. Though their theoretical mechanisms differ, they yield the same results.

1926   Heisenberg proposes the uncertainty principle.

1927   Georges Lemaître proposes that the universe began in a "big bang."

1927   Niels Bohr posits the theory of complementarity, which holds, among other things, that results of measuring objects in the quantum realm vary with the measuring method used.

1929   Hubble discovers the universe is expanding.

1930   E. O. Lawrence and Milton S. Livingston invent the cyclotron particle accelerator.

1930   Paul A. M. Dirac completes his theory of the electron; one of his equations predicts a particle that would be the mirror image of an electron, but with a positive charge. Carl Anderson discovers that particle, the positron, in 1932.

1932   Jan Oort postulates the existence of "dark matter."

1932   James Chadwick identifies the neutron.

1933   Wolfgang Pauli proposes the existence of an electrically neutral, weak particle—the neutrino; 25 years later its existence is confirmed.

1934   Irène and Frédéric Joliot-Curie show that elements can be made radioactive.

1934   Engineers in Germany perfect the electron microscope.

1937   The first radio telescope is built.

1938   Otto Hahn, Lise Meitner, and Fritz Straussmann demonstrate that the uranium nucleus can split, producing fission, when struck with a neutron.

1938   Hans Bethe and Carl von Weizsäcker theorize that in the intense heat and pressure in the interior of a star, hydrogen nuclei combine with each other to form helium. In 1952 this theorizing bears fruit in the form of the fusion (hydrogen) bomb.

1939   Einstein writes a letter advising President Franklin D. Roosevelt that recent research points to the possibility of creating "powerful bombs of a new type." In 1942, the Manhattan Project to build an atomic bomb begins.

1942	Enrico Fermi obtains the first self-sustaining nuclear fission (chain) reaction.
1945	The Manhattan Project produces atomic weapons. The first bomb is exploded at the Trinity Site in New Mexico; the United States drops two more on Japan. World War II ends.
1948	Shinichiro Tomonaga, Julian Schwinger and Richard Feynman develop the quantum electrodynamics theory of how light and matter interact.
1948	200-inch Hale Telescope begins operating at Mount Palomar, California.
1948	Researchers at Bell Labs invent the transistor.
1955	Canadian experiments indicate subatomic particles called neutrinos have mass.
1955	The first patent is issued for a nuclear reactor.
1958	Charles Townes invents the laser.
1960s	Astronomers using radio telescopes discover "quasi-stellar radio sources," or quasars.
1963	Murray Gell-Mann and George Zweig propose the quark model for subatomic particles.
1965	Arno Penzias and Robert Wilson measure cosmic background radiation.
1965–73	Theorists develop quantum chromodynamics—the study of quarks and the gluons that act on them.
1967	The term *black hole* comes into wide use to describe celestial bodies long theorized to exist.
1970s	Theorists devise string theory.
1970s	Theorists devise the Standard Model of elementary particles. Subsequently, researchers using particle accelerators detect the model's components.
1971	The Uhuru X-ray satellite detects a black hole.
1977	The Very Long Array of radio telescopes begins operating in New Mexico.
1979	Researchers discover evidence of the "gluon," the theorized particle that holds the atom's nucleus together.

1983   Carlo Rubbia and his colleagues detect W and Z particles—particles that produce the weak force.

1986   Materials that conduct electricity with no resistance at relatively high temperatures are invented.

1990   Hubble Space Telescope is launched; its mirror has problems that shuttle astronauts correct later.

1993   Congress kills funding for the Supercolliding Superconductor, ceding the cutting edge of particle physics to Europe.

1994   The "top" quark is found, completing experimental proof for the Standard Model.

1996   The age of the universe is "settled" at 15 billion years. In 1999 it is settled again, at 13.5 billion years.

1998   Astronomers discern that the expansion of the universe is accelerating—some theorize that "dark energy" drives the process.

1998   Neutrinos, thought to be mass-less, turn out to have mass.

2003   Astronomers report signs of "dark energy."

2007   The Large Hadron Collider at CERN, the European center for particle research, the world's most powerful particle accelerator, opens underground in Switzerland.

2012   Researchers discover the Higgs boson—the particle that endows everything in the universe with mass.

What's next? In physics and astrophysics, as with all of science, each new finding seems to open doors to new research. Among the big issues tormenting researchers today are the nature of dark matter, the nature of dark energy, and the quest to mesh the theory of relativity, Einstein's theorizing of the very big, with quantum mechanics, Planck's theorizing about the very small.

CONTRIBUTORS' BIOGRAPHIES

Hanson W. Baldwin (1903–1991), a longtime military editor for *The New York Times,* won a Pulitzer Prize in 1942 for his coverage of World War II.

William J. Broad is a science writer for *The New York Times* who has won two Pulitzer Prizes as well as an Emmy Award. He is the author of *The Science of Yoga: The Risks and the Rewards* (Simon and Schuster, 2012).

Malcolm W. Browne (1931–2112), a longtime science writer for *The New York Times,* won a Pulitzer Prize in 1964 for his coverage of the Vietnam War for the Associated Press. His 1963 photograph of the self-immolation of a Buddhist monk became one of the most memorable images of the conflict.

Lawrence E. Davies (1900–1971) was a *New York Times* reporter for forty-four years, covering politics, industry, agriculture, science and sports.

Clinton J. Davisson (1881–1958) was an American physicist who won the 1937 Nobel Prize in physics for his discovery of electron diffraction.

Lee Edson (1918–2008) contributed articles to *The New York Times* on science and technology.

John W. Finney (1923–2004) was a senior correspondent and editor in the Washington bureau of *The New York Times* for thirty years.

Ben A. Franklin (1927–2005) was a national correspondent and reporter in the Washington bureau for *The New York Times* for thirty years.

James Glanz is the former Baghdad bureau chief for *The New York Times* and is now one of its investigative reporters. He holds a PhD in astrophysical sciences.

Barry James is a reporter for the *International Herald Tribune.*

George Johnson is a science writer and the author of a number of books, including *The Ten Most Beautiful Experiments* (Random House, Inc., 2008) and *Strange Beauty: Murray Gell-Mann and the Revolution in Twentieth-Century Physics* (Random House, Inc., 1999).

Waldemar Kaempffert (1877–1956) was a science writer and museum director who worked for *The New York Times* from 1922 until 1953.

Ferdinand Kuhn Jr. (1905–1978) was *The New York Times* chief London correspondent from 1936 to 1940.

William L. Laurence (1888–1977) was a science writer for *The New York Times*. He received two Pulitzer Prizes, and, as the official historian of the Manhattan Project, was the only journalist to witness the 1945 Trinity test of the atom bomb in New Mexico and, shortly thereafter, the nuclear bombing of Japan.

William H. Lawrence (1916–1972) was a journalist who worked for *The New York Times* for more than twenty years (1941–1961).

W. J. Luyten (1899–1994) was a Dutch-American astronomer who worked at the Harvard College Observatory (1923–1930) and taught at the University of Minnesota (1931–1967).

John Markoff is a science correspondent for *The New York Times* and the author of *What the Dormouse Said: How the Sixties Counterculture Shaped the Personal Computer Industry* (Penguin Group, 2006).

Dennis Overbye is a science reporter for *The New York Times* and the author of *Lonely Hearts of the Cosmos: The Scientific Search for the Secret of the Universe* (1991) and *Einstein in Love* (Penguin Group, 2000).

Maya Pines is a science writer and reporter who has worked for several papers and magazines.

Simon Romero is the Brazil bureau chief for *The New York Times*. He's been working at *The Times* since 1999.

Garrett P. Serviss (1851–1929) was an astronomer and science fiction writer who specialized in explaining scientific details that made them clear to the ordinary reader.

Walter Sullivan (1918–1996) was an award-winning science reporter and editor for *The New York Times*. During his fifty-year career, he was considered the dean of science writers.

Laurie Tarkan writes about health issues for *The New York Times* and national magazines.

John Noble Wilford is an author and award-winning journalist for *The New York Times*. He won two Pulitzer Prizes (1984 and 1987).

IMAGE CREDITS

INDEX

BOOKS IN THE
NEW YORK TIMES SERIES:

The New York Times Book of Math

The New York Times Book of Physics and Astronomy

The New York Times Book of Wine

Thanks to Phyllis Collazo for her assistance with photos and
to Peter Morance for his assistance with illustrations.